И. И.

★

ИСТОРИЯ ЭКОНОМИЧЕСКОЙ МЫСЛИ

ИЗДАНИЕ ТРЕТЬЕ
СО ВТОРОГО ДОПОЛНЕННОГО

Научно-политической секцией Государственного Ученого Совета допущено в качестве учебного пособия для вузов

ГОСУДАРСТВЕННОЕ ИЗДАТЕЛЬСТВО
МОСКВА ★ 1929 ★ ЛЕНИНГРАД

A History of Economic Thought

by

Isaac Ilych Rubin

Translated and edited by Donald Filtzer

Afterword
by
Catherine Colliot-Thélène

First published as Istoriya ekonomicheskoi mysli (Gosizdat RSFSR). This translation taken, with the permission of the New York Public Library, from a copy of the second printing of the second, revised Russian edition (1929) in that library's possession.

This edition first published in 1979
Second impression 1989

Translation by Don Filtzer

345 Archway Road
London N6 5AA

Afterword by Catherine Colliot-Thélène

The Frontispiece is taken from the New York Public Library photocopy of the 1929 Russian edition of Rubin's book.

ISBN 0 7453 0301 3

CONTENTS

Part Five
The Decline of the Classical School 289

Part Six
Conclusion: A Brief Review of the Course 363

Afterword
By Catherine Colliot-Thélène 385

Editor's Preface

This English edition of Isaac Rubin's *A History of Economic Thought* has been prepared from the New York Public Library's copy of the 1929 reprinting of the second, revised Russian edition. As the reader will learn from Rubin's Preface, the book is made up of a series of lectures and was used as a university text. The book must have been in fairly general use, because the reprint of the second edition ran to 5,000 copies. The lectures were intended to be used alongside two other texts, Marx's *Theories of Surplus Value* and an anthology compiled by Rubin of extracts from pre-Classical and Classical political economy, *Classics of Political Economy From the Seventeenth to the Mid-Nineteenth Century* [*Klassiki politicheskoi ekonomii ot XVII do srediny XIX veka*] (Gosizdat RSFSR, 1926).

The design of Rubin's book has presented certain difficulties in translating and editing an English edition. Because it was to be used together with the above-mentioned collection, *A History of Economic Thought* contains no references for any of its quotations. Thus we have had to go through the laborious task of tracking down the standard English editions of the works of the many philosphers and economists from whom Rubin quotes. In most cases this was relatively straightforward; in others, such as the Physiocrats or Sismondi, whose works are translated either only partially into English or not at all, we have on occasion had to be satisfied with re-translating Rubin's own Russian rendering of the passages in question. The reader will see from the Editor's notes that these represent only a very small minority of the quotations, and that most passages are from the English original (in the case of French authors, most quotes are either from the standard English translation or have been translated directly from the French).

In editing the work we have provided copious notes directing the reader to the original sources; very often we have also given quotations fuller than those provided by Rubin, so as to allow the reader to gain a better sense of the arguments of Petty, Smith, Ricardo, etc. We have also used the notes to guide the reader to other secondary sources that she or he might find useful and to explain historical and conceptual references that might be unclear in the main text.

As for the terminology used, we have in general followed this rule:

where Rubin is paraphrasing a particular author we have tried to retain that author's own usage, whereas when translating Rubin's discussions of these texts we have opted for the terminology accepted in modern usage. There are certain exceptions, e.g., in the section on Adam Smith where we have replaced Smith's term 'commandable labour' with the more modern 'purchasable labour'. We have also followed the standard practice of not modernizing the spelling or syntax of the passages quoted.

In a small number of cases we have deleted certain sentences or phrases in which Rubin is recapitulating a doctrine that he has already discussed on several occasions. These repetitive summaries, e.g., of the Physiocrats' views on productive labour or Smith's theory of profit, whilst perhaps of value in maintaining the continuity of Rubin's classroom lectures, are a genuine obstacle to someone trying to read the text straight through. In no case have we cut more than one or two sentences at a time, and the sum of these elisions amounts to no more than two or three printed pages: thus the reader need have no fear about whether she or he is receiving a genuine 'original edition'.

Finally, I should like to acknowledge the assistance of the reference staff of the main library of the University of Glasgow and of the staff of the Sidney Jones Library, University of Liverpool, who gave me invaluable help in locating and using many of the original editions from which I had to take quotations. I should like also to thank Prof. D.P. O'Brien of the University of Durham and Prof. Andrew S. Skinner of the University of Glasgow for their help in tracking down certain highly elusive passages. Needless to say, all of these people are blameless for any remaining shortcomings in this volume.

Donald Filtzer
Birmingham, England
April 1979

Author's Preface to the Second Edition*

The study of the history of economic thought holds immense historical and theoretical interest. As a science it is closely tied, on the one hand, to the history of economic development and the struggle between the classes and, on the other, to theoretical political economy.

From an historical point of view, economic doctrines and ideas can be seen to have been amongst the most important and influential forms of ideology. As with other forms of ideology, the evolution of economic ideas depends directly upon the evolution of economic forms and the class struggle. Economic ideas are not born in a vacuum. Often they arise directly out of the stir and strife of social conflicts, upon the battleground between different social classes. In these circumstances, economists have acted as arms-bearers for these classes, forging the ideological weapons needed to defend the interests of particular social groups—often not concerning themselves any longer with developing their own work and giving it greater theoretical foundation. This was the lot that befell the economists of the mercantilist period (16th and 17th centuries), who devoted countless topical pamphlets to the ardent defence of the interests of merchant capital. Yet even if we look at the Physiocrats and the economists of the Classical school, whose works conform far more to the demands of theoretical clarity and logical coherence, we have little difficulty in identifying the social and class forces behind the different currents of economic thought. Though it occurs less openly and with greater complexity, we still find that the requirements of economic policy exert a powerful impact upon the orientation of economic ideas. In the most abstract constructs of the Physiocrats or Ricardo—those that seem farthest removed from real life—we shall discover a reflection of

*The present edition contains the following additions to the first edition of this work: 1) a concluding chapter, Chapter Forty, giving a brief review of the material covered; 2) a name index; 3) a subject index, to make it easier to situate individual problems historically; 4) certain additions to the bibliography. Other than the additional chapter already referred to, the text of the book has in no way been altered.

contemporary economic conditions and an expression of the interests of particular social classes and groups.

While being thorough in tracing the influence of economic development and the changing forms of class struggle upon the general direction of economic thought, we nevertheless must not lose sight of our other task. Once we arrive at the more advanced stages of social development, the systems constructed by economists no longer represent a loose aggregation of isolated practical demands and theoretical propositions; instead, they appear as more or less logically coherent theoretical systems, whose separate parts are to a greater or lesser extent in harmony both with one another and with the overall character of the ideology appropriate to a particular social class during a given historical epoch. The Physiocratic system, for example, when taken as a whole can only be correctly understood against the background of socio-economic conditions in eighteenth-century France and the struggles which these generated between different social classes. We cannot, however, limit ourselves to studying the social and economic roots of the Physiocrats' system. We must examine the latter as a system: as an organic totality of logically interconnected concepts and propositions. The first thing we must uncover is the close connection between the Physiocrats' economic theory and their overall world view, especially their social philosophy (i.e., their views on the nature of society, economy, and state). Secondly—and this is where the most important of our tasks begins—we must reveal the logical connection which binds together the different parts of the system or, conversely, identify those places where such connection is absent and the system contains logical contradictions.

What makes an account of the history of economic thought particularly difficult is this two-sided nature of our task: the necessity to impart to the reader at one and the same time an exposition of both the *historical* conditions out of which the different economic doctrines arose and developed, and their *theoretical meaning*, i.e., of the internal logical relationship of ideas. We have tried to allocate sufficient space to the historical and theoretical parts of our exposition. Each section of our book (with the exception of the first) is prefaced by a general historical study which depicts the economic conditions and class relations which were to find expression in the ideas put forward by the economists concerned. However, we have allocated even greater space to our theoretical analysis of these doctrines, especially where, as in the sections devoted to Adam Smith and David Ricardo, we are dealing with grandiose theoretical systems permeated by a single idea. In these sections our theoretical analysis has received first priority, since our main task was, in our view, to provide readers with a thread to guide them through the complex and entangled maze of these economists' theoretical ideas.

Without this type of detailed theoretical analysis no history of economic thought could ever perform the service we have the right to expect of it, namely to act as a faithful companion and guide facilitating our study of the theory of political economy. For we do not analyze the doctrines of Smith simply to gaze at a vivid page from the history of social ideology, but because it permits us to gain a deeper understanding of theoretical problems. Familiarity with Smith's theories can provide the reader with one of the best introductions to a more serious study of the problem of value, just as a knowledge of Ricardo's theories facilitates the study of the problem of rent. These are difficult problems: in theoretical political economy they stand before us in their full magnitude and in their most complicated and involved form; but for a reader acquainted with the historical process through which they were built upon and acquired their complexity the difficulties are in large measure removed. The ideas and problems of the early economists will be more easily understood by the reader if they are posed and formulated more simply; an analysis of the contradictions so often encountered in their works (even of such intellectual giants as Smith and Ricardo) is of tremendous intellectual and pedagogical value.

If the knowledge of the history of economic thought is on the whole essential for a deeper understanding of theoretical political economy, this is all the more true when it comes to understanding Marx's theoretical system. To construct his system Marx first laboriously and conscientiously studied a wealth of economic literature, itself the product of the labours of several generations of English, French, and Italian economists from the 17th to the mid-19th centuries. Marx was the leading expert of his time on the economic literature of the 17th and 18th centuries, and probably no one has surpassed him even to this day. On the very first page of *Capital* the reader encounters the names of the elders, Barbon and Locke. And at every step in his subsequent exposition, both in his text and in his footnotes, Marx stops to select with evident enjoyment a particularly valuable thought that he has discovered in the early economists. No matter how rudimentarily or naively this idea may originally have been expressed, Marx nonetheless gives it his full attention and diligently analyzes it, so as to prize out the valuable hidden kernel that went unnoticed at first sight.

Marx's attentive and painstaking treatment of his forerunners is not to be taken as the whim of a dilettante, of an expert and connoisseur in old economic writings. Its cause is far more profound and serious. Ever since publication of his *Theories of Surplus Value* we have had substantial access to the laboratory of Marx's thought and have glimpsed at first hand with what profound seriousness and intellectual effort Marx carried out his study of those who had preceded him. We cannot but admire and marvel at the tirelessness with which he tracked down the twists and turns and the most subtle offshoots of the ideas of the economists he was investigating. We now know that the abundance of brief remarks on Smith, Ricardo, and other economists which Marx scattered throughout the footnotes to *Capital* are the abbreviated, not to say parsimonious, resumés of the highly detailed—and on occasion tiresome—researches contained in *Theories of Surplus Value*. It is only in the light of the *Theories* that we can fully appreciate how much these footnotes—made almost as if in passing—are an organic part of the text of *Capital*, and how inseparable Marx saw the tasks of studying his predecessors and constructing his own system. Every step that permitted Marx to penetrate more deeply into the works of his predecessors brought him closer to this construction. And each success gained in resolving this latter problem opened up to Marx new treasure chests of ideas which had lain buried in the long-known and partly forgotten writings of past economists. In his own system Marx made full use of the intellectual skills deployed by economists over the preceding centuries; thanks to him the ideas and knowledge that his forerunners had accumulated were brought together into a grand synthesis. Here is why the study of the history of economic thought is so essential both to an elucidation of the historical background to Marx's economic system and to the acquisition of a more profound understanding of his theory.

From what we have said, the reader can draw certain conclusions about what method is most desirable for studying the history of economic thought. In our view the most efficacious method is for the reader to combine this study with a parallel study of theoretical political economy. This does not mean that readers of *A History of Economic Thought* can take up the book without first being familiar with a general course in political economy. Our book is intended for those readers who, after taking an introductory course in political economy, would like to acquire an understanding of the evolution of basic economic ideas and at the same time undertake a more serious and detailed investigation into theoretical problems. For these readers our book can serve both as a systematic course in the history of economic thought and as an historical introduction to a more thorough study of Marx's system. One way in which the reader could familiarize himself simultaneously with the historical and theoretical material would be as follows. In the course of going through *A History of Economic Thought* the reader can mark off certain sections for more thorough study, e.g. on how the labour theory of value evolved through Petty, Smith, and Ricardo. By dividing up the material according to specific problems, readers will immediately find themselves faced with the need to combine their historical study with a theoretical one. From Petty's first, brilliant sketches to the agonizing contradictions which Ricardo's ideas consistently came up against, the history of the labour theory of value is one of the gradual accumulation of problems and contradictions. Readers can correctly understand this process only if their own thought proceeds in parallel with the historical exposition and critically analyzes and surmounts those problems and contradictions which in the course of history have confronted economists. To conduct such a critical analysis successfully the reader has no recourse but to turn to theoretical political economy.

Readers will draw maximum benefit from their endeavors if, instead of limiting themselves to reading and studying the present course, they turn directly to the works of the economists we are analyzing. In our view, readers would draw particular advantage from familiarizing themselves with the works of Smith and Ricardo, even if this is limited to only a few selected chapters.* For those readers who would like to acquaint themselves more thoroughly and in greater detail with the economic doctrines of Smith and Ricardo, Marx's most important predecessors, we would recommend that they order their studies as follows. After studying those parts of our book devoted to Smith and Ricardo, it is then necessary to become acquainted at the very least with the chapters of their works that we have already indicated. Parallel to reading the chapters in Smith and Ricardo on value, wages,

* We recommend that the reader refer to Chapters I, V, VI, VII, and VIII of Book One of Smith's *An Inquiry Into the Nature and Causes of the Wealth of Nations*, and Chapters I, II, IV, V, and XX of Ricardo's *Principles of Political Economy and Taxation*. (For the reader's convenience we have prepared a collection of extracts from the works of the economists of the 17th to 19th centuries, entitled *Klassiki politicheskoi ekonomii* [*Classics of Political Economy*] (Gosizdat RSFSR, 1926). The excerpts in this collection have been arranged in an order roughly corresponding to that in which we discuss the economists in the present work.)

etc., the reader can then turn to those sections of *Theories of Surplus value* where Marx presents his own critical analysis of their views on these questions. Readers will be well rewarded for the effort expended on a careful study of these critical remarks: they will learn to probe more deeply both into the works of these economists and into Marx's own theoretical system.

It remains for us to say a few words about the scope of the material covered by our book. We begin our account with the English mercantilists of the 16th and 17th centuries, and conclude with the mid-19th century, i.e., with the period when Marx was in the process of laying down the basis of his new economic doctrine, which supplanted the classical theory of Smith and Ricardo. Some historians of economic ideas begin their account with the ancient philosophers (Plato, Aristotle), in whose work are to be found some penetrating reflections and observations on various economic problems. But their economic considerations were themselves reflections of the slave economy of antiquity, just as the writings of the medieval church reflected the feudal economy. We cannot include them in our book since it is our task to provide the reader with an idea of how *contemporary* political economy—a science whose object of study is capitalist economy—came into being and evolved. This science arose and developed only with the appearance and development of its object of study, i.e. capitalist economy itself. We therefore begin our account with the age of mercantilism, the epoch when capitalism, in the rudimentary form of merchant capital, first sprang into existence.

On the other hand, we do not see that it is possible to limit our study any more narrowly than we have already done. There are historians who take up their account from the era of the Physiocrats or Adam Smith, when economic enquiry had already taken shape as more or less coherent, finished theoretical systems. But if we begin from this point, when contemporary political economy had already emerged in its essentially finished form, we will not have made accessible that critically important process through which this science *came into being*. Just as a complete understanding of the capitalist economy is impossible without knowledge of the epoch of primitive capitalist accumulation, so, too, there can be no proper comprehension of the evolution of contemporary political economy without a general acquaintance with the economists of the mercantilist age. This obviously does not mean that we can include all of the more or less distinguished economists from that period in our course. Mercantilist literature had no shortage of representatives populating the most diverse countries of Europe. Our priority however was not comprehensiveness of material otherwise our book would inevitably have been dry and condensed, overburdened with facts and boring for the reader. To avoid this we have limited the first section of the book in two respects: first, we have included only the English mercantilist literature, as this was the most developed and played the most important role in preparing the way for the emergence of the Classical school; second, we have chosen only those of the English mercantilists who most clearly spoke for their particular historical age, in order to concentrate as far as possible upon their specific contribution. We have tried to follow this same principle in the other sections of the book, concentrating our exposition only upon the most important themes. Our preference has been to limit our selection to the most prominent and brilliant representatives of the different currents of economic thought, and to accord them greater attention than is usually the case with courses designed for a wider circle of readers. We hope that by *limiting the number of themes* and analyzing each of them *in greater detail* we will more readily arouse within the reader a lively interest in our science.

I. I. Rubin

Part One

Mercantilism and its Decline

CHAPTER ONE

The Age of Merchant Capital

The age of *merchant capital* (or early capitalism) covers the 16th and 17th centuries. This was an era of major transformations in the economic life of Western Europe, with the extensive development of seafaring trade and the emerging predominance of commercial capital.

The economy of the later middle ages (the 12th to 15th centuries) can be characterized as a *town* or *regional* economy. Each town, together with its surrounding agricultural district, comprised a single economic region, within whose confines all exchange between town and countryside took place. A substantial portion of what the peasants produced went for their own consumption. A further part was given over as quickrent to the feudal lord, and what meagre surpluses were left were taken to the neighboring town for sale on market days. Any money received went to purchase goods fashioned by urban craftsmen (textiles, metalwares, etc.). The lord received a quickrent—established by custom—from the peasant serfs who lived on his estates. Over and above this, he also received the produce from his manor's own tillage, which was worked by these same peasants doing compulsory labour service (the *barshchina*, or *corvée*). A large part of these products were for the lord's own consumption, or for that of his innumerable household servants and retainers. Anything left over was sold in the town, so that the receipts could be used to buy either articles made by local craftsmen or luxuries brought in by traders from far away countries, primarily from the East. What therefore distinguished rural feudal economy was its overwhelmingly *natural* character and the feeble development of money exchange.

If the rural economy was organized around the *feudal demesne*, the industry of the towns was organized into *guild handicrafts* where production was carried out by small *master craftsmen*. Each master owned the simple tools and instruments necessary for his trade, and worked personally in his own shop with the help of a small number of assistants and apprentices. His products were made either on special order from individual consumers or were held in stock for sale to local inhabitants, or peasants who had journeyed in to market. Because the local market was limited, the craftsman knew in advance the potential volume of demand for his product, while the backward, static technique of craft production allowed him to tailor the volume of production to exactly what the market would bear. The craftsmen of each profession all belonged to a single union, or *guild*, whose strict rules permitted them to regulate production and to take whatever measures were necessary to *eliminate competition*—whether between individual masters of a given guild or from persons who were not guild members. This right to a monopoly over producing and selling within a given region was accorded only to members of the guild, who were bound by the guild's strict code of rules: no master could arbitrarily expand his output or take on more than the statutory number of assistants and apprentices; he was obliged to turn out products of an agreed quality and to sell them only at an established price. The removal of competition meant that craftsmen could market their wares at high prices and be assured of a relatively prosperous existence, in spite of the limited size of their sales.

By the late middle ages there were already signs that the regional, or town economy which we have just described was in a *state of decline*. However, it was not until the epoch of merchant capital (the 16th and 17th centuries) that the break up of the old regional economy and the transition to a more extensive national economy became in any way widespread. As we have seen, regional economy was based on a combination of the rural feudal demesne with the guild handicrafts in the towns; it was, therefore, only with the decomposition of both of these that the disintegration of the regional economy could occur. In both cases their decomposition was brought about by one and the same set of basic causes: the rapid development of a *money economy*, the expansion of the *market*, and the growing strength of *merchant capital*.

With the end of the crusades in the late middle ages trade expanded *between the countries of Western Europe and the East* (the Levantine trade). The European countries acquired, firstly, raw materials from the tropical countries (spices, dyestuffs, perfumes) and, secondly, finished goods from the highly-developed Eastern craft industries (silk and cotton textiles, velvet, carpets, and the like). Such luxury articles, imported into Europe from so far away, were very dear, and were purchased overwhelmingly by the feudal aristocracy. In the main it was the Italian trading cities, Venice and Genoa, which carried on this commerce with the East, dispatching their fleets across the Mediterranean Sea to Constantinople, Asia Minor, and Egypt, where they bought up Eastern commodities that had in large part been delivered from India. From Italy these commodities were transported to other European countries, some in the commercial convoys of these same Italians, others overland to the North, through the South German towns (Nuremberg, Augsburg, and others) and on to the towns of Northern Germany which had formed themselves into the *Hanseatic League* and controlled the Baltic and North Sea trade.

The military conquests of the Turks in the 15th century cut the Italians off from direct contact with the countries of the East. But the fledgeling interests of commercial capital demanded the continuation of so profitable a source of trade, and consequently Europe undertook an intense search for direct, oceanic routes to India—efforts which were crowned with brilliant success. In 1498, the Portuguese Vasco da Gama rounded the Southern tip of Africa and found a direct route to India. In 1492, Columbus, whose mainly Spanish expedition was also seeking a direct path to India, accidentally discovered America. From this point onwards, the old Levantine trade with the East across the Mediterranean gave way to an ocean going commerce in two directions: eastwards to *India*, and westwards to *America*. International commercial hegemony passed out of the hands of the Italians and the Hanseatic cities to those countries situated along the *Atlantic Ocean*: first to *Spain* and *Portugal*, afterwards to *Holland*, and finally to *England*.

The colonial trade brought enormous *profits* to European merchants, and enabled them to accumulate sizable *money capitals*. They would purchase colonial commodities for next to nothing and sell them in Europe at an enormous markup. Colonial trade was *monopoly* trade: each government would attempt to establish a monopoly over the trade with its own colonies, and block foreign ships and traders access to them. Thus the riches of the American colonies, for example, could only be exported to Spain, while only Spanish merchants had the right to supply these colonies with European commodities. The Portuguese did exactly the same with India, as did the Dutch, once they had ousted the Portuguese from that part of the world. The Dutch entrusted their India trade to the Dutch East India Company, a special joint stock company set up by them in 1602, which received a trading monopoly for this purpose. Similar *'companies'* (i.e., joint stock companies) were founded by the French and English, and each received a commercial monopoly with their respective colonies. It was out of the far flung activities of these societies that the English East India Company, founded in 1600, later developed.

As a consequence of the colonial trade, huge quantities of *precious metals* (mainly silver at first) were shipped into Europe, thus increasing the quantity of money in circulation. In America (Mexico, Peru) the Europeans came upon rich silver mines, which could be worked with far less labour than the poor and exhausted mines of Europe. On top of this the mid-16th century saw the introduction of a significant improvement in the technology of silver extraction—the amalgamation of silver with mercury—and copious streams of cheap American silver and gold flowed into Europe. Its first point of arrival was Spain, which owned the American colonies. But it did not stay there: backward, feudal Spain was compelled to purchase industrial goods, both for its own consumption and for export. And so Spain's negative balance of trade resulted in an outflow of its precious metals to all the countries of Europe, the largest masses being accumulated in Holland and England, the nations where the development of merchant and industrial capital was most advanced.

If trade with the colonies prompted a flow of precious metals into Europe, this flow in its turn brought with it a growth in commercial exchange and a money economy. The stocks of precious metals in Europe grew by three to three and a half times during the 16th century alone. Such an enormous rise in the mass of precious metals, whose value had fallen as a consequence of the greater ease with which they could now be extracted, produced as an inevitable consequence a universal rise in prices. Indeed, 16th century Europe experienced a '*price revolution*.' Prices of everything rose sharply, two to three times on average, but sometimes even more. Thus in England, for example, prices of wheat, which for several centuries had held constant at five to six shillings per quarter had reached twenty-two shillings by 1574 and forty shillings by the end of the same century. While wages also went up, they lagged appreciably behind the rise in prices: whereas provisions were now twice as expensive (i.e., their prices had risen by 100%), the growth in wages was only between 30 and 40%. By the close of the 17th century real wages had fallen to approximately half of what they had stood at at the start of the 16th century. The rapid *enrichment of the commercial bourgeoisie* in the 16th and 17th centuries was accompanied by a drastic decline in the standard of living of the lower classes of the population, *the peasantry, craftsmen, and workers*. The impoverishment of the peasantry and craftsmen appeared as the inevitable result of the break up of the feudal order in the countryside and the guild crafts in the towns.

The rise of the money economy heightened *the feudal lords' demand for money* and at the same time opened up the potential for an extensive market in agricultural produce. The feudal lords of the most advanced commercial nations (England and Italy) began to replace the *in natura* obligations of their peasants with a money quickrent.* The peasant serfs whose previous obligations had been precisely fixed by long-standing custom were gradually turned into free tenants who rented the land by agreement of the lord. Though they had acquired their freedom, its embodiment, the rent, proved more of a burden as time went on. Often the lord preferred to lease his land not to small-scale peasants, but to larger, better-off farmers who had it within their means to make improvements to their holdings. The English landowners of the end of the 15th and beginning of the 16th centuries often *cleared the small-scale peasant-tenants off their land*, or 'enclosed' the communal lands which the peasants had previously used for grazing their cattle, since the areas thus made free could be put to better use raising sheep. As English and Flemish cloth manufacturers increased their demand for wool, so prices shot up and sheep breeding became a more profitable undertaking than cultivating the soil. 'Sheep swallow down the very men themselves,' said Thomas More at the beginning of the 16th century. Another of his contemporaries wrote: 'Gentlemen do not consider it a crime to drive poor people off their property. On the contrary, they insist that the land belongs to them and throw the poor out from their shelter, like curs. In England at the moment thousands of people, previously decent householders, now go begging, staggering from door to door.'[1]

If in the countryside the feudal order was in a process of decomposition, in the towns the growth of merchant capital was causing a simultaneous *decline of guild handicrafts*. The petty craftsman could preserve his independence only so long as he was producing for the local market with exchange taking place between the town and its immediate environs. But side by side with the growth of international trade there was also the development of trade between the different regions and towns within a given country. Certain towns specialized in the manufacture of particular items (e.g., textiles or armaments), which they produced in too large a quantity for their sale to be

* In the backward countries of Europe (Germany, Russia), the growth of monetary exchange led to a completely different development: the landlords transferred their peasants onto a *corvée* system and expanded the area subject to this type of tillage. In this way they were able to obtain a greater quantity of grain for selling.

limited to the local surroundings; hence *markets further afield* had to be sought. This was particularly true of the *cloth industry*, which had started to flourish in the towns of Italy and Flanders (and later on, in England) even by the end of the middle ages. Even then the master weaver could no longer depend on the immediate consumption of the local market for sales, and so he sold his cloth to middlemen, who transported large consignments to areas where demand existed. The *buyer up* now occupied an intermediary position between consumer and producer, gradually asserting his domination over the latter. At first he purchased individual batches of commodities from the craftsman as the occasion arose; later he bought up everything the craftsman produced. With the passage of time he began to give the craftsman a money advance; and in the end he came to provide the raw materials at his own expense (e.g., thread or wool), farming them out to individual craftsmen (spinners, weavers, etc.) who were then paid a remuneration for their labour. From this moment the *independent craftsman* was turned into a *dependent handicraft worker*, and the *merchant* into *a buyer up-putter out*. In this way the merchant capitalist, moving from the sphere of trade, worked his way into the production process, organized it and gained control over the labour of large numbers of handicraft workers working in their own homes. The independent guild crafts, which had so dominated the economy of the towns in the late middle ages, gave way in the 16th and 17th centuries to the rapid rise of *cottage industry* (the so-called domestic system of capitalist industry). It made especially rapid headway in those branches of production, such as cloth manufacturing, which worked for specific markets or for export to other countries.

Peasants dispossessed of their land and ruined craftsmen swelled the already numerous ranks of beggars and vagabonds.The measures adopted by the state against vagabondage were harsh: able-bodied vagabonds were lashed or had their chests branded with red-hot irons; persistent vagrants were liable to execution. At the same time maximum wage rates payable to workers were established by law. The brutal moves against vagabondage, and the laws setting maximum wages were attempts by governments of the day to turn these declassed social elements into a disciplined obedient class of wage workers who, for a pittance, would offer up their labour to a youthful and growing capitalism.

What thus took place in the age of merchant capital (the 16th and 17th centuries) was the accumulation of huge capitals in the hands of the commercial bourgeoisie, and a process of separation of the direct producers (handicraftsmen and in part the peasantry) from the means of production—i.e., the formation of a class of wage labourers. Having gained domination in the field of foreign trade the merchant bourgeoisie penetrated from there into those branches of industry which worked for export. The handicraft workers who laboured in these industries were subordinated to the merchant-exporter and buyer up-putter out. With *foreign trade* and the imposition of the latter's control over *cottage industry*, capitalism celebrated its first victories.

This transition from feudal to capitalist economy enjoyed the active promotion of the *state authorities*, whose increasing centralization ran parallel with the growing strength of merchant capital. The commercial bourgeoisie suffered greatly at the hands of the antiquated feudal regime: firstly, because the fragmentation of the country into seperate feudal estates made commercial relations between them difficult (agressions from the lords and their knights, the levying of duties, and the like) and secondly, because the rights of access to the individual towns was refused to traders from other cities. To smash through the privileges of the estate holders and towns, a strong crown was essential. But the bourgeoisie also needed a powerful state to protect its international trade, to conquer colonies, and to fight for hegemony over the world market. And so the youthful bourgeoisie came out as a partisan of the strong royal houses in the latter's struggle against the feudal lords. The transition from the closed off town and regional economy to a truly national one demanded the transformation of the weak feudal monarchy into a centralized state which could rely on its own bureaucracy, army, and navy. Thus the age of *merchant capital* was also the age of *absolute monarchy*.

But if the young bourgeoisie supported the crown, the latter, for its part, took measures to nurture and develop the burgeoning capitalist economy. There were political as well as economic and financial considerations which made this alliance with the bourgeoisie essential for the crown. In the first place, the maintenance of a bureaucracy and an army demanded enormous expenditures, and only a wealthy bourgeoisie could provide the means to cover these through taxes, commercial (customs) duties, state loans (both compulsory and voluntary), and, lastly, through the fees paid to the state for the right to exact state revenues from the population [tax farming]. Secondly, the crown needed the support of the 'third estate' (the bourgeoisie) in its struggle with the feudal lords. It was, therefore, during the age of merchant capitalism that *a close alliance was formed between the state and the commercial bourgeoisie*, an alliance which found expression in *mercantilist policy*.

The basic feature of mercantilist policy is that the state actively uses its powers to help implant and develop a young capitalist trade and industry and, through the use of protectionist measures, diligently defends it from foreign competition. While mercantilist policy served the interests of both these social forces, it was dependent upon which partner in this union proved the stronger—the state or the merchant bourgeoisie—as to whether its *fiscal* or its *economic* aspect gained the upper hand. In its opening phase mercantilism had above all to foster the fiscal aims of enriching the state coffers and augmenting state revenues, and this it did by making the population bear a heavier tax burden and by attracting precious metals into the country (early mercantilism, or the *monetary balance system*). But as the bourgeoisie grew in strength mercantilism became increasingly transformed into a means of bolstering capitalist trade and industry and defending it through protectionism. Here we have developed mercantilism, or the *balance of trade system*.

1 The statement by More is from *Utopia*. The quotation immediately following it is unattributed and thus translated from the Russian.

CHAPTER TWO

Merchant Capital and Mercantilist Policy in England in the 16th and 17th Centuries

Although practically all the countries of Europe practiced a mercantilist policy during the early capitalist period, it is through the example of *England* that its evolution can be traced out most clearly.

Compared to some other European nations, such as Italy and Holland, England was relatively late in embarking upon the pursuit of colonies and the development of its industry. At the start of the 16th century England was still overwhelmingly agricultural and commercially underdeveloped. Its exports were *raw materials*, e.g., hides, metals, fish, and above all, wool, which was purchased by the highly developed cloth industry of Flanders. From abroad came manufactured articles, such as Flemish cloth, copperware, etc. This import and export trade was in the main in the hands of *foreign merchants* from Italy and the Hanse. The Hanseatic traders had a large factory[1] in London; as it was their ships which conveyed commodities in and out of England, the latter was hampered from developing her own shipping. When English merchants ventured onto the Continent (which was not often) it was primarily to purchase wool in Flemish towns—first at Bruge and later, from the 16th century onwards, at Antwerp, where they had their own factory.

Under these conditions, there was no wealthy native merchant class, and the country was poor in money capital. The English government—at least to the end of the 16th century—regarded foreign trade with the wealthier nations primarily from a *fiscal* perspective. Duties were levied upon imports and exports alike, especially the export of wool. Every single transaction between English and foreign merchants was subject to strict state control, first to assure that the treasury received the appropriate *duties*, and second to guarantee that *no money was sent out of the country*. With the government always short of funds, constantly having either to debase the coinage or to resort to loans in order to keep the treasury solvent, the outflow of precious metals was a source of deep apprehension given the state's shortage of money capital. The export of gold and silver was strictly forbidden. According to the *'Statutes of Employment'* foreign merchants who brought commodities into England were obliged to spend all moneys received from selling them upon the purchase of other commodities inside the country. As soon as a foreign trader journeyed into England he was put under the control of a respected local resident who acted as his 'host'. The 'host' kept a sharp watch over all transactions carried out by the journeying 'guest' and entered them into a special book. The 'guest' had a maximum of eight months to sell all his stocks and use his receipts to buy English commodities. Any attempt by a foreign merchant to evade the 'host's' control resulted in imprisonment. During the second half of the fifteenth century the system of 'hosts' gave way to one of control exercised by special government inspectors and overseers.[2]

It was one thing to put an embargo upon the export of precious metals out of England. Care still had to be taken to attract these metals into the country from abroad. To this end the law obliged English traders exporting commodities to repatriate a specified portion of their receipts in hard cash. In order that the government would be able to maintain control over the foreign transactions of its merchants it allowed them to export their commodities only to certain continental towns(the so-called *'staples'*.)[3] For instance, in the early part of the 14th century English wool could be exported only to Bruges, Antwerp, Saint-Omer, and Lille. In these 'staples' the English government installed special officials whose job it was to oversee all transactions between English and foreign traders and to see to it first, that the correct amount of duty was paid to the English treasury, and second, that a portion of the receipts taken in from the sale of English commodities was designated for despatch back to England, either as metal or as foreign coinage.

Early mercantilist policy was, therefore, primarily *fiscal* policy, whose over-riding aim was to *enrich the treasury*, either directly, through the collection of import and export duties, or indirectly, by increasing the quantity of precious metals present within the country (here, too, the intention was to make possible a rise in state revenues in the future). On the one hand, the 'Statutes of Employment' forbade foreigners from exporting hard currency out of England; on the other, the creation of the 'staples' inevitably promoted the inflow of money from abroad. To see that its laws were complied with the state had to regulate the activities of both English and foreign traders strictly and rigidly, and exercize meticulous supervision over each and every commercial transaction, be they conducted inside or outside England's borders. By blocking gold and silver from going out of the country and by attracting these metals in from abroad, early mercantilist policy was directed towards improving the nation's *monetary balance* and can therefore be designated as a *monetary balance system*.

As commerce and industry developed, this policy began to hinder the turnover of trade. The controls that it entailed could be maintained only so long as foreign commercial deals were not overly numerous, were done in hard cash, and were confined in their majority to transactions with foreign traders who had come to England. While England's principal export was its wool—famous for its superior quality and enjoying a monopoly position on the market—the ban on exporting commodities other than to the 'staples' imposed little sacrifice on English merchants. The money balance system corresponded to a level of foreign trade that was poorly developed, concentrated in the hands of foreign merchants, and limited overwhelmingly to the export of raw materials. The future development of English trade and industry during the 16th and 17th centuries led inevitably (as we will see later on) to a break with the outmoded money balance system and to its replacement with a more advanced mercantilist policy, the so-called *balance of trade system*.

Over the course of the 16th and 17th centuries the basis of England's exports gradually shifted from *raw materials* (wool) to the *export of finished products* (cloth). England's *cloth industry* had started to enjoy a rapid development as far back as the 14th century, when rural weavers in Flanders, prevented from pursuing their craft by the urban guilds of their own country, moved to England. Weaving established itself there as cottage industry, situated in rural localities and free from any subordination to guild regulations. The English wool that had heretofore been exported to Flanders for working up now began to be processed partly in its country of origin. In the 16th century there was a reduction in the export of raw English wool and a sharp growth in the export of unfinished cloth.* Deprived of English wool the Flemish cloth industry now began to fall into decline, and by the start of the 17th century had already ceded first place to England. While in earlier times the main item of English exports had been wool, that role now passed to cloth.

*Up until the middle of the 17th century English cloth was exported unfinished; finishing and dyeing were carried out in Holland and France.

The export of English cloth abroad became the province of a special trading company, the Merchant Adventurers, whose activity expanded throughout the 16th century. English cloth required new markets, to which end the Merchant Adventurers were granted the right to conclude independent trade agreements and to export cloth to new foreign markets. The old monopoly of the 'staples' was thereby broken. By the close of the 16th century the English merchants no longer sat at home with their commodities, or in the continental 'staple' towns, awaiting the arrival of foreign buyers. No longer could they sell simply the raw materials (i.e. wool) which they monopolized; they had to sell finished goods (cloth), and for this they had to maintain a strong competitive position on the world market against the cloth of other countries, especially that of Flanders. What now began was a struggle for *domination over the world market* and the elimination of foreign competition. To win out, English traders abandoned their passive role in commerce for an active one—they started conveying their own commodities in their own boats to far flung markets returning with the goods they had purchased—particularly from the colonies. English ships were now dispatched across the Mediterranean in search of Eastern products; factories were established in Venice and Hamburg. The Italian and Hanseatic merchants in England had their monopoly broken up: in 1598 the Hanse traders' factory was shut down by the English government and the merchants themselves expelled from the country.

As English merchants now ventured forth onto the world's markets, the country was forced to *pursue an active colonial policy*. The wealthiest colonies had already been seized by other states, namely Spain and Portugal. With time Holland, and to some extent France acquired sizable colonial possessions. The entire history of England from the 16th to the 18th centuries is a history of its struggles with these nations for commercial and colonial superiority. Its weapons in this struggle were *the founding of its own colonies, commercial treaties, and wars*. The English fitted out their own expeditions to India, where they established the factories that were to mark the beginning of their domination over that country. At the end of the 16th century they founded a number of colonies in North America which were eventually to form the United States of America. England forced her way into the colonies already held by other countries, partly through illegal contraband, partly by means of commercial agreements. It was the latter that gave the English the right to send their ships into the Portuguese colonies in India and to export their cloth to

Portugal. With her more dangerous adversaries England waged war after bloody war. The end of the 16th century saw England emerge victorious from her war with Spain, whose navy, the indomitable Armada, was completely and utterly routed in 1588. England's main rival in the 17th century was Holland, who possessed the world's strongest merchant fleet and a flourishing commerce and industry. The 17th century for England was the century of its struggle against the Dutch while the 18th was taken up with its struggle against the French. Of the years extending from 1653 to 1797, England spent 66 of them engaged in naval wars. The outcome was that England emerged as the world's mightiest seafaring and colonial commercial power.

Thus the second half of the 16th century brought with it profound changes in England's domestic economy: raw materials (wool) began to lose their dominant position in English exports to finished products (cloth); the importance of foreign trade in the national economy grew immensely. England developed her own wealthy commercial bourgeoisie who, as buyers up, partially penetrated industry. The prosperity enjoyed by foreign trade was accompanied by the rise of shipping and industry, as cottage industry replaced the guild crafts. Compared to *commerce*, however, the role of *industrial capital* was still extremely modest: it had not yet outgrown the primitive form of the capital of the buyer up, and its penetration of production was primarily limited to those branches of production which either worked directly for *export* or were tied closely to the export trade. The aggrandizement of *bourgeois* moneyed interest at the expense of the *landowners* inevitably found itself reflected in state policy. The bourgeoisie increasingly tried to extend its influence over the state and use it to accelerate the transition from a feudal to a capitalist economy. The two English revolutions of the 17th century were themselves graphic expression of the bourgeoisie's aspirations. For its part the state had an interest in rapidly developing trade and industry as a means of enhancing its own power and enriching the treasury. And so the money balance system, that old, outmoded set of restrictive, essentially fiscal measures, gradually gave way to the state's intervention on a broad front, as it actively fostered the growth of *capitalist trade, shipping, and export industry* with the aim of consolidating England's position on the world market and doing away with her foreign competitors.

Fully-fledged mercantilism was above all a policy of *protectionism*, i.e., the use of customs policies to stimulate the growth of native industry. It was protectionism which was to speed up England's transformation from an agricultural to a commercial and industril nation. Customs duties now started to be used to further *economic* as well as fiscal ends. Previously, the government had, for fiscal reasons, levied duties indiscriminately upon every type of export item; now, however, the state began to differentiate between *raw materials* and *finished products*. To provide English industry with the cheap raw materials it required the government either raised their duties or forbade their export altogether. In the years when corn prices went up neither corn nor other agricultural products could be sent out of the country. On the other hand when it came to finished goods, the state encouraged their export by every possible means, exempting them from duties or even offering an export subsidy. The same type of discrimination—though in the reverse direction—applied to imports. The import of wool, cotton, linen, dyestuffs, leather, and other raw materials was not only freed of customs levies, but even subsidized, and otherwise encouraged. Conversely, the import of foreign finished products was either banned or subjected to high tariffs. Such a customs policy meant that *native industry was to be shielded to the detriment of agriculture, which produced raw materials*. It must be added that in England, where capitalism was quick to penetrate into agriculture and where part of the landowning class formed a bloc with the bourgeoisie, the government endeavoured to pursue policies favourable to farming. But in France, where agriculture was still feudal, the crown (especially under Colbert) often utilized mercantilist policy to win to its side the merchant and industrial bourgeoisie as allies in its struggle against the feudal aristocracy.

Insulated from foreign competition, English commercial and industrial capital was able to acquire *a monopoly hold* not merely over the home market, but over *the colonies*, as well. A law entitled the 'Navigation Act,' issued by Cromwell in 1651, prohibited the export of colonial products from Britain's colonies to any country other than England; in like manner, commodities could be delivered to the colonies only by English traders using either English or colonial ships. The same act established that all commodities imported into England had to be carried either by English ships or by ships belonging to the country where the commodities were produced. This latter provision was directed against the Dutch, whose shipping, at that time, serviced a large share of the world's transport and had earned that country the title of 'The Carriers of Europe'. The Navigation Act dealt a staggering blow to Dutch shipping and did much to stimulate the growth of England's *merchant navy*.

The policies of the late mercantilist period, geared as they were to expanding foreign trade and to promoting the development of shipping and the export-oriented industries upon which that trade depended, corresponded to a higher level of merchant capitalist development than did the policies of mercantilism's first phase. In contrast to early mercantilism, where exports were limited to a small number of '*staples*', developed mercantilism was *expansionist,* aiming at the maximum extension of foreign trade, the seizure of colonies, and hegemony within the world market. Early mercantilism exercized rigid control over each *individual commercial transaction*; late mercantilism restricted its regulation of trade and industry (both of which were growing rapidly) to a broader, *national scale*. Early mercantilism was concerned to regulate directly the movement of *precious metals* in and out of the country; late mercantilism sought to achieve this same end by regulating *the exchange of commodities* between the home country and other nations. The late mercantilists in no way relinquished the desire to attract the maximum volume of precious metals into the country: *the state* aspired first and foremost to improve the condition of government finances; the *merchant class* looked upon a greater mass of precious metals as a necessary condition for the stimulation of commercial turnover; and, finally, the *landlords* hoped that an abundance of money would raise prices on agricultural produce and lower the rate of interest payable on loans. All these different class interests helped nurture and sustain mercantilist belief in the need to attract money into the country. But the late mercantilists understood that the inward and outward flows of money from one country to another are the consequence of commodity exchange between them, and ready money comes into a nation when its commodity exports exceed its imports. And so they saw a positive *balance of trade*—guaranteed by the forced export of commodities and curtailment of their import—as the best means for achieving a favourable monetary balance. The entire protectionist system was directed at improving this balance of trade: it limited imports of foreign goods and, through its colonial policy and its ability to provide cheap raw materials and cheap labour, etc., it helped make native industry competitive on the world market. By way of distinguishing it from the 'money balance system' of early mercantilism, late mercantilist policy can be termed '*a balance of trade system*'.

Although this transition from early mercantilism to the system based on the balance of trade testifies to the rise of commercial and industrial capital, the latter were still not strong enough to give up the state's tutelage and do without its assistance. Mercantilist policy went hand in hand with state *regulation* of all aspects of national economic life. The state interfered in trade and industry with a barrage of measures designed to steer these in the desired direction (duties or prohibitions on imports and exports, subsidies, commercial treaties, navigation acts, etc.). It imposed *fixed prices paid to working hands and on articles of subsistence* and forbade the consumption of articles of luxury. It granted specific individuals or trading companies *monopoly right* over trade or industrial production. It offered *subsidies and tax concessions* to entrepeneurs and sought out experienced master craftsmen for them from abroad. Later on, at the end of the 18th century, this policy of comprehensively regulating economic life was to elicit violent opposition from the rising and newly-consolidated industrial bourgeoisie, but during the epoch of early capitalism, when it corresponded to the interests of the *commercial bourgeoisie*, it found complete and total support amongst the ideologues of that class—the *mercantilists*.

1 The 'factories' were walled, self sufficient trading settlements where foreign merchants would be housed and from which they would transact their business. Quite often all of the merchants coming from outside the town in question would reside within the same physical settlement. At the same time, however, they became the starting point for many of the new merchant associations that were coming into being at this time.

2 Much of this control fell to the Justices of the Peace, who had a wide range of powers to regulate commerce, wage rates, etc., on a country-wide, and not simply a guild-town basis.

3 The Staple policy was more than just a means of channelling and restricting trade; since it provided a monopoly over the local market for any merchant company to which it was granted, it became a right jealously sought after from the crown. Rubin's usage here obscures the actual origin of the institution, as a means by which particular towns would attempt to establish themselves as trading centres by becoming the principle 'place of contract' (as the Italians termed it) for trade in various commodities. Once accomplished, as in Bruges and Antwerp, which used their staple policy to build themselves into large market centres, attempt would be made to use this concentration of trade to foster local crafts and commerce.

CHAPTER THREE

The General Features of Mercantilist Literature

The age of early capitalism also saw the birth of *modern economic science*. Admittedly, among the thinkers of *antiquity and the middle ages* one can already find reflections on a range of economic questions. But the economic considerations of such ancient philosophers as Plato or Aristotle were themselves a reflection of the ancient slave economy, just as those of the medieval scholastics reflected the economy of feudalism. For both, the economic ideal was the self-sufficient, consumer economy, where exchange was confined to the surpluses produced by individual economies and was carried out *in natura*. For Aristotle, professional commerce conducted with the aim of earning a profit was a calling that went 'against nature'; to the medieval scholastics it was 'immoral'. Thomas Aquinas, the well-known Canonist writer of the 13th century, cited the words of Gratian about the sinfulness of trade: 'Whosoever buys a thing ... in order that he may gain by selling it again unchanged and as he bought it, that man is of the buyers and sellers who are cast forth from God's temple.'[1] Thus it was with great abhorrence that ancient and medieval thinkers looked upon usurer's capital, under the impact of which the break up of natural economy was to take on an even faster tempo. During the latter half of the middle ages a number of church decrees were issued which totally proscribed the levying of interest upon loans, and which threatened usurers with excommunication.

As capitalism developed these medieval attitudes towards economic activity became obsolete. The early ideal had been the self-sufficient *natural economy*; now the nascent bourgeoisie and the crown were seized by a passionate *thirst for money*. Formerly *professional* commerce had been considered a sin; now foreign trade was looked upon as the *main source of a nation's wealth*, and all measures were applied in the effort to expand it. In previous times the collection of interest had been banned; now the need to develop trade and the growth of the money economy meant that either ways were found to elude these proscriptions or they were done away with.

The new economic views, corresponding as they did to the interests of an infant capital and commercial bourgeoisie, found their proponents in the *mercantilists*. This appellation is used to designate a vast number of writers of the 16th to 18th centuries who lived in the different countries of Europe and dealt with economic themes. The volume of their writings was enormous, although many were of only topical interest and are no longer remembered. Nor can it be said that all mercantilists professed to 'mercantilist theory:' firstly because they were by no means in agreement on all issues, and secondly because nowhere in their works is there to be found a unified 'theory' that embraces all economic phenomena. The tenor of mercantilist literature was more *practical* than theoretical, being overwhelmingly devoted to those *specific questions of the day* that had been thrown up by the development of early capitalism and which urgently demanded a practical solution. The enclosure of common lands and the export of wool; the privileges of foreign traders and the monopolies granted to trading companies; the prohibitions on the export of precious metals and the limits placed upon interest rate levels; the standing of the English currency regarding that of other countries and fluctuations in its rate of exchange—these were all issues of vital practical concern to the English merchant bourgeoisie of the time and formed the main preoccupation of English mercantilist literature—the most advanced in Europe.

Like the topics themselves, the conclusions arrived at in mercantilist writings were primarily practical in their orientation. These were not armchair scholars, divorced from real life and dedicated to the discussion of abstract theoretical problems. Many amongst them took an active part in practical affairs, as merchants, as board members of trading companies (e.g., the East India Company), or as trade or customs officials. They approached the problems with which they were concerned not as theoreticians seeking to uncover the laws of economic phenomena, but as practical men who sought to *influence the course of economic life by enlisting the active assistance of the state*. Much mercantilist writing consisted of partisan pamphlets, urgently defending or refuting particular state measures from the standpoint of the interests of the merchant bourgeoisie. But to do this, to be able to justify a particular practical policy, they had to prove that what they advocated was in the interests of the economy, and hence they were compelled to trace the causal connection between different economic phenomena. And so in this gradual, halting fashion, there grew up, in the form of auxiliary tools to assist in the resolution of *issues related to economic policy*, the first frail shoots of a theoretical investigation into the phenomena of capitalist economy—the shoots of what was to become the contemporary science of political economy.

We noted earlier that mercantilist policy was the expression of the union between the crown and the developing merchant bourgeoisie, and that it depended upon the relative strengths of the two social forces involved in this temporary bloc as to whether mercantilism became bureaucratic or bourgeois-capitalist in character. In backward countries such as Germany, where the bourgeoisie was weak, it was the bureaucratic side which predominated; in the advanced countries, of which England was the most notable, its capitalist side won out. In correspondence with this state of affairs German mercantilist literature primarily bore the outlook of *bureaucratic officialdom*, while in England it reflected that of *commerce and trade*. To use the highly apt description given by one economist, German mercantilist works were in the main written by officials for officials; those in England, by traders for traders. In backward Germany, where the guild system hung on tenaciously, there was a splendid flowering of 'Cameralist' literature, dedicated mostly to questions of financial management and the administrative control over economic life. In England there grew up out of the discussions around questions of economic policy the precursors of those theoretical ideas that were later to be taken up and developed by the Classical school. It will always be the literature of the commercial-merchant school, which was the most advanced and characteristic body of mercantilist literature, that we have in mind. Receiving its clearest formulation in England,* it exerted the greatest influence upon the future evolution of economic thought.

The '*merchant*' character of mercantilist literature was manifested in its consistent defense of rising merchant capital, whose interests were identified with those of the state as a whole. The mercantilists strenuously emphasized that the growth of commerce was of benefit to all sections of the population. 'When trade flourishes *the income to the crown* is augmented, *lands and rents* are improved, navigation increases and the *poor people* find work. If trade declines, all these decline with it.'[2] This formula of Misselden's (from the first third of the 17th century) was intended to affirm that the interests of the commercial bourgeoisie coincided with those of the other social forces of the time: *the crown, the landlords, and the working class*. The

*Besides the works of the English mercantilists, *Italian* mercantilist literature of the 16th to 18th centuries is also of considerable interest, especially in its discussion of monetary circulation.

attitude taken by mercantilist literature to these different social groups reveals clearly how closely tied it was to the class interests of the merchant bourgeoisie.

Thus the mercantilists came out as advocates of a close alliance between the *commercial bourgeoisie and the crown*. The object of their concern was to increase 'the wealth of king and state,' and to foster the growth of 'trade, navigation, stocks of precious metals, and royal taxes'; they asserted that if the country had a favourable balance of trade this would make it possible for the royal treasury to accumulate greater sums of money. Along with this they insistently repeated that the crown could increase its revenue only where foreign trade grew—i.e. where there was a growth of bourgeois incomes. '... A King who desires to lay up much mony must endeavour by all good means to maintain and encrease his forraign trade, because it is the sole way not only to lead him to his own ends, but also to enrich his Subjects to his farther benefit' (Thomas Mun, writing in the first third of the 17th century). The money accumulated by the state treasury must not exceed that level which corresponds to the volume of foreign trade and the nation's income. Otherwise, 'all the mony in such a State, would suddenly be drawn into the Princes treasure, whereby the life of lands and arts must fail and fall to the ruin both of the publick and private wealth.' An economic collapse deprives the crown of the ability to pursue the profitable undertaking of 'fleecing its subjects.'[3] Thus the crown itself has every interest in actively employing measures to assist the growth of commerce, even where this acts to the temporary detriment of its fiscal interests, for instance when it is a case of lowering customs duties. 'It is needful also not to charge the native commodities with too great customes, lest by indearing them to the strangers use, it hinder their vent' [4] (Mun).

While the mercantilists wanted to make the crown an active ally of the merchant bourgeoisie, they could entertain no such hopes towards the landowners. They knew that the measures they were advocating often provoked the disatisfaction of the landlords; they nevertheless endeavoured to allay this discontent by pointing out that the growth of trade brings with it a rise in the prices of agricultural produce and thus also in rents and the price of land. 'For when the Merchant hath a good dispatch beyond the Seas for his Cloth and other wares, he doth presently return to buy up the greater quantity, which raiseth the price of our Woolls and other commodities, and consequently doth improve the Landlords Rents... And also by this means money being gained, and brought more abundantly into the

Kingdom, it doth enable many men to buy Lands, which will make them the dearer' (Mun).[5] With arguments such as these the plenipotentiaries of the young bourgeoisie attempted to interest *the landlord class* in the successes of commerce; this did not, however, mean that they turned a blind eye to the conflict of interests that lay between them. The mercantilists had already given the landowners advance warning that the interests of trade and export industries would have to be placed before those of agriculture and the production of raw materials. 'And forasmuch as the people which live by the Arts are far more in number than they who are masters of the fruits, we ought the more carefully to maintain those endeavours of the multitude, in whom doth consist the greatest strength and riches both of King and Kingdom: for where the people are many, and the arts good, there the traffique must be great, and the Countrey rich' [6] (Mun). In later mercantilist literature one can find a sharp polemic between representatives of the financial bourgeoisie and those of the landowners over the level of interest that should be charged on loans.*

There was one question, however, over which the interests of both classes still coincided and showed no sign of divergence: the exploitation of the *working class*. The throngs of landless peasants and ruined craftsmen, the declassed vagabonds and homeless beggars thrown up by the break up of the rural economy and guild handicrafts were a welcome object of exploitation for both industry and agriculture. *The legal limit placed upon wages* on the whole won the lively approval of landlord and bourgeois alike. The mercantilists never ceased to moan about the 'indolence' of the workers, or about their lack of discipline and slow adaptation to the routine of industrial labour. If bread is cheap, the worker works only two days a week, or however long it takes to assure the necessities of life, and the rest of the time is free for carousing and drunkenness. To get him to labour on a constant basis and without interruption he must, over and above state compulsion, be prompted by the biting lash of famine and necessity—in short, by the compulsion of the high price of corn. At the start of the 19th century the English bourgeoisie battled with the landlords to drive down the price of corn and, in so doing, the price of labour power as well. But in the 17th century many English mercantilists found themselves in complete agreement with the landowners in advocating high corn prices as a means of compelling the workers to toil. They even advanced the paradoxical claim that *dear corn makes*

* See Chapter Six, below.

labour cheap, and vice versa, since dear corn would cause the worker to apply himself with greater exertion.

According to Petty, writing in the second half of the 17th century: 'It is observed by Clothiers, and others who employ great numbers of poor people, that when Corn is extremely plentiful, the labour of the poor is proportionably dear: And scarce to be had at all (so licentious are they who labour only to eat, or rather to drink).'[7] It follows from this that 'the Law that appoints such Wages...should allow the Labourer but just wherewithall to live; for if you allow double, then he works but half so much as he could have done, and otherwise would; which is a loss to the Publick of the fruit of so much labour.'[8] For Petty there is nothing unjust about 'limiting the wages of the poor, so as they can lay up nothing against the time of their impotency and want of work'.[9] The public, in Petty's view, must undertake to provide for those unfit for work; as for the unemployed, they should be set to work in the mines, on the construction of roads and buildings, etc., a policy to be recommended because it will 'keep their mindes to discipline and obedience, and their bodies to a patience of more profitable labour when need shall require it'.[10] In their advocacy of the interests of youthful capitalism and their concern for the conquest of foreign markets for English traders and exporters, the mercantilists were naturally preoccupied with the mobilization of an adequate core of disciplined and inexpensive working hands. The mercantilists advocated something akin to *the iron law of wages*—albeit only in embryonic form. However, consistent with the general nature of their doctrine, this law does not as yet appear as a theoretical proposition, but as a practical prescription: the mercantilist view was that a worker's wages *must not* exceed the minimum necessary means of subsistence.

The commercial-merchant standpoint of English mercantilist literature, which so clearly emerges in its attitude towards the different social classes, also left its imprint upon the body of problems—and their solutions—with which it was concerned. The view is often expressed that mercantilist doctrine is reducible to the declaration that the precious metals are the sole form of wealth. Adam Smith sharply criticizes 'the absurd notion of the mercantilists that wealth consists of money'. And yet such a characterization is quite unjust. They looked upon increases in the quantity of precious metals not as a source of a nation's wealth, but as one of the signs that this wealth was growing. It is only the early mercantilists whose intellectual horizons remained naively confined to the sphere of monetary circulation. The late mercantilists, in putting forth the doctrine of the 'balance of trade,' uncovered the connection between the movement of the precious metals and the overall development of trade and industry. There was still much about their analysis of the interconnection between different economic phenomena that was superficial, but it was nevertheless free of the naive notions of their forerunners and opened the way for future scientific development. We must now move on to describe the content and evolution of the mercantilists' views.

1 Cited in R.H. Tawney, *Religion and the Rise of Capitalism* (London, John Murray, 1964), pp. 34-35.
2 Translated from the Russian.
3 Thomas Mun, *England's Treasure by Forraign Trade*, in *Early English Tracts on Commerce*, edited by J.R. McCulloch, originally published by the Political Economy Club, London, 1856, reprinted for the Economic History Society by Cambridge University Press, 1954, pp. 188-89.
4 Mun, *England's Treasure*, McCulloch edition, p. 133.
5 *Ibid*, p. 142.
6 *Ibid*, p. 133.
7 Sir William Petty, *Political Arithmetick*, in *The Economic Writings of Sir William Petty*, two volumes, edited by Charles Henry Hull, reprinted by Augustus M Kelly (New York, 1963), Vol. 1, p. 274.
8 Petty, *A Treatise of Taxes and Contributions*, in *Economic Writings* (Hull edition), p. 87.
9 *Ibid*, p. 20.
10 *Ibid*, p. 31.

CHAPTER FOUR

The Early English Mercantilists

The attention of the first English mercantilists of the 16th and early 17th centuries was drawn to the *circulation of money*. Here decisive changes were taking place that were working much to the disadvantage of broad layers of the population, the merchant class in particular. To begin with, the influx of American gold and silver into Europe had brought with it, as a matter of course, a *revolution in prices*: as commodity prices went up so a general wail of disatisfaction and complaint arose over the inadequacy of the supply of money. Secondly, as England was relatively more backward than Holland, the *rate of exchange* used in the trade between the two countries often worked *to the detriment of England*, so that one unit of Dutch currency would exchange for a greater sum of English shillings. It thus became profitable to send English gold and silver coins of standard value into Holland to be reminted. An export of ready money *out of England* was observed, and with it spread the conviction that this was the primary factor behind the universal complaint of a money shortage.

To the early mercantilists the interrelation between the circulation of money and the circulation of commodities was still unknown: they had yet to comprehend that the deterioration of England's rate of exchange and the ensuing flight of ready money out of the country was the inevitable result of an unfavourable balance of trade. When these men debated a topical issue they did so as practicioners with little disposition towards seeking the ultimate causes behind it; and so it was most often to the realm of monetary circulation itself that they looked to find the reasons behind the outflow of money namely, to the debasement of the coinage. At the beginning of the 16th century this was a common practice for monarchs all over Europe, with the English crown one of the worst offenders. The Crown would issue new coins having the same face value as the old coins, but containing a smaller amount of metal. But since these new coins, though lighter in weight than the old coins of standard value, were legally pegged at their value, it became profitable to send the old money out of the country, either to be reminted or exchanged for foreign coin. The fact that bad money drives out good from domestic circulation and forces it abroad was noted by *Thomas Gresham*, one of the early mercantilists, in the middle of the 16th century, and has since become known as 'Gresham's Law'. It was this debasement of English coins that the early mercantilists were prone to see as the reason for the English currency's depreciation against the Dutch (as indicated by the shilling's deteriorating rate of exchange), and for the fact that precious metals were being exported. To do away with the evils of the debasement of the coinage, the worsening rate of exchange, and the steady seepage of money from the country the mercantilists advocated that the state use coercion and intervene directly into the sphere of monetary circulation. They demanded that the government issue coins of *standard weight*, and recommended that the rate of exchange be *compulsorily regulated* (in other words, private individuals would be prohibited from buying foreign coins at more than the fixed number of English shillings). But what they clamoured for with still greater insistence was *a ban on the export of money* from England and the adoption of strict measures to stop the drain of precious metals. Their advice had no effect. The state had neither the ability nor the inclination to issue coins of standard weight. As for the rest of the mercantilists' recommendations, these were merely an attempt to reinforce or revitalize traditional government practices that had already become outdated. The state had previously imposed a rigid prohibition upon exporting money from England. Similarly, it had endeavoured to peg the rate of exchange and to regulate it through 'royal money changers' who would exchange foreign for English currency at a fixed rate. But these efforts were powerless in the face of the elemental laws of commodity and monetary circulation, laws which still lay beyond the perceptions of the early mercantilists.

One of the outstanding relics of mercantilist ideas from this early period is a work entitled *A Compendious or Briefe Examination of Certayne Ordinary Complaints, of Divers of Our Country Men in These Our Days*, which appeared in 1581 under the initials, 'W.S.' At one time the author was thought to be none other than William Shakespeare, but general opinion came to ascribe it to William Stafford. More recent studies have established that the book, though published in 1581, was actually written in 1549 by *John Hales*, and in our discussion we will designate it as the work of Hales (Stafford).[1]

The work is written in the form of a conversation between the representatives of different classes of the population: a knight (or landowner), a farmer (or husbandman), a merchant, a craftsman, and a theologian. It is obvious that the latter expresses the views of the author in his attempts to reconcile the interests of these social classes. All those engaged in the debate bemoan the high level of prices, with each attempting to pass the blame off onto the representative of another class. From the knight we hear that merchants have raised prices on commodities so high that the landowners are left with the stark choice between abandoning their holdings or shifting from cultivating the soil to the more profitable line of raising sheep. The farmer-husbandman complains about the enclosure of grazing lands and the higher rent that he must pay to the landlords. The merchant and the craftsman are agrieved by the upward trend in workers' wages and by the fall off in trade.

The theologian, seeking to bridge the interests of the various parties, lays before them the general causes of the growing impoverishment of the realm: the debasement and *deterioration of the English coinage* and the *export of ready money* consequent upon it. Old English coins of standard value are rapidly moving out of the country: 'every thing will go where it is most esteemed; and therefore our treasure thus goeth over in ships.' What is more, this deterioration in the worth of the coinage has caused imported commodities to become vastly more expensive—their prices having risen by a full third. Foreign merchants assert that they are selling their commodities at no greater profit than before, but that they are compelled to raise their prices by virtue of the erosion in the value of English coin. Our coin, as is known, is priced not 'by its name but esteemed both [by the]value and quantity of the stuff it was made of.'[3] On the other hand, the prices on commodities which foreigners purchase within England have risen to a lesser extent. We sell our own produce, in the main our raw materials, cheaply, and foreigners work them up into industrial wares which are sold back to us dear. Thus, from English wool, foreigners fashion cloth, coats, shawls, and the like; from English leather they make belts and gloves, and from English tin, spoons and dishes, all of which articles are then imported back into England. 'What grossness be we of, that see it and suffer such a continual spoil[age] to be made of our goods and treasure ... '. 'They do make us pay at the end for our stuff again; for the stranger custom,[4] for the workmanship, and colours, and lastly for the second custom in the return of the wares into the realm again; whereas with working the same within our Realm, our own men should be set on work at the charges of the strangers;[5] the custom should be borne all by strangers to the kinge; and the clear gains to remain within the Realm.'[6]

Thus where foreign trade is characterized by the export of raw materials and the import of expensive finished products, it will become *a pump for wringing money out of the country*. This applies for the most part to the import trade. In dealing with industrialists and merchants one has to differentiate between three types: vintners, milliners, and traders in imported commodities (e.g., those from the colonies), who send money out of the country; a second group, comprising butchers, tailors, bakers, and other such entrepreneurs, who both receive and spend their money inside the country; and finally, a third group, who work up wool into cloth and process leather. As this third category works for the export market and attracts money into the country, it warrants the patronage and encouragement of the Crown. It is necessary to encourage the domestic processing of English raw materials, to which end it is advisable either to prohibit or inhibit the export of unprocessed raw materials, and to put a ban on the import of finished products manufactured abroad. It is more profitable to purchase our own manufactures, even though they be dearer in price, than to buy foreign ones. Hales (Stafford) gives the following example to illustrate his view that native industry demands protective tariffs to gain an implantation: One day I asked a book-binder 'why we had no white and brown paper made within the Realm, as well as they had made beyond the Sea. Then he answered me that there was paper made a while [back] within the Realm. At the last, said he, the man perceived that made it that he could not [af]ford his paper as good cheap as that came from beyond the seas, and so he was forced to lay down [the] making of paper. And no blame to the man; for men would give never the more for his paper because it was made here; but I would have either the paper stayed from coming in, or else so burdened with custom that, by the time it came hither, our men might afford their paper better cheap than strangers might do theirs ... '.[7]

Hales (Stafford) is a typical representative of that early mercantilism which grew out of the backward economic conditions of 16th-century England. Through the pages of his book we gain a glimpse of a relatively underdeveloped country, which *exports primarily raw materials and imports finished manufactures*, and suffers under the weight of the foreign merchant. More than anything else it is monetary phenomena that attract his attention: for him the source of all evil is the *debasement of coin and the export of money*. As he sees it, England grows poor because foreigners ship its money out of the country, while other nations grow rich from its influx. The reason for this outflow of money is the unfavourable rate of exchange of England's currency; thus to halt it requires first, that coins be issued of *standard value* (so as to give stability to the exchange rate), and second, that there be a *reduction in the import of foreign-made finished goods*. And so Hales (Stafford) advocates that appreciation of the coinage be accompanied by measures leading to an improvement in the balance of trade: 'We must at all times endeavour to purchase from strangers no more than what we sell to them; for otherwise we make ourselves poor and them rich'. But unlike the later mercantilists, who were to discover that fluctuations in the rate of exchange have a regular and law-determined dependence upon a positive or negative balance of trade, Hales's (Stafford's) 'Monetarist' ideas led him to reverse the conceptual connection between these two phenomena: in his understanding, *debasement of the coinage* produces *a deterioration in England's rate of exchange*; from this ensues a *general rise in the prices of foreign commodities* which then aggravates England's negative balance of trade. What separates him further from the later mercantilists is that he looks not so much to a stimulation of English exports to improve the balance of trade (he even demands their curtailment, insofar as they consist of raw materials), but rather to a contraction in the number of foreign goods brought into the country. Such a conception corresponded well to a period when English capitalism was undeveloped and in transition, and when the English bourgeoisie, already demanding a cutback in the export of raw materials, could not yet hope to find ample markets abroad for the products of its own industry. This was an age of *defensive*, rather than *aggressive* protectionism: Hales (Stafford), for whom the dream of the mititant acquisition of foreign markets for English manufactures did not yet exist, had as his ideal a native industry that would gain sufficient implantation to work up the raw materials of its own country and push the wares of foreign industry out of the English market.

One can find these early mercantilist ideas even amongst such figures as *Misselden, Malynes, and Mills,* all of whom wrote during the first part of the 17th century. Lacking any understanding of the dependence of the rate of exchange upon the balance of trade, they hoped to improve the former through direct measures of state compulsion. Misselden counseled the government to fix the rate of exchange by relying on treaties with other states. According to Malynes the rate could be bolstered and the export of money halted by resurrecting the rigid restrictions of early mercantilism—for example, the office of 'royal money changer' and its right to compulsorily fix the rate paid for foreign coins (i.e., the rate of exchange), or the prohibition against traders paying foreigners in gold. Mills even went so far as to protest against the abolition of the ancient monopoly of the 'staples'. And while Misselden acknowledged these types of restrictions as outmoded, he levelled his own categorical objection *to any and all exportation of money abroad*. This was the one point on which all the early mercantilists concurred, and it exemplified the distinction between themselves and the mercantilists of the later period.

1 The work actually attributed to Hales bears the title *A Discourse of the Common Weal of the Realm of England*, reprinted by Cambridge University Press, 1893. The book attributed to Stafford, published in 1581 under the title listed in Rubin's text (the same edition to which Marx refers in Volume One of *Capital*), differs slightly from the original, and it is now assumed that 'W.S.,' be it Stafford or not, was the editor. All quotations are from the Cambridge edition of *A Discourse of the Common Weal*. Unlike the later mercantilist works we have altered the spelling in the text to conform to modern usage, for the text would have been quite incomprehensible otherwise. We have not, however, 'modernized' the language. Where insertions or changes have been necessary they are placed in brackets [].

2 Hales, *A Discourse of the Common Weal*, p. 79.

3 *Ibid*, p. 102.

4 I.e., the customs levied by foreign countries upon the importation of English raw materials.

5 I.e., at foreigners' expense.

6 Hales, *A Discourse of the Common Weal*, pp. 64-65.

7 *Ibid*, pp. 65-66.

CHAPTER FIVE

Mercantilist Doctrine At Its Height

THOMAS MUN

As English commerce and industry developed, the cumbersome restrictions of the early mercantilist period proved, as we have already seen,* more and more archaic, and were either done away with or allowed to maintain a mere formal existence, devoid of any of their former practical content. As soon as English merchants went in active quest of new foreign markets for their commodities the 'staples' were abolished. On the other hand, once the English traders had managed to displace those of the Hanse and Italy, they established their own direct connections with the East, where they purchased the produce of the colonies. But for this they had to *send ready money out of England*; the old laws placing absolute proscription upon such activities fell into disuse (although they were to remain officially in force until 1663). Where this especially applied was with the English *East India Company*, which had built up a vast trade with India. The company carried out of India spices, indigo, textiles, and silks, some of which remained within English borders, but much of which was subsequently resold—at great profit—to other European countries. This carrying trade, whereby England acted as the middleman for foreign produce, was exceptionally lucrative and necessitated the export from England of large quantities of ready money. The total mass of imports into England from India was greater than the exports to India from England; the difference had to be made up through the export of hard cash from England. Without it there was no way that the East India Company could have sustained its commercial activities. Naturally, the Company was subjected, in its turn, to furious attack by defenders of the old restrictive regime. Even at the close of the 17th century the view was expressed that 'the East India trade will be the ruin of a large part of our industry unless something is done to prevent it'; in the early part of the century this conviction was nearly

*See Chapter Two, above.

universal. It was inevitable that if the partisans of the East India trade were to marshall their arguments against the blanket prohibition on exporting money they would have to develop a critique of the antiquated views of the early mercantilists. To counter the old '*monetary system*' they came up with a new theory of the '*balance of trade.*' The new views received their most brilliant expression in a book by a member of the East India Company's board of directors, *Thomas Mun* (1571-1641), entitled *England's Treasure by Forraign Trade.* Mun's work, which though written in 1630 was not published until 1664 after the author's death, typifies mercantilist literature more than any other, and was to become, in Engels's words, 'the mercantilist gospel.'[1]

Mun did not contest the previous doctrine about the benefits accruing to a nation from the acquisition of precious metals or, as he called it, the multiplication of its 'treasures.' What he argues, however, is that such 'treasures' cannot be multiplied by *the state taking coercive measures to regulate monetary circulation directly* (prohibition on the export of money, a fixed exchange rate, changes in the metallic content of coins, etc.). Whether or not there is an influx or outflow of precious metals depends exclusively upon whether the balance of trade is positive or negative. 'The ordinary means therefore to increase our wealth and treasure is by *Forraign Trade*, wherein wee must ever observe this rule; to sell more to strangers yearly than wee consume of theirs in value. For suppose that when this Kingdom is plentifully served with the Cloth, Lead, Tinn, Iron, Fish and other native commodities, we doe yearly export the overplus to forraign Countreys to the value of twenty two hundred thousand pounds; by which means we are enabled beyond the Seas to buy and bring in forraign wares for our use and Consumption, to the value of twenty hundred thousand pounds: By this order duly kept in our trading, we may rest assured that the Kingdom shall be enriched yearly two hundred thousand pounds, which must be brought to us in so much Treasure.'[2]

In other words, *money will flow into the country* as the result of a *positive balance of trade*. It follows from this premise that if money is to be drawn into the country it will not be because of the cumbersome regulations of early mercantilism, but the result of a comprehensive economic policy directed towards the promotion of exports, shipping and export-oriented industries, as a means of improving the balance of trade. Clearly, the balance of trade can be bolstered either by cutting back on the import of commodities or by expanding their export. Here again, we note the fundamental difference between Mun and his predecessors. The mercantilists of the early period called for a ban on the export of money and a reduction in the import of foreign commodities; Mun, on the other hand, pins his hopes on the development of *England's commodity exports*. This difference in their respective points of view was itself a reflection of England's gradual transition from a nation which imports foreign manufactures to one which exports its own. Mun, then, appears on the scene as the representative of a rising merchant capital which is in the process of acquiring new markets and is aspiring to expand its exports. Whereas the concern of Hales (Stafford) was to guard the domestic market from the flood of foreign wares, Mun's thoughts now centre on the conquest of foreign markets for England. It is, of course, true that Mun has nothing against reducing the importation of foreign commodities; but he does object to the previous methods by which this was achieved, namely direct prohibition. Measures such as these will only provoke other countries to do likewise, much to the detriment of English exports: and it is after all the expansion of these that for Mun is the primary objective.

Mun urgently demands that export trade, shipping, and export industries be encouraged and expanded. England must draw benefit not simply from its *'natural'* produce, i.e. its raw materials surpluses, but also from *'artificial'* produce, that is, industrial articles of its own production and the commodities imported from other countries (e.g. India). To this end there must be incentive, first to have *raw materials worked up by domestic industry* and exported as finished products, and second to develop the *carrying trade*, whereby the produce of nations such as India is imported in order to be resold to other countries at a higher price. This 'reworking' of raw materials and 'resale' of foreign commodities are extolled by Mun as the main sources of a nation's enrichment:

'We know that our own natural wares doe not yield us so much profit as our industry,' since the value of the cannon and rifles, nails and ploughs is so much greater than that of the iron from which they are wrought, and just as the price of cloth is higher than that of wool. In consequence, 'we shall find these Arts more profitable than the natural wealth', [3] and it is essential that they be strongly encouraged. What is needed is to win markets for the export of our industrial wares, but this is possible only if we can cheapen their price. 'We may ... gain so much of the manufacture as we can, and also endeavour to sell them dear, so far forth as the high price cause not a less vent in the quantity. But the superfluity of our commodities which strangers use, and may also have the same from other Nations, or may abate their vent by the use of some such like wares from other places, and with little inconvenience; we must in this case strive to sell as cheap as possible we can, rather than to lose the utterance of such wares.'[4] We have found from experience that by selling our cloth cheaply in Turkey we have been able to greatly increase its sale at the expense of the Venetians. On the other hand, a few years back, when the excessive price of our wool caused our cloth to become very dear, we momentarily lost half our foreign sales. A cheapening of our cloth by 25% can raise our sales by more than half, and though the individual merchant incurs a loss because of lower prices, this is more than compensated for by the gain to the nation as a whole.[5] The arguments here advanced by Mun as to the *benefits derived from low prices* demonstrate the extent to which the English economy had been transformed from the mid-16th to mid-17th centuries. The complaint of the early mercantilists had been that the selling price of English cloth was too low; some amongst them had advocated steps to see that prices on exported commodities were raised. By the time we come to Mun the situation had changed: the export of raw materials had given way to that of finished industrial products, and England was now faced with the task of expanding its export potential and displacing its numerous competitors. Wherever it was not possible to get a monopoly hold on the market foreign competitors would have to be crushed through recourse to low prices.*

The working up of raw materials and the export of domestic manufactures could not be the only source of profit for the country: there must also be *resale of foreign produce*. Here Mun's primary concern is to defend the carrying trade—especially that with the East Indies—against the attacks of its opponents. The import of foreign commodities and their subsequent export and resale to other countries, argues Mun, brings wealth to both the Kingdom at large and the royal treasury. Especially lucrative is the transport of merchandise from such far away places as the East Indies. These colonial commodities

*The need to lower prices in order to compete successfully on foreign markets was advanced by the late-17th century mercantilists. *Child* wrote: 'If it were a question of trade alone we could, as the proverb states, command any market we pleased. But in the conditions that we currently find ourselves in, where each nation endeavours to seize the greatest possible share of trade, it is another proverb that holds good: whoever wishes to profit too much loses everything.' [Translated from the Russian—D.F.] *D'Avenant*, too, states that only with a low price of labour and manufactured commodities can one maintain a competitive position on foreign markets. All of these arguments clearly and unmistakably express the standpoint of the merchant-exporter.

can be acquired for a pittance: a pound of pepper, for instance, acquired for three pence will fetch twenty-four pence on the markets of Europe. Not all of this twenty-one pence margin goes to the merchant, of course, since the outlays on long-distance sailing are enormous, what with the costs of conveyance, hiring and maintaining seamen, insurance, customs levies, taxes, and the like. But when the transport is done on English vessels these sums are totally spent on English shores, thus enriching that country at the expense of others.[6] 'We make a far greater stock by gain upon these *Indian* Commodities, than those Nations doe where they grow, and to whom they properly appertain, being the natural wealth of their Countries.'[7] In this instance the development of commerce will bring greater profit to a country than will its 'natural' riches if the latter have not been fructified by trade and industry.

What aroused objection to the East India trade was the fact that, as we have already seen, it necessitated the exportation of money as payment for the commodities being imported from India. And so Mun addresses himself in detail to the pros and cons surrounding the export of money to India: We noted earlier that the overall excess of England's exports over its imports came to 200,000 pounds sterling, and that this sum entered the country as ready money. The question arose, what to do with it? Those in favour of putting a blanket ban on exporting money counselled that it should remain there in England, a position that Mun vigourously opposes: 'If... having gained some store of mony by trade with resolution to keep it still in the Realm; shall this cause other Nations to spend more of our commodities than formerly they have done, whereby we might say that our trade is Quickned and Enlarged? no verily, it will produce no such good effect: but rather according to the alteration of times by their true causes wee may expect the contrary.'[8] In such a case the money will lay within the country as lifeless treasure, and would only prove a source of gain if it were again put back into commercial circulation. Suppose, for instance, that out of this sum, £100,000 is exported to the East Indies, and that the commodities which are purchased with it are then resold in other countries at a far greater price (say, for £300,000). Evidently, a sizable profit accrues to the nation as a result of this operation. And though it is true that the number of commodities being imported is increased, this is solely to produce an even greater rise in exports later on. Opponents of the East India trade object that while it is money that we issue out only commodities are received back in return. But if these commodities are not for our own consumption but for future resale, the entire difference between their purchase price and sale price must necessarily accrue 'either in mony or in such wares as we must export again.' 'They that have Wares cannot want mony,' for by selling them one earns a profit.[9] Each quantity of money that we export to India comes back to us augmented by a profit. 'And thus we see that the current of Merchandize which carries away their Treasure, becomes a flowing stream to fill them again in a greater measure with mony.'[10] *There is enormous profit to be had by a country when it exports its money* to meet the needs of the carrying trade. 'For if we only behold the actions of the husbandman in the seed-time when he casteth away much good corn into the ground, we will rather accompt him a mad man than husbandman: but when we consider his labours in the harvest which is the end of his endeavours, we find the worth and plentiful encrease of his actions.'[11]

Mun's book brilliantly exemplifies *mercantilist literature* at its height. Mun writes as a man of action: the problems which he confronts are practical ones, as are the solutions which he advances. Seeking to marshall arguments against the old restrictions and the direct regulation of monetary circulation, Mun arrived at a theory of *the dependence of monetary movements and the rate of exchange upon the balance of trade.* He did not object to the importance of bringing money into the country, but held the view that the only means by which this could be profitably accomplished was if the development of foreign trade, shipping and export industries brought about an improvement in the balance of trade. Here, then, was a fusion of the early mercantilist view, to the effect that money comprises the principal component of a nation's wealth, with the later mercantilist notion that foreign trade is that wealth's primary source. The discussions within mercantilist literature centre mainly around these two basic themes: first, the importance of *money* and of the means by which a nation can acquire it, and second, *foreign commerce and the balance of trade*.

The conceptual errors for which the mercantilists were upbraided by the free traders, beginning with the Physiocrats and Adam Smith, were first, that the true wealth of a nation resides in its *produce*, and not its *money*, and second, that its real source is *production*, and not *foreign trade*. But such a critique fails to grasp that, for all their theoretical naivety, the formulae advanced by the mercantilists represented an attempt to resolve the basic problems of their age and of their social class namely, on the one hand, those of the transformation of a natural economy into a money economy, and on the other, those of acquiring primary accumulations of capital in the hands of the merchant bourgeoisie. As spokesmen for this class, their concern lay first and foremost in drawing a substantial slice of the economy into the orbit of monetary exchange. Their concern for augmenting the nation's wealth centred not on whether there was a growth in production for use, or use values, but rather on whether there was an increase in the number of products capable of being sold or converted into money, in short, in the growth of *exchange value*. The mercantilists understood perfectly well, of course, that people subsist off bread and meat, and not off gold. But in an economy where the development of monetary circulation was weak and the bulk of its bread and meat was still produced for direct consumption, rather than for realization on the market, exchange value, in the perceptions of the mercantilists, lay not in the products themselves, but in *money*. Since not all products of labour constitute exchange values, i.e. commodities transformable into money, *exchange value* became naturally confused with the physical form of that product which functions as *money*, i.e., gold and silver. Though such a confusion was theoretically naive, this furious chase after precious metals so characteristic of the early mercantilists was itself a reflection of the painful transition from a natural to a commodity-money economy. The influx of precious metals was to serve as a tool for speeding up this process in the interests of the commercial bourgeoisie. Since foreign trade was at that time both the arena within which the circulation of money was most extensively developed, and the sole means by which countries deprived of their own gold and silver mines could draw in precious metals, it followed that the intensified drive to acquire these metals would be combined (as in the balance-of-trade doctrine) with a policy of promoting foreign trade and forcibly developing exports.

The disproportionate value accorded to foreign trade by the mercantilists is to be explained not simply by its great potential for transforming products into money and attracting precious metals: *the enormous profits derived from foreign trade* helped foster *primitive capital accumulation* by the merchant class. It was not to the growth of money economy in general that the commercial bourgeoisie aspired, but of a money-capitalist economy. The process of transforming products into money was to be accompanied by the accumulation of the latter and its own conversion into profit-bearing money, that is, into capital. But for the most part, really large profits were only to be had in this period through foreign commerce, in particular through trade with the colonies. By buying commodities cheaply on some markets (where, as with the colonies, the merchants and trading companies of a particular government often enjoyed a monopoly) and selling them dear on others, wealth and capital could be rapidly accumulated—not to mention the direct plundering of the colonies and the forcible appropriation of the produce of their inhabitants. In an age when the merchant occupied a near monopoly position between producers (e.g., colonial subjects or craft workers) and consumers (e.g., landlords or peasants), even 'peaceful' foreign trade afforded him the chance to exploit both to his own benefit. Merchants grew rich by purchasing producers' commodities below their value and selling them to consumers at prices where their value was exceeded. In this period the basic source of commercial profit was *non-equivalent exchange*. It was, then, natural that the mercantilists saw profit only in the net profit of trade, or '*profit upon alienation*,' which had its source in the mark up that the merchant added to the price of the commodity.

It stands to reason that when the origin of profit is non-equivalent exchange the advantages falling to one party in the exchange are equal to the losses incurred by the other—one person's gain is the other's loss. *Internal trade* of this kind leads merely to a redistribution of wealth amongst a country's individual inhabitants, but does nothing to *enrich the country as a whole*. This can come only from *foreign commerce*, where one nation is enriched at the expense of another. '... By what is Consum'd at Home, one loseth only what another gets, and the Nation in General is not at all the Richer; but all Foreign Consumption is a clear and certain Profit.'[12] With these words *D'Avenant*, writing at the end of the 16th century summarized the general mercantilist belief that foreign commerce and those sections of industry working for the foreign market yield the greatest profit. 'There is much more to be gained by Manufacture than Husbandry, and by Merchandize than Manufacture.'[13] '... A Seaman is in effect three Husbandmen.'[14] One should not conclude from this that Petty (who wrote these words) had forgotten the importance of agriculture as the source of a country's foodstuffs. Petty simply meant that with capitalism totally absent in agriculture and having only weakly penetrated into industry, the sphere within which the capitalist economy would enjoy extensive development and allow for the vigorous accumulation of capital would be commerce, particularly foreign trade.

As we have seen, the exaggerated importance which the mercantilists attached to *money* had its roots in the conditions of *transition from a natural to a commodity-money economy*; similarly, the overemphasis which they placed upon *foreign trade* was the logical result of the role of the latter as a source of *immense profits* and a sphere of activity that allowed for the rapid *accumulation of capitals*. And though both these mercantilist ideas were later to be cruelly ridiculed as absurd, they nonetheless reflected the historical conditions of the merchant-capitalist age and the real interests of those social classes for whom the mercantilists acted as spokesmen. As the mercantilists' overwhelming concern was with questions of economic policy, and as economic theory was only in its infancy, they remained content with ill-developed and naive theoretical formulas, provided that these answered the practical demands of their time. Our legacy from the mercantilists is not a comprehensive economic theory embracing the totality of capitalist economic phenomena, but a body of work containing only rudimentary theoretical conceptions whose development and substantiation were left to later economists. Thus the separate strands of mercantilist doctrine—the one dealing with exchange value and money, the other with profit and foreign trade—suffered different fates. As commercial conditions altered and industrial capitalism developed, the falaciousness of the theory of foreign trade as the sole source of profit became obvious. Further evolution of economic thought at the hands of the Physiocrats and the Classical school was to dismiss the mercantilist interpretation of *foreign trade and profit*. The embryonic theories of *exchange value and money* in mercantilist literature proved, on the contrary, capable of additional theoretical development: appropriated by subsequent schools of economists, and freed from the naive confusion of exchange value with money and of money with gold and silver, these embryonic theories were worked upon and advanced. Their profound interest in the problem of trade and the process whereby commodities are exchanged for money permitted the mercantilists to put forward a substantial number of correct ideas on the nature of exchange value and its monetary form. There is, in particular, within mercantilist literature, the beginnings of a *labour theory of value*, which was to play a part of great significance in the subsequent evolution of our science.

1 Engels, *Anti-Dühring*, (Progress Publishers edition, Moscow, 1969). The phrase is actually Marx's, and not that of Engels, as it was Marx who wrote the chapter on the historical development of political economy from which it comes (Part Two, Chapter X, 'From the *Critical History*'). On Mun's book Marx had this to say: 'The particular significance of this book was that, even in its first edition [*A Discourse of Trade from*

England unto the East Indies, 1609; the 1621 edition is reprinted in McCulloch, *op cit*, pp. 1-47—D.F.], it was directed against the original *monetary system* which was then still defended in England as being the policy of the state; hence it represented the conscious *self-separation* of the mercantile system from the system which gave it birth. Even in the form in which it first appeared the book had several editions and exercised a direct influence on legislation. In the edition of 1664 (*England's Treasure, etc.*), which had been completely rewritten by the author and was published after his death, it continued to be the mercantilist gospel for another hundred years. If mercantilism therefore has an epoch-making work ... it is this book ...' (*Anti-Dühring*, p. 274).

2 Mun, *England's Treasure*, McCulloch edition, p. 125.
3 *Ibid*, pp. 133-34.
4 *Ibid*, p. 128.
5 *Ibid*, p. 128. The passage here is Rubin's paraphrase of Mun's text.
6 *Ibid*, pp. 130-31 & 136.
7 *Ibid*, p. 131.
8 *Ibid*, p. 138.
9 *Ibid*, p. 137.
10 *Ibid*, p. 139.
11 *Ibid*, p. 141.
12 Charles D'Avenant, *An Essay on the East-India Trade*, etc., London 1697, in D'Avenant, *Discourses on the Publick Revenues, and on the Trade of England* ..., Part II, London, 1968, p. 31. Cited in Marx, *Theories of Surplus Value*, Part One (Progress Publishers edition, Moscow, 1969), p. 179.
13 Petty, *Political Arithmetick*, in *Economic Writings*, Hull edition, Volume One, p.256.
14 *Ibid*, p. 258.

CHAPTER SIX

The Reaction Against Mercantilism

DUDLEY NORTH

Although he stood opposed to the out of date prohibitions against the export of money, Thomas Mun had nevertheless acknowledged the need for the government to exercize control over foreign trade as a means of improving the balance of trade and drawing money into the country. The first person to develop a critique of the principals behind mercantilist policy was *Dudley North*, whose *Discourses upon Trade* appeared in 1691. A prominent merchant, and, later on, a Commissioner of the Customs, North comes out in defense of merchant and money capital, which had already become sufficiently developed to feel the constraints of excessive state tutelage. North is the first of the early prophets of the *idea of free trade*. He dedicates his tract to a discussion around two central themes: first, the restrictions which the state, in its desire to attract money into the country, has imposed *upon foreign trade*, and second, the legal limitation placed upon the *level of interest*. On both these issues North consistently demands that the state cease its interference into economic life.

To the mercantilists, for whom the aim of foreign trade was to increase the nation's stock of money, trade was understood to be above all the exchange of a product, or use value, for money, or exchange value. With North the concept of trade is something different, being *an exchange of certain products for others*; foreign trade, then, is an exchange of the produce of one nation for that of another, to their common benefit. In this exchange, money functions simply as a medium. '...Gold and Silver, and, out of them, Money, are nothing but the Weights and Measures, by which Traffick is more conveniently carried on than could be done without them.'[1] If trade prospers or declines the cause is not to be found in the inflows and outflows of money; to the contrary, an increase in the quantity of money is consequent upon a growth of trade.

This is not the idea held by public opinion, which is prone to ascribe any stagnation in commerce to a *shortage of money*. When a merchant cannot find a market for his commodities he sees as the cause an insufficient amount of money within the country, a view, however, which is deeply mistaken. 'But to examine the matter closer, what do these People want, who cry out for Money? I will begin with the Beggar; he wants, and importunes for Money: What would he do with it if he had it? buy Bread, &c. Then in truth it is not Money, but Bread, and other Necessaries for Life that he wants. Well then, the Farmer complains, for the want of Money; surely it is not for the Beggar's Reason, to sustain Life, or pay Debts; but he thinks that were more Money in the Country, he should have a Price for his Goods. Then it seems Money is not his want, but a Price for his Corn, and Cattel, which he would sell, but cannot.'[2] Such a failure to sell is the result either of an excessive supply of corn or cattle or of a shortfall in the demand for them, owing to the poverty of the consumers or to the blockage of exports abroad.

Commerce, therefore, suffers not from a *shortage of money*, but from a break in the steady flow of *commodity exchange*. Generally speaking there can be no such thing as a shortage of money, since a country is always in possession of as much money as is needed for the purpose of commerce, that is, for the exchange of commodities. '... If your are a rich People, and have Trade, you cannot want Specifick Coyn, to serve your occassions in dealing.'[3] For even if a country does not mint its own coinage it will be supplied in sufficient quantity by the coins of other nations. On the other hand, 'when Money grows up to a greater quantity than Commerce requires, it comes to be of no greater value than uncoyned Silver, and will occasionally be melted down again.'[4] North is thus led to the conclusion that the *circulation of money will regulate itself* to correspond with the demands of commodity circulation. And while a country has nothing to fear from a *shortage* of money, it is equally of no avail for the state to resort to compulsory measures to *augment* it.

Measures designed to retain money within the country will only *retard commerce*. 'Let a law be made, and what is more, be observ'd that no Man whatsoever shall carry any Money out of a particular Town, County, or Division, with liberty to carry goods of any sort: so that all the Money which every one brings with him, must be left behind, and none be carried out. The consequence of this would be, that such Town or County were cut off from the rest of the Nation; and no Man would dare to come to Market with his Money there;

because he must buy, whether he likes, or not: and on the other side, the People of that place could not go to other Markets as Buyers, but only as Sellers, being not permitted to carry any Money out with them. Now would not such a Constitution as this, soon bring a Town or County to a miserable Condition, with respect to their Neighbours, who have free Commerce?'[5] The same sorry fate would befall an entire nation were it to introduce similar restrictions upon trade, for 'a Nation in the World, as to Trade, is in all respects like a City to a Kingdom, or Family in a City.'[6] North's ideal is that world commerce be as free and unfettered as possible.

A country which by its acts and decrees holds onto its money, turning it thereby into idle hoards, directly inflicts upon itself *a loss*. 'No Man is richer for having his Estate all in Money, Plate, &c. lying by him, but on the contrary, he is for that reason the poorer. That man is richest, whose Estate is in a growing condition, either in Land at Farm, Money at Interest, or Goods in Trade: If any man, out of an humour, should turn all his Estate into Money, and keep it dead, he would soon be sensible of Poverty growing upon him.'[7] Whether it be an individual or an entire nation, enrichment will come not by accumulating ready money, but only by continuously throwing it back into circulation as money capital—as profit-bearing money. In North's eyes the road to prosperity lies not in the accumulation of *money hoards*, but in the growth of *trade* and a rise in the general mass of *profit and capitals*. In his polemic against mercantilist policy North overcomes the theoretical error that the mercantilists had made in confusing *money* (precious metals) with *exchange value* in general, with *capital*. By recognizing that money is a medium of exchange and a measure of value for actual commodities, North comes very close to arriving at a correct understanding of the distinction between money and exchange value. With even greater clarity he explores the difference between money and capital, developing ideas already tentatively advanced by Mun. Mun had seen a positive balance of trade as more than a means of attracting and accumulating precious metals: it was a sign that greater capital was being invested in commerce and that profits were flowing into the country. But he also advocated that the state keep close watch over the balance of trade and take steps to improve it. For North, too, it was a conscious goal that commercial capitals and profits should be built up, but the best means to this end was *free trade* and not the restraining interference of the state.

North extended this same principle of government noninterference to another question, that of the *level of interest*, an issue which generated furious debate—and a mass of literature—throughout the 17th and 18th centuries. This was an issue over which *the interests of the landowning class and the money capitalists came into sharp conflict.* The medieval laws forbidding the exaction of interest payments were repealed in England by Henry VIII in 1545. Interest could now be charged on loans, although it could not exceed 10% per annum. At the beginning of the 17th century the legal ceiling was lowered to 8%, and to 6% in 1652. Especially persistent in their pressure for further reductions in the rate of interest were the landed aristocracy, whose prodigal living and constant borrowings dropped them straight into the usurers' clutches. A fall in the interest rate would benefit the landlords in two ways: first, the interest payments owing to money lenders would be reduced, and second, the price of land would go up and with it the prospect of selling it at greater profit.

In 1621, Culpeper, an early partisan of landed interests, wrote: 'Wherever money is dear, land is cheap, and where money is cheap, land is dear.' 'The high interest on loans compels the sale of land at a cheap price.'[8]

Support for the landlords' demands for a lower rate of interest came also from certain sections of the industrial and commercial bourgeoisie, especially those holding an interest in the affairs of the East India Company. The lower the interest on loans the more willing rentiers would be to invest their disposable funds in the Company's shares, and the higher these shares would be quoted. *Child* wrote in 1668, that if the rate of interest earned from lending was high (6%), no-one would wish to invest their money in precarious transoceanic commerce which itself could offer only 8%-9%. Basing themselves upon the example of Holland, where the rate of interest was low, it was deemed by Child and other writers that keeping interest rates down would assure the stimulation and profitability of commerce, and so they demanded that rates be legally reduced.

Against this, arguing that government regulation over the level of interest was primarily of service to an idle aristocracy and not to the merchant class, the defenders of moneyed capital demanded that these controls be totally repealed. In reality, to broad sections of merchants these laws were of little use, for despite a legal ceiling of 6% their quest for credits compelled them to pay rates of interest far in excess of what the law allowed, rates which could at times go as high as 33%. Thus, quite a number of the writers who upheld the interests of money and commercial capital demanded *the repeal of the legal limit upon the rate of interest*, arguing that it contravened the 'natural' laws of capitalist economy. Amongst these writers were *Petty, Locke*, and *North*.

It was North's view that a reduction in the interest rate would benefit the gentry far more than it would the trader: '...the Moneys imployed at Interest in this Nation, are not near the Tenth part disposed to Trading People, wherewith to manage their Trades; but are for the most part lent for the supplying of Luxury, and to support the Expence of Persons, who though great Owners of Lands, yet spend faster than their Lands bring in.'[9] A legal limit upon interest will only create a difficult and precarious situation for those merchants seeking credit, and will exercize a retarding influence on trade. 'It is not low Interest makes Trade, but Trade increasing...makes Interest low,' by augmenting the number of capitals being accumulated and put towards investment.[10] If the level of interest is to fall it will come from the unhindered expansion of trade, and not out of compulsory regulation. Hence : 'it will be found best for the Nation to leave the Borrowers and the Lender to make their own Bargains, according to the Circumstances they lie under.'[11]

It is characteristic that in order to justify the earning of interest upon capital North attempts to equate this form of revenue with *agricultural rent*. 'But as the Landed Man letts his Land, so these [traders] still lett their Stock; this latter is call'd Interest, but is only Rent for Stock, as the other is for Land.'[12] And so the state can as little legislate a reduction of interest from 5% to 4% as it can bring down the rent paid on an acre of land from ten shillings to eight. We find Petty and Locke equating in similar fashion the interest upon capital with land rent. For the former was still at this time a new form of revenue, and could only be theoretically explained and justified in practice by drawing an equation between it and the traditional source of income, the rent on land.

For its time North's book was a remarkable phenomenon, containing as it does the *first formulation of the ideas of free trade* that were developed in full by Hume and Smith. A man who transcended his age, North was one of the earliest prophets of mercantilism's decline. To the mercantilists, international commerce was like a chess match in which the gain of one is the loss of another. To North, this trade was mutually profitable to all nations who took part. The mercantilists differentiated between '*profitable*' and '*unprofitable*' branches of trade, depending upon what effect they had upon the balance of trade. For North 'there can be no Trade unprofitable to the Publick;

for if any prove so, men leave it off.'[13] The mercantilists upheld the strict tutelage by the state over economic life; North demanded *free trade and government nonintervention,* since it is impossible 'to force Men to deal in any prescrib'd manner.' We also find in North a deeper analysis of theoretical questions, namely the distinction that he makes between *capital* and *money*, and his observation that *the circulation of money will regulate itself* in accord with the requirements of the circulation of commodities.

Still, this theoretical analysis is for North a subordinate instrument, a means for making his criticism of mercantilist policy more incisive. Questions of economic policy still predominate: where the literature deals with theoretical arguments they are fragmentary and incomplete. To gain a proper understanding of the theoretical heritage of mercantilism we must now step back in time somewhat, to *William Petty*, so that we then may pass to the mid-18th century, which forms a period of transition from mercantilist to Classical literature.

1 Dudley North, *Discourses upon Trade*, in McCulloch, *Early English Tracts on Commerce*, pp. 529-30.
2 *Ibid*, p. 525.
3 *Ibid*, p. 531.
4 *Ibid*, p. 531.
5 *Ibid*, pp. 527-28.
6 *Ibid*, p. 528.
7 *Ibid*, p. 525.
8 Thomas Culpeper, *A Tract against Usurie*, London, 1621. Quotations translated from the Russian.
9 North, *Discourses*, in McCulloch, *op cit*, p.520.
10 *Ibid*, p. 518.
11 *Ibid*, p. 521.
12 *Ibid*, p. 518.
13 *Ibid*, p. 513.

CHAPTER SEVEN

The Evolution of the Theory of Value

WILLIAM PETTY

We have already noted that, in their majority, mercantilist writers were concerned overwhelmingly with questions of economic policy, and showed little inclination toward theoretical study. However, the need to justify various practical measures increasingly compelled them to fall back upon arguments of a theoretical nature. Thus the struggle against restrictions on the circulation of money, for example (the prohibitions on exporting coins, etc.), provided the impetus for developing the theory of the balance of trade. But, influenced by the sweeping, generalizing character of 17th century mathematics and empirical philosophy (*Bacon, Hobbes*), and aware of the need to conduct a broader and more radical re-examination of mercantilist doctrine in order to meet the new, increasingly complex demands of economic development, English mercantilist literature, beginning with the mid-17th century, displayed a growing *concern with theory*.

Along with its basic,'*merchant*' tendency, a '*philosophical*' current now appeared in mercantilist literature that was more disposed towards theoretical generalizations. Alongside the narrow practitioners debating the practical questions of their day, there now appeared among mercantilist writers people with a broad scientific outlook (*Petty*) and the most eminent philosophers of the age (*Locke, Berkeley, Hume*). Even the authors who were practical men of action displayed in their works a deeper concern for theoretical issues (*North, Barbon, Cantillon*). As a theoretical movement, although its ideas were still rudimentary and unripened, it left a most valuable legacy in its theory of *value* and its theory of *money*.

In its modern form the *problem of value* could only be posed once the guild handicrafts had begun to give way to capitalist economy. During the age of medieval crafts, prices on goods were regulated by the guild and town authorities. The fixed prices that the guilds placed upon hand-crafted articles were intended to assure craftsmen a 'decent subsistence' or reward for their labour. Thus it is not surprising that the most prominent Canonist writers of the 13th century, *Albert the Great* and *Thomas Aquinas*, taught that the value of a product depends upon 'the quantity of labour and the outlays' expended upon its production. Although outwardly this formula resembles the later labour theory of value, there is a substantial difference between them. The economic soil from which this formula grew was that of *craft production*, rather than *capitalism*. What the authors had in mind by it were the outlays which the craftsman made for raw materials and implements and a 'decent' reward for his labour. The price that they were concerned with was not the one that was *actually* established through the process of *market competition*, but the '*just price*' (*justum pretium*) that *had* to be set by the authorities in order to accord with the traditional conditions of the medieval crafts. Thus the problem of value was posed '*normatively*'.

With the appearance of capitalist economy this situation altered, as the price fixing of the guilds increasingly gave way to a process of competition between buyer and seller. *Price formation via regulation* was replaced by the *spontaneous price formation of the market*. What had been a magnitude fixed in advance and compulsorily established was now the result of a complex process of competition about which there could be no prior knowledge. For the writers of the 13th century the discussion had been about what price *ought* to be established out of considerations of *justice*; the economists of the 17th century approached the problem from the other direction: they wanted to discover the *law-determined regularity* which governed the process of price formation *as it actually occurred on the market*. The *normative* formulation of the problem of value had given way to that of *scientific* theory.

Nevertheless, during the epoch of early capitalism it was no easy feat to find any definite regularity behind the phenomena of price formation. Free competition had not yet seized hold of all sectors of the economy, nor had it fully displayed its law-governed regularities. Its workings were still limited to a great extent by the survivals of the price fixing of the *guilds*, by *mercantilist* regulation over trade and industry, and by the *monopoly* rights of the trading companies. The mercantilists themselves continued to hold faith in the possibility of *regulating* economic life through recourse to state enactments. For them *the idea of a market spontaneously regulated by certain laws* was an alien concept, to be developed only later on by the Physiocrats, and especially by Adam Smith.*

The economists who observed the chaos of multitudinous determinants that together made up the process of price formation under early capitalism very often gave up any attempt to discover the *law-determined regularity* which lay beneath it. The constant and sharp fluctuations in market prices suggested that the prices on commodities depended exclusively upon what accidental relationship between supply and demand existed at any given moment. From this idea arose the first rudiments of the *theory of supply and demand*, a theory which gained wide currency amongst the mercantilists and which the famous philosopher, *John Locke*,** formulated in these words: 'All things that are bought and sold, raise and fall their price in proportion, as there are more buyers or sellers. Where there are a great many sellers to a few buyers, there use what art you will, the thing to be sold will be cheap. On the other side, turn the tables, and raise up a great many buyers for a few sellers, and the same thing will immediately grow dear.' [1] If one talks about exchange value it can only be 'in a given place and at a given moment in time'; there can be no question of exchange value having any fixed and objectively determined level.

This denial of any law-determined regularity behind price formation was a position also arrived at by the early advocates of *the theory of subjective utility*. *Nicholas Barbon*,† an English contemporary of Locke's, was an active participant in the fever of promotional speculation that gripped England at the end of the 17th century. The spectacle of prices dancing about from such speculative activity was one that could readily lend itself to the idea that 'no commodity possesses a precisely determined price or value'.

'The value of all commodities derives from their *usefulness*' (that is, their ability to 'satisfy human wants and needs'), and it changes with

*See below, Chapter 11 and Chapter 20.

**Born in 1632, Locke died in 1704. Besides his famous philosophical and sociological investigations he wrote a purely economic study, *Some Considerations of the Consequences of the Lowering of Interest, and Raising the Value of Money*. On Locke see also the end of the present chapter.

†Born in 1640, died in 1698. His major work is his *A Discourse Concerning Coining the New Money Lighter. In answer to Mr Locke's Considerations about Raising the Value of Money*. See in addition the beginning of the next chapter.

changes in the 'humour and the whims of the people who make use of them'.

This theory outlined by Barbon found little success amongst the mercantilists. Its further development came only in the mid-18th century, in the works of the late mercantilist, *Galiani*,* those of the famous Physiocrat, *Turgot*,** and, especially, in those of *Condillac*,† an opponent of the Physiocrats who had nevertheless been greatly influenced by them. Condillac is justifiably regarded as the forefather of modern psychological theories of value. He differentiates between the *abstract* utility of a given type of thing, for example, corn, and the *concrete* utility contained in a given unit of that corn. The value of a thing is determined by its concrete utility, which in turn depends above all upon its *scarcity*, i.e., upon the quantity of it that is currently available.

The respective partisans of the theory of supply and demand and of the theory of subjective utility virtually *renounced* the task of discovering the law-determined regularities behind price formation. Yet as economic life developed economists were peremptorily confronted with this problem. The successful beginnings and then the diffusion of free competition rendered theoretically unsatisfactory for economists the notion that the phenomena of price formation were accidental in nature. In earlier times, the trading companies, which then held a monopoly, used to dictate prices arbitrarily to the consumer, and would often destroy parts of their commodity stocks to keep price levels high (in so doing they gave an illustration of just how powerful was the law of supply and demand). With the appearance of industrial capitalism this situation had changed. In his advance calculations, the industrialist was determined to see that the sale price of a commodity at least compensated him for his *costs of production*. Out of the seemingly aleatoric dance of prices economists found a stable base point that prices must necessarily conform to—the costs of production incurred in the manufacture of a commodity. And so there arose the *theory of production costs*.

James Steuart,‡ one of the last mercantilists (1712-1780),

*Galiani (1728-1780), an Italian who lived many years in Paris, was the author of *Della moneta* [*On Money*] (1750) and *Dialogues sur le commerce des blés* (1770). See also Chapter 10 below.

**See below, Chapter 10.

†*Condillac*, Frenchman, famous philosopher and representative of sensualism, wrote his economic work *Le Commerce et le gouvernement* in 1776.

‡ His *An Inquiry into the Principles of Political Oeconomy* appeared in 1767. See also the end of the present chapter and the beginning of Chapter 8.

divided the price of a commodity into two different parts: '*the real value* of the commodity, and the *profit* upon alientation'.[2] A commodity's 'real value' represents a precisely determined magnitude, equal to its *costs of production*. To calculate these production costs it is necessary to know, first, the number of units of the commodity produced by the worker in the course of a day, a week, or a month; second, the value of the workman's means of subsistence (i.e., the sum of his wages) and of the implements that he employs in his labour; and third, the value of his raw materials. 'These three articles being known, the price of manufacture is determined. It cannot be *lower* than the amount of all the three, that is, than the real value; whatever is *higher*, is the manufacturer's *profit*.'[3] And how is the size of this profit determined? This is a question Steuart cannot answer. Here we see the fundamental flaw in the theory of production costs, which to this day it has not managed to escape from: its inability to explain the origin and magnitude of *surplus value* or *profit* (in the broad sense of the term), i.e., the surplus of a product's price *over its costs of production*. Being a true mercantilist, Steuart supposes that the sale price of a commodity will exceed its 'real value', and that the capitalist's enrichment derives from 'profit upon alientation', 'which will ever be in proportion to demand, and therefore will fluctuate according to circumstances'.[4] As a result, Steuart passes up the opportunity to find the law-determined regularity which determines the magnitude of surplus value, or profit. It would only have been possible to discover this using the *labour theory of value*. Like the theories we have just discussed, this one, too, arose during the mercantilist epoch. To trace its roots we must go back to William Petty.

A man of rare gifts and versatility, *William Petty* (1623-1687), though a physician by profession, simultaneously devoted himself to mathematics, geodesy, music, and ship-building. Born the son of a small craftsman, he died an English peer and a millionaire, having acquired his fortune by the shameless methods of an adventurer by taking part in the partition of the lands of Irish rebels. Being a true son of the 17th century, with its brilliant flowering of mathematics and its desire to transcribe every picture of the world into mathematical formulae, Petty was primarily concerned with the quantitative side of economic phenomena. In keeping with the spirit of 17th century empirical philosophy, Petty aspired to the observation and precise, *quantitative description* of real phenomena. In the preface to one of his works, which he entitled a *Political Arithmetick*,* he describes his method as follows: 'The Method I take to do this, is not very usual; for instead of using only comparative and superlative Words, and intellectual Arguments, I have taken the course ... to express myself in Terms of *Number, Weight*, or *Measure*; to use only Arguments of Sense, and to consider only such Causes, as have visible Foundations in Nature.'

Petty shared this interest in the statistical description of economic phenomena with a number of other economists of his age: *Graunt*, who compiled tables on mortality; *D'Avenant*, who concerned himself with statistics on trade; and *King*, the author of the well-known 'King's Law', which states that fluctuations in the supply of corn call forth far sharper fluctuations in its price (if, for instance, the quantity of corn available falls by half because of poor harvest, the rise in price which follows will be four or five-fold). Unlike these other writers, however, Petty's interest in statistical observations was not for their own sake, but because they afforded material for *theoretical analysis*. He not only compiled facts on population growth, movements of commodity prices, wages, rents, the price of land, and so on, but having made these observations, he endeavoured to penetrate into what it was that bound them all together. It is true that Petty was not fully conscious of the difficulties involved in moving from individual statistical data to broad theoretical generalizations, and that his boldness led him to make hasty generalizations and derivative constructs that were often in error. Yet his conjectures and hypotheses invariably displayed the great sweep of a mind of genius and earned him a reputation as one of the founders of modern political economy and forebears of the labour theory of value.

As a mercantilist, for whom the exchange of products for money had utmost importance, Petty was especially concerned with the problem of price, meaning by this not a product's market price, accidentally determined by 'extrinsic' causes, but its '*natural price*', which depends upon 'intrinsic' factors. In keeping with the mercantilist identification of money and the precious metals, Petty poses this problem of 'natural price' or *value* in the form of a question: why is a certain quantity of silver offered for a given product? In his answer Petty sketches out with ingenious simplicity the basic ideas of the

*This work was issued in 1690, after Petty's death. His other works include *A Treatise of Taxes and Contributions*, published in 1662, and *The Political Anatomy of Ireland*, published in 1672.

labour theory of value. 'If a man can bring to *London* an ounce of Silver out of the Earth in *Peru*, in the same time that he can produce a bushel of Corn, then one is the natural price of the other; now if by reason of new and more easie Mines a man can get two ounces of Silver as easily as formerly he did one, then Corn will be as cheap at ten shillings the bushel, as it was before at five shilings *caeteris paribus*.'[5] '...Corn is cheaper where one man produces Corn for ten, then where he can do the like but for six...Corn will be twice as dear where there are two hundred Husbandmen to do the same work which an hundred could perform.'[6] Corn and silver will have equal value provided that equal quantities of labour have been expended on their production. The magnitude of a product's value depends upon *the quantity of labour expended on its production*.

From the magnitude of a product's value Petty moves on to analyze its individual components. He distinguishes two parts to the value of any product (which by way of example he usually takes to be corn): *wages* and *the rent on land*. Before, when we were discussing the general features of mercantilist literature,* we noted that Petty had deemed it essential to place a legal limit upon wages, to what was necessary for the worker's provision. By assuming in his theoretical discourses that this is the level of wages which prevails, Petty is thereby able to determine the size of agricultural rent *in natura*, i.e., in corn: 'Suppose a man could with his own hands plant a certain scope of Land with Corn, that is, could Digg, or Plough, Harrow, Weed, Reap, Carry home, Thresh, and Winnow so much as the Husbandry of his Land requires; and had withal Seed wherewith to sowe the same. I say, that when this man hath subducted his seed out of the proceed of his Harvest, and also, what himself hath both eaten and given to others in exchange for Clothes, and other natural necessaries; that the remainder of Corn is the natural and true Rent of the Land for that year.'[7] The *in natura* size of rent is determined by deducting the articles of the worker's consumption (his wages) and the costs of his means of production (his seed) from the total product. Thus what Petty has in mind, and presents in the guise of *the rent on land*, is *total surplus value*, including profit.

Having determined the rent *in natura*, Petty then asks what its *price will be in money*, i.e., for what quantity of silver can it be exchanged. 'But a further, though collateral question may be, how much English money this Corn or Rent is worth? I answer, so much as the money,

*See above, Chapter Three.

which another single man can save, within the same time, over and above his expence., if he imployed himself wholly to produce and make it; *viz.* Let another man go travel into a Countrey where is Silver, there Dig it, Refine it, bring it to the same place where the other man planted his Corn; Coyne it, &c. the same person, all the while of his working for silver, gathering also food for his necessary livelihood, and procuring himself covering, &c. I say, the Silver of the one, must be esteemed of equal value with the Corn of the other: the one being perhaps twenty Ounces and the other twenty Bushels. From whence it follows, that the price of a Bushel of this Corn to be an Ounce of Silver.'[8] Once the price of a bushel of corn is known this can be used to determine the price of the corn that makes up the rent—i.e., the total *money rent.*

Petty follows this discussion with a very bold attempt to deduce the *price of land* from the total money rent. By Petty's time land in England had already become an object of buying and selling, with a determinate price approximately equal to the total annual rent multiplied by 20 (or, more precisely, by 21). Petty knew from business experience that a parcel of land yielding an annual rent of £50 would sell for approximately £1,000. Petty asks, why is the price of land equal to 20 times its annual rent? Taking his investigation into rent as his starting point, but without knowledge of the laws governing the formation of profit and interest, Petty could not know that this relationship between the annual rent and the price of land depends upon the average rate of interest prevailing at the time (in England this was around 5%), and that the former changes together with the latter (for example, if the rate of interest falls from 5% to 4% the price on the same tract of land would rise to £1,250, or to 25 times the annual rent). Thus Petty resorts to the following, artificial argument: The buyer reckons that by purchasing land he will be guaranteeing a set annual income for himself, his son, and his grandson; people's concern for posterity usually did not extend any farther. Suppose that the buyer is around fifty years old, the son about twenty-eight and the grandson seven. According to the statistics from Graunt's mortality tables these three people can count on living on average a further twenty-one years. Thus, reckoning on the land providing an annual income for twenty-one years, the buyer agrees to pay a sum twenty-one times greater than the overall yearly rent.

However erroneous this argument of Petty's may be, it contains within it one fertile idea of profound truth: 'the value of land' is none other than the sum of a definite *number of annual rents*. Since the size of money rent depends upon the value of a bushel of corn, and this in turn is determined by the quantity of labour expended on its production, it follows that labour is the source not only of the value of corn, but in the final accounting also of the 'value of land'. Petty's argument represents an early and daring attempt to subject agricultural phenomena to the law of labour value. The other side of this, however, is that Petty's concentration upon land rent testifies to the still overwhelming predominance of agriculture. Economic theory, although turning to new concepts and ideas in order to generalize the phenomena of the new capitalist economy, often dresses them up in the clothing of concepts and ideas inherited from the era when agriculture and feudal forms of land tenure were dominant. Within economic theory the basic category of capitalist economy—profit—has still not detached itself from land rent, but is dissolved into it: *all surplus value, including profit, appears under the heading of rent.* In part this theoretical disregard for the category of profit is explained by the difficulties involved in working out new categories that correspond to the reality of new phenomena; but it is also explained by the fact that manufacturing profit at this time still played but a secondary role, while commercial profit was viewed by the mercantilists as a mark up on the price of a commodity. Petty singled out only one form of profit as special, and that was the *interest on loan capital*. This was a necessary distinction, in view both of the huge importance that loan capital had at the time, and of the sharp class antagonism that existed between money capital and landed interests.* But having specially singled out money interest, Petty all the same regarded it as a derivative form of revenue, as though it were a substitute for rent. Because he did not understand that fluctuations in the price of land dutifully follow upon fluctuations in the level of interest. Petty imagined that the relation between these two phenomena was in fact the other way round: he explained the *level of interest* from the *level of land prices*. If a parcel of land could be purchased for £1,000 and yielded an annual rent of £50, the owner of a capital of £1,000 would naturally agree to loan it out only on condition that the money received in interest was no less than the £50 received per year in rent: thus, given the price of land, the rate of interest was established at 5%.

As we see, Petty was the first to sketch in the outlines of the labour theory of value, and attempted on the basis of this to explain

*See the preceding chapter.

the quantitative relationships between different phenomena: between the quantity of a product and the quantity of silver for which it would exchange; between natural wages and natural rent; between natural rent and money rent; between money rent and the price of land; and between the price of land and the rate of interest. Yet along with these rudiments of a correct understanding of the relationship between value and labour we often find in Petty a different concept of value, in which the source of the latter is ascribed to *labour and nature*. Petty gave brilliant expression to this idea in his famous phrase, 'Labour is the Father and active principle of Wealth, as Lands are the Mother'.[9] It is clear that he is speaking here about material wealth, or *use values*, whose production indeed requires the active union of the forces of nature with human activity. However, once the value of a product (which he does not differentiate from the product itself) is created by labour and land, the determination of the magnitude of this value necessitates that there first be found a general measure by which the action of the forces of nature and the labouring activity of man can be compared. So the problem arises of a '*measure of value*', which itself rests upon the problem of the '*par between land and labour*'. '...all things ought to be valued by two natural Denominations, which is Land and Labour; that is, we ought to say, a Ship or garment is worth such a measure of Land, with such another measure of Labour; forasmuch as both Ships and Garments were the creatures of Lands and mens Labours thereupon: This being true, we should be glad to finde out a natural Par between Land and Labour, so as we might express the value of either of them alone as well or better than by both, and reduce one into the other as easily and certainly as we reduce pence into pounds.'[10]

How, then, do we resolve this 'most important Consideration in Political Oeconomies,' how do we make 'a *Par* and *Equation* between Lands and Labour'?[11] Both land and labour participate in the process of creating use values; let us examine the *proportion* in which each of them does so. Suppose that a calf is set out to pasture on two acres of uncultivated land, and that the weight which it puts on in the course of a year represents a quantity of meat sufficient to feed a man for fifty days. It is obvious that there having been no assistance from human labour the land has produced fifty 'days food'; the sum of these daily rations comprises the yearly 'rent' from this particular plot of land. If one man now cultivates this same land and in a year produces a greater number of daily food rations, the excess over and above the original 50 rations will constitute his 'wages'; in this way the shares of land (rent) and of labour (wages) are both expressed in *one and the same unit,* in *'days food'*. Thus 'the days food of an adult Man, at a Medium, and not the days labour, is the common measure of Value ... Wherefore I valued an *Irish* Cabbin at the number of days food, which the Maker spent in building of it.' In other words, its value is determined by the sum of the wages paid to the builder.[12]

We therefore see that there is a sharp disparity and contradiction between these two constructs of Petty. Previously he was speaking about *exchange value*; now he is concerned with *use value*. Before it was labour that he considered as the source of value; now it is labour and land. Previously he deduced the value of land (or, to be more exact, the price of land) from *labour*; now he seeks *'a par between land and labour'*. Before, he took as his measure of value *the quantity* of labour; now he takes the 'value of labour', i.e., *wages*. Previously Petty determined the magnitude of rent on the land by deducting *the worker's means of consumption* (i.e., wages) from the total product; now he finds wages by deducting *rent* from this product. If Petty can be justifiably acknowledged as the father of the labour theory of value he can also be recognized as the forebear of those basic errors and contradictions in its formulation that it has taken economic thought two hundred years to overcome. In one variant or another these fundamental mistakes—the confusion of exchange value with use value, the search for an equation between land and labour, and the confusion between the quantity of labour and the 'value of labour'—were repeated in the ensuing literature, including that of the English economists whose writings filled the near 100-year period that separated the activity of Petty from the works of Adam Smith. Let us now dwell briefly on Locke, Cantillon, and James Steuart.

For *Locke* the source of value is labour, value being understood, however, as material wealth, or *use value*. 'Nature and the earth furnished only the most worthless materials as in themselves',[13] i.e., without the assistance of human labour. How great the contrast between these natural works of nature and those products modified by human labour! Labour is the source of the powerful increase in the wealth of modern nations. 'I think it will be but a very modest computation to say, that of the products of the earth useful to the life of man, nine-tenths are the effects of labour'.[14] 'For whatever bread is more worth than acorns, wine than water and cloth or silk than leaves, skins, or moss, that is wholly owing to labour and industry'.[15] Labour is the primary source of a commodity's *use* value; as we have already seen, however, its *exchange* value, in Locke's view, is determined by the law of supply and demand.

With *Cantillon** (who died in 1734) we again find confusion between exchange value and use value, as well as a further attempt to deduce value from land and labour. 'Land is the Source of Matter from whence all Wealth is produced. The Labour of man is the form which produces it: and Wealth in itself is nothing but the Maintenance, Conveniences, and Superfluities of Life.'[16] Once a thing is created by *land and labour* 'the price or intrinsic value of a thing is the measure of the quantity of Land and of Labour entering into its production'.[17] Cantillon was clearly influenced by Petty, and rather than stopping at the simple determination of value by land and labour looks to find an *equation* between these two elements. Nor was he satisfied with Petty's solution, who at one point, as we have seen, reduces 'the value of land' to labour, and at another designates a man's daily subsistence (the food ration) as the common denominator between 'the value of land' (rent) and 'the value of labour' (wages). Cantillon, as a forerunner of the Physiocrats, awards the palm of superiority to *land*, and endeavours to reduce the value of the worker's labour to the value of that plot of land that would be sufficient to feed him and his family. Thus 'the intrinsic value of any thing may be measured by the quantity of Land used in its production and the quantity of Labour which enters into it, in other words by the quantity of land of which the produce is allotted to those who have worked upon it.'[18] Proceeding from Petty's mistaken ideas, Cantillon moves even farther away from a correct formulation of the labour theory of value. What is more, in order to reduce 'the value of land' to labour he, on the contrary, establishes an equality between human labour and a determinate plot of land.

Finally to *James Steuart*;** we find with him, too, this same confusion between exchange value and use value. Within a concrete product of labour (i.e., a use value) Steuart differentiates between the *material substratum*, which is given by nature, and the modification

*His *Essai sur la nature du commerce en général* appeared in 1755, after the author's death. [The French edition of Cantillon's work appeared under the name of Richard Cantillon, and was reprinted in an Amsterdam edition in 1756 (the edition that Marx quotes in Volume I of *Capital*). An English edition, published as *The Analysis of Trade, Commerce, etc., by Philip Cantillon, late of the City of London, Merchant*, appeared in 1759. Although the French edition claims to be a translation of an English original, Marx notes that both the date of the English edition and the fact that it contained substantial revisions from the French make this impossible. See *Capital*, Vol. I (Penguin edition), p. 697.—*Trans.*]

**See the beginning of the present chapter. On his theory of money see the end of Chapter Eight.

made upon it by *human labour*. Though it may seem odd, he calls the natural material out of which the product is created its 'intrinsic worth'. The 'intrinsic worth' of a silver vase is the raw material (the silver) out of which it was fashioned. Its modification by the labour of the worker who made the vase makes up its 'useful value'. 'Here two things deserve our attention. First, the simple substance, or the production of nature; the other, the modification [*preobrazovanie*—transformation, *I.R.*], or the work of man. The first I shall call the *intrinsic worth*, the other, the *useful value*... The value of the second must be estimated according to the labour it has cost to produce it.'[19] What Steuart has in mind, therefore, is the *concrete useful labour* which creates use value and gives 'form to some substance which has rendered it useful, ornamental, or, in short, fit for man, mediately or immediately.'[20]

It was, therefore, during the mercantilist age that there appeared in embryo the main theories of value that were to play an important part in the subsequent history of economic thought: the theory of *supply and demand*, the theory of *subjective utility*, the theory of *production costs*, and the *labour theory of value*. Of these, it was not until the appearance of the Austrian School that the theory of subjective utility was employed in economic science with any kind of success. Of the others, it was the *labour theory of value* which had the greatest impact upon the further evolution of economic thought. At the hands of Petty and his followers the labour theory of value suffered a multitude of glaring contradictions, being pushed by Locke onto the same level as the theory of supply and demand, and by Steuart onto the plane of the theory of production costs. The labour theory of value owed its future progress to the Classical School and to scientific socialism. Petty's heirs were *Smith, Ricardo, Rodbertus, and Marx*.

1 John Locke, *Some Considerations of the Consequences of the Lowering of Interest, and Raising the Value of Money* (1691)., published as an *Essay on Interest and Value of Money* by Alex. Murray & Son, London, 1870, p. 245.

2 Sir James Steuart, *An Inquiry into the Principles of Political Oeconomy*, (abridged edition in two volumes, edited by Andrew S. Skinner, published for the Scottish Economic Society by Oliver & Boyd, Edinburgh, 1966), Vol. One, p. 159. Rubin's emphasis.

3 *Ibid*, Vol. One, pp. 160-61. Rubin's emphasis.

4 *Ibid*, Vol. One, p. 161.

5 Petty, *A Treatise of Taxes and Contributions*, in *Economic Writings*, Hull edition pp. 50-51.

6 *Ibid*, p. 90.

7 *Ibid*, p. 43.

8 *Ibid*, p. 43.
9 *Ibid*, p. 68.
10 *Ibid*, pp. 44-45.
11 Petty, *The Political Anatomy of Ireland,* in *Economic Writings* (Hull edition), p. 181 (Petty's italics).
12 *Ibid*, pp. 181-82. By 'days food' Petty means the food necessary for one day's subsistence. The words 'at a Medium' were not included in Rubin's quotation from this passage, but have been reinserted here because of their importance for Petty's argument and as evidence of Petty's genuine insight into the question of socially necessary labour. In the passage immediately preceding the one to which Rubin refers he says: 'That some Men will eat more than others, is not material, since by a days food we understand 1/100 part of what 100 of all Sorts and Sizes will eat, so as to Live, Labour, and Generate. And that a days food of one sort, may require more labour to produce, than another sort, is also not material, since we understand the easiest-gotten food of the respective Countries of the World.'
13 Locke, *Two Treatises of Civil Government*, Everyman edition (London, J.M. Dent &Sons, 1962), p.138.
14 *Ibid*, p. 136.
15 *Ibid*, p. 137.
16 Richard Cantillon, *Essai sur la nature du commerce en général*, edited with an English translation and other material by Henry Higgs (London, Macmillan & Co., for the Royal Economic Society, 1931), p. 3.
17 *Ibid*, p. 29.
18 *Ibid*, p. 41.
19 Steuart, *Principles*, Skinner edition, Vol. One, p. 312. Steuart's italics.
20 *Ibid*, Vol. One, p. 312.

CHAPTER EIGHT

The Evolution Of The Theory of Money

DAVID HUME

Within the theoretical heritage of the mercantilist period we also find, along with the rudiments of a labour theory of value, attempts to develop a theory of money. Together with questions surrounding the balance of trade, it was *the problem of money* which most attracted and gave birth to an extensive literature; this was especially the case in the Italian cities, where there was a developed moneyed bourgeoisie and where monetary circulation was in a state of constant confusion. While in England a multitude of writings appeared entitled 'A Discourse Upon Trade', in Italy the traditional title was 'A Discourse on Money'. All of these works focussed upon questions of economic policy: prohibitions upon the export of money, the debasement of coins, and the like. The incessant debasement of coinage by rulers provoked furious debates. Those who defended the power of the kings and princes upheld their right to reduce the metallic content of coins, arguing that the value of coins is determined not by the quantity of metal they contain, but by state edict.'Money is value created by law', wrote *Nicholas Barbon*,* a partisan of the 'legal', or state 'theory of money. Defenders of the *commercial bourgeoisie* (which suffered from fluctuations in the value of coinage) demanded the minting of coins of standard weight. These forefathers of the 'metallic' theory of money argued that the constant decline in the metallic content of the coinage inevitably led to a drop in its value. Finally, there were certain writers who proposed a compromise solution, most clearly expressed by the well-known *John Law* at the beginning of the 18th century. According to Law's doctrine, the value of coins is composed of two parts: one, its 'real value'[1] is determined by the value of the metal that it contains; beyond this, however, it possesses also an 'additional

*On Barbon, see the preceding chapter.

value', which stems from the use of the metal in question as specie and from the additional demand for this metal that the manufacture of coins thus produces.

Because the aims were practical, the arguments and discussions in mercantilist writings on money are haphazard and disconnected. It is only in the mid-18th century, when mercantilist literature was in the last days of its decline, that we find more or less finished statements of the two theories which have come to play such an important role in ensuing writings on money, right up to the present day: The well-known 'quantity' theory of money, as put forth by *David Hume*; and an opposing theory advanced by *James Steuart*.

David Hume (1711-1776) was both a celebrated philosopher and an outstanding economist. His *Essays*, which appeared in 1753, levelled, with their ingenious and brilliant critique, the final blow to mercantilist ideas. As Hume was in general a clear-cut defender of free trade he cannot, of course, be counted as a mercantilist in any exact sense of the term. Usually in the history of economic thought Hume is accorded a place somewhere between the Physiocrats and Adam Smith, of whom he was both a direct predecessor and a close friend. Nevertheless, in the interests of giving greater clarity to our presentation we think it permissable to consider Hume's works in the present section, which covers not only the age when mercantilist ideas were at their zenith, but the period of their decline as well.

The issues around which Hume's ideas centered were the very same as had been constantly debated within mercantilist writings, namely *the balance of trade, the rate of interest,* and *money*. In his discussion of the first two of these Hume's power lies not so much in his originality as in the brilliant development and decisive formulation that he gave to ideas already expressed before him, by North in particular. If, at the end of the 17th century, North's voice had stood practically alone, Hume, with his critique of mercantilism, was by the mid-18th century expressing the general thinking of his age.

Hume's sharply critical stance towards the idea of a *balance of trade* flows from his general conception of *commerce*. For the mercantilists, the object of foreign trade was to bring advantage to the trading nation at the expense of others; foreign trade for Hume, however, consists of a mutual exchange of the different material products that the separate nations have produced by virtue of the diversity of their 'geniuses, climates, and soils'. It therefore follows that one nation can sell its surplus products to another only if the latter itself possesses excess produce which it can offer in exchange. '...If our neighbours have no art or cultivation, they cannot take [our commodities]; because they will have nothing to give in exchange.'[2] Thus 'an encrease of riches and commerce in any one nation, instead of hurting, commonly promotes the riches and commerce of all its neighbours'.[3] Every nation has an interest in the more extensive development of international commerce and in doing away with those 'numberless bars, obstructions, and imposts, which all nations of Europe, and none more than *England*, have put upon trade from an exorbitant desire of amassing money...or from an ill-grounded apprehension of losing their specie'.[4] In the course of commerce the precious metals will be distributed between individual countries proportionately to their 'trade, industry, and people'.[5] Should the quantity of a nation's ready money exceed this *normal level* it will *flow out* of the country; in the reverse situation there will be an *influx*. Compulsory measures to increase the quantity of money in a country are unnecessary.

The mercantilists had maintained that a rise in the quantity of money lowers a country's rate of interest and thereby stimulates commerce. Hume's essay, '*Of Interest*', was dedicated to refuting these ideas. The level of interest depends not upon an abundance of precious metals, but upon the following three factors: the volume of *demand for credit*, the number of *capitals* that are free and seeking investment, and the *size of commercial profits*. '...The greater or less stock of labour and commodities [i.e., capital—*I.R.*] must have a great influence; since we really and in effect borrow these, when we take money upon interest.' Interest falls not because of a 'great abundance of the precious metals', but because of an increase in the number of lenders having 'property or command' over them. Out of the growth of trade free capitals are accumulated and the number of lenders rises, at the same time as there is a decline in commercial profit. Both of these call forth a drop in the rate of interest. Since the same growth of commerce that 'sinks the interest, commonly acquires great abundance of the precious metals', people are mistakenly disposed to take the latter as the cause of the rate of interest's decline. The fact is, however, that both of these phenomena—the abundance of money and a low interest rate—are conditioned by one and the same factor: the expansion of trade and industry.[6] On the question of interest Hume is developing ideas outlined by North; his service consists in the insistence with which he differentiates *capital* from *money* and in his correct idea that the rate of interest depends upon the *level of profits*. Hume has in view a more developed system of credit relations than did the mercantilists: the latter often spoke about consumer credit, to which the landlords, especially, took recourse; Hume, however, has in mind productive credit going to merchants and manufacturers.

The most original part of Hume's economic doctrine is his *quantity theory of money*; this, too, is closely tied to his polemic against the mercantilists, who saw an increasing quantity of ready money as a powerful stimulus to the expansion of trade and industry. Hume's express goal was to show that even a protracted rise in the overall volume of money could in no way increase a nation's wealth, but would have as its sole result a corresponding and universal rise in the nominal prices of commodities. Thus Hume's polemic with the mercantilists led him to a 'quantity' theory of money, according to which *the value* (or purchasing power) *of money* is determined by the latter's *overall* quantity.

Suppose, argues Hume, that the quantity of money within a nation doubles. Does this mean an increase in its riches? Not at all, since it is products and labour that make up a nation's wealth. 'Money is nothing but the representation of labour and commodities, and serves only as a method of rating and estimating them.'[7] It is a conditional unit of account, an 'instrument which men have agreed upon to facilitate the exchange of one commodity for another', and, as such, has no value of its own.[8] Following Locke, and asserting that money has 'chiefly a fictitious value',[9] Hume sets himself firmly upon the ground of a *nominalist* theory of money and in opposition to the mercantilist doctrine that money alone (i.e., gold and silver) possesses true value.

Obviously, then, once the monetary unit becomes but a representative for a determinate number of commodities, any *rise* in the overall quantity of money (or decrease in the general mass of commodities) will mean that each unit of the country's money commands *fewer commodities*. 'It seems a maxim almost self-evident, that the prices of everything depend on the proportion between commodities and money, and that any considerable alteration on either has the same effect, either of heightening or lowering the price. Encrease the commodities, they become cheaper; encrease the money, they rise in their value',[10] and vice versa. An increase in the quantity of money—whose sole result is to universally raise commodity prices—is incapable of bringing the slightest benefit to a country; from the standpoint of international commerce it can even prove harmful, since when commodities become more expensive a nation is less competitive on the world market. If foreign trade is left out of account, the effect of raising the quantity of money is neither good nor bad, any more than it would matter to the merchant whether he kept his books with arabic numerals or Roman ones, the latter simply requiring more symbols to register the same number. 'Money having chiefly a fictitious value, the greater or less plenty of it is of no consequence, if we consider a nation within itself; and the quantity of specie, when once fixed, though ever so large, has no other effect, than to oblige every one to tell out a greater number of those shining bits of metal, for clothes, furniture or equipage.'[11]

Hume's forerunner in the development of the quantity theory was the famous French writer *Montesquieu* (1689-1755), author of the work *De l'esprit des lois.** Montesquieu proposed a purely mechanical relationship between a country's quantity of money and its level of commodity prices: double the amount of money, for example, and the result is a two-fold jump in prices. The task that Hume set himself was to trace the *economic process* by which changes in the quantity of money exercized their effect upon commodity prices. He depicts this process as follows: 'Here are a set of manufacturers or merchants, we shall suppose, who have received returns of gold and silver for goods which they sent to *Cadiz***. They are thereby enabled to employ more workmen than formerly, who never dream of demanding higher wages, but are glad of employment from such good paymasters. If workmen become scarce, the manufacturer gives higher wages, but at first requires an encrease of labour; and this is willingly submitted to by the artisan, who can now eat and drink better, to compensate his additional toil and fatigue. He carries his money to market, where he finds every thing at the same price as formerly, but returns with greater quantity and of better kinds, for the use of his family. The farmer and gardener, finding that all their commodities are taken off, apply themselves with alacrity to the raising more; and at the same time can afford to take better and more cloths from their tradesmen, whose price is the same as formerly, and their industry only whetted by so much new gain. It is easy to trace the money in its progress through the whole commonwealth; where we shall find, that it must first quicken the diligence of every individual, before it encrease the price of labour.'[12]

*An embryonic version of the quantity theory of money is to be found as early as the 16th century in the works of the Frenchman *Bodin* and the Italian *Davanzati*. Bodin was the first to point out that the fall in the value of money was to be accounted for not simply by the debasement of coin, but also by the inflow of large masses of gold and silver from America. [J. Bodin, *Discours sur le rehaussement et diminution des monnoyes*, Paris 1578—*Trans.*]

**I.e., in Spain, which owned the rich silver and gold mines of America.

Thus, if a group of traders comes into possession of a greater sum of money this will raise their demand for specific commodities and gradually cause the latter's price to rise. The dealers in this last group of commodities will in turn manifest a heightened demand for other commodities, whose price will also eventually go up. In this way the *greater demand* which is stimulated by a *growth in the quantity of money* will spread *from one group of commodities to another* and will lead gradually to a *general rise in the level of prices*, or to a fall in the value of a unit of money. 'At first, no alteration is perceived; by degrees the price rises, first of one commodity, then of another; till the whole at last reaches a just proportion with the new quantity of specie which is in the kingdom.'[13]

By endeavouring to describe the influence that a growing quantity of money and increased demand exert upon the motivation and behaviour of producers (on the one hand, encouraging them to expand production, on the other, raising their demand for other commodities), Hume liberated the quantity theory of money from the *naively mechanical* way that it had been formulated by Montesquieu, and paved the way for newer, *psychological* variants of the theory. In doing so, however, Hume introduced into it one major *qualification*: the rise in commodity prices that follows an increase in the quantity of money is not a rapid phenomenon, but proceeds over what, on occasion, can be extremely protracted periods, affecting different commodities at different points in time. There was one other important limitation that Hume imposed upon his theory: '... Prices do not so much depend on the absolute quantity of commodities and that of money, which are in a nation, as on that of the commodities which come or may come to market, and of the money which circulates. If the coin be locked up in chests, it is the same thing with regard to prices, as if it were annihilated; if the commodities be hoarded in magazines and granaries, a like effect follows. As the money and commodities, in these cases, never meet, they cannot affect each other.[14]

Hume's theory of money is in turn a reaction against the mercantilist concept of money and a *theoretical generalization from the phenomena of universal price rises* that Europe experienced during the 'price revolution' of the 16th-17th centuries (when there had been a massive influx of silver and gold from America). Yet Hume failed to take account of one crucial circumstance: side by side with the huge increase in Europe's quantity of precious metals, there was also a sharp fall in their value, as the richer American mines were opened up and as major technical improvements in extraction and processing were introduced (the discovery in the middle of the 16th century of the process of amalgamating silver with mercury lowered production costs considerably). The fall that took place in the value of precious metals, and the simultaneous and rapid growth of money economy and of the mass of commodities being thrown onto the market together demanded a far greater mass of money than before—a demand that was met by the inflow of American silver and gold. The 'price revolution' of the 16th and 17th centuries could not, therefore, be explained simply as the product of an increase in the quantity of money: the fact that the prices of commodities were rising reflected a fall in the value of the precious metals themselves. The nominalist conception of money as a simple token, with no value of its own but rather with a 'fictitious' value that derives from, and alters with fluctuations in the amount of money, proved to be profoundly mistaken when applied to metallic money.

Without dwelling on the other inadequacies of the quantity theory (that it ignores the velocity of money turnover, the role of credit money, etc.) it should be noted that Hume himself introduced corrections into the theory that opened the way for its supersession. For it is Hume, we see, who recognizes that when the quantity of money in a country doubles from one to two million rubles, the additional million might be accumulated in 'chests' as a hoard; in that case 'the quantity of money in circulation' will remain as before, at one million rubles, and no rise in commodity prices will ensue. The doubling of the nation's money gives rise to no surge of commodity prices, since part of this mass of money will lie outside of circulation. But if that is so, the question will arise, what determines the quantity of money that *enters into circulation?* Obviously it is the demands of commodity circulation, which in turn depends upon the mass of commodities and their prices (the latter depending on the value of commodities and the value of the precious metals that function as money). It is impossible, therefore, to assert that the *quantity of money in circulation* determines the *prices of commodities*; to the contrary, it is the demands of commodity circulation—which includes the *prices of commodities*—that determines the *quantity of money in circulation*.

Such was the position put forward in the middle of the 18th century by *James Steuart*, whom we have already encountered above.*

*See Chapter Seven, above.

On questions of economic policy, Steuart (whose work appeared in 1767) was a belated spokesman for the views of the mercantilists, and in this regard cedes a great deal to Hume when it comes to his grasp of the needs of his own epoch. His attachment to mercantilist ideas, however, protected him from the nominalist error of seeing money as a simple token. In his objection to the quantity theory of money Steuart argues that *the level of commodity prices* depends upon other causes than *the quantity of money within a country*. '"The standard price of every thing" is determined by "the complicated operations of demand and competition", which "bear no determined proportion whatsoever to the quantity of gold and silver in the country"'.[15] 'Let the specie of a country, therefore, be augmented or diminished, in ever so great a proportion, commodities will still rise and fall according to the principles of demand and competition; and these will constantly depend upon the inclinations of those who have *property* or any kind of *equivalent* whatsoever to give; but never upon the quantity of *coin* they are possessed of.'[16] The volume of commodity circulation and the prices of commodities are what determine what money is demanded in circulation. 'Now the state of trade, manufactures, modes of living, and the customary expence of the inhabitants, when taken all together, regulate and determine what we may call the mass of ready-money demand...'[17] 'The circulation of every country... must ever be *in proportion to the industry of the inhabitants, producing the commodities which come to market* ... If the coin of a country, therefore, fall below the *proportion* of the produce of industry *offered to sale*... inventions, such as symbolical money, will be fallen upon to provide an equivalent for it. But if the specie be found above the proportion of the industry, it will have no effect in raising prices, nor will it enter into circulation: it will be hoarded up in treasures ... Whatever be the quantity of money in any nation, in correspondence with the rest of the world, there never can remain *in circulation*, but a quantity nearly proportional to the consumption of the rich, and to the labour and industry of [its] poor inhabitants'.[18]

Steuart, therefore, denies that commodity prices are dependent upon the quantity of money in circulation; to the contrary, it is the quantity of money in circulation which is determined by *the demands of commodity circulation*, including *the level of commodity prices*. Taking the total mass of money in the country, one part enters into *circulation*; what remains over and above the money that commodity circulation requires lies outside the latter, either to be accumulated as a hoard (*reserves*) or as articles of luxury. If commodity circulation's demand for money expands, part of this hoard is put into circulation; in the opposite situation coin will flow out of circulation. The ideas that Steuart had put forward in contraposition to the quantity theory were extended in the 19th century by *Tooke*,* and then later on by *Marx*. These two theories—Hume's quantity theory, on the one hand, and Steuart's doctrine, on the other—represent in brilliant fashion the two basic tendencies in the theory of monetary circulation that even to this day are vying for supremacy in economic science.

*Tooke's main work was his *History of Prices, and of the State of the Circulation, from 1839 to 1847 inclusive*. [Thomas Tooke, 1774-1858. Rubin gives the work as *A History of Prices (1838-1857)*. There was an earlier edition of Tooke's treatise (London, 1838), with the same title, only covering the period 1793 to 1837. There was also a later edition, co-authored with William Newmarch (1820-1822), *A History of Prices, and of the State of the Circulation during the Nine Years 1848-1856*; its two volumes formed Volumes Five and Six of the *History of Prices from 1792 to the Present Time* (London, 1857). We have found no reference to an edition for the years cited by Rubin. The edition listed here was published in London in 1848—*Trans.*]

---

1 The term used by Rubin is 'intrinsic value' (*vnutrenyaya stoimost'*). Law's own term is that given here: 'Silver was exchanged in proportion to the use-value it possessed, consequently in proportion to its real value. By its adoption as money it received an additional value.' John Law, *Considérations sur le numéraire et le commerce* (1705), cited by Marx in *Capital*, vol. I, p. 185.
2 David Hume, 'Of the Jealousy of Trade,' in *David Hume, Writings on Economics*, edited with an introduction by Eugene Rotwein (Madison, Wisconsin, University of Wisconsin Press, 1970), p. 79.
3 *Ibid*, p. 78.
4 Hume, 'Of the Balance of Trade,' in Rotwein, *op cit*, p. 75. Rubin's rendition of this passage in the Russian is little more than a paraphrase; the original is given here.
5 'Of the Balance of Trade,' *ibid*, p. 76.
6 Hume, 'Of Interest,' in *ibid*, pp. 50-56.
7 Hume,'Of Money,' in *ibid*, p. 37.
8 'Of Money,' in *ibid*, p. 33.
9 'Of Interest,' in *ibid*, p. 48.
10 'Of Money,' in *ibid*, pp. 41-42.
11 'Of Interest,' in *ibid*, p. 48.
12 'Of Money,' in *ibid*, p. 38.
13 'Of Money,' in *ibid*, p. 38.
14 'Of Money,' in *ibid*, p. 42.
15 Rubin presents this sentence as if he has quoted it directly from Steuart. In fact, he has quoted it from Marx, *A Contribution to the Critique of Political Economy* (London, Lawrence & Wishart, 1970), p. 166, where Marx was paraphrasing Steuart by combining elements of separate sentences from Chapter xxviii of Book II of the *Principles* (Skinner edition, p. 344 and pp. 341-42, respectively): 'I have laid it down as a principle, that it is the complicated operations of demand and

competition, which determines the standard price of everything' (p. 344). 'From this I still conclude, that it is in countries of industry only where the standard prices of articles of the first necessity can be determined; and since in these, many circumstances concur to render them either higher or lower than in other places, it follows, that in themselves they bear no determinate proportion whatsoever, to the quantity of gold and silver in the country...' (pp. 341-42). The sentence we have given in the English text is quoted from Marx; the phrases quoted from Steuart are in double quotations.

16 Steuart, *Principles* (Skinner edition), Vol. Two, p. 345. Steuart's italics.

17 Steuart, *Principles*, cited in Marx, *Critique*, pp. 165-66.

18 Steuart, *Principles* (Skinner edition), Vol. Two, p. 350. Steuart's italics.

Part Two

The Physiocrats

CHAPTER NINE

The Economic situation in Mid-18th-Century France[1]

Before taking up the history of the Physiocratic school, we must first outline, in its general features, the state of the French economy in the middle of the 18th century. The Physiocratic school captured the attention of a broad circle of society above all by virtue of its programme for the regeneration of agriculture and its protest against mercantilist policy. To understand how this school emerged we must acquaint ourselves with the *condition of agriculture* and the fortunes that befell mercantilist policy in France during the 18th century.

France had consistently pursued a mercantilist policy beginning with the administration of *Colbert* (1661-1682), Louis XIV's famous minister. Colbert is regarded as the classical representative of mercantilism, and has sometimes been mistakenly taken as the founder of mercantilist policy itself (which thus became known as 'Colbertism'). In reality, Colbert was merely pursuing with a dogged consistency a policy that was generally typical of the early capitalist period, that of using the state to give implantation to trade, shipping, and industry. Colbert hoped by these means first of all to make the country more wealthy and replenish the state treasury (which suffered from constant deficits,) and, secondly, to politically weaken the feudal aristocracy. In order to develop domestic trade, Colbert wanted to do away with the customs posts that existed between the separate provinces, and with the stationing of guards along the roads and bridges that belonged to individual feudal lords. But opposition from these same provinces and lords meant that Colbert could form his single customs union over a part of the country only; it needed the Great French Revolution to finish the job of uniting all of France together into a single customs union. To develop foreign trade, Colbert took care to build up shipping, constructed a sizable fleet, encouraged trade with India, and founded colonies in America. He placed foreign trade upon the so-called 'balance of trade' system: the import of foreign-made industrial goods was forbidden or impeded, while the export of French manufactures was stimulated by the use of bonuses. Colbert spared no expense in his

his efforts to implant new branches of industry in France, especially those which worked for export. He assisted in the setting up of workshops for cloth, linen, silk, lace, carpets, stockings, mirrors, and the like, handed out subsidies, premiums, and interest free loans to their organizers, whom he freed from tax obligations while granting many of them monopoly rights over manufacturing. To ensure that industry would have cheap hands and inexpensive raw materials, he put a ban on the export of corn and primary materials, much to the detriment of agriculture.

The industry thus implanted at state expense was subjected by Colbert to the strictest *state control*. As a means of guaranteeing that French commodities would win out over foreign competition, the state took care to see that they were of high quality. There were countless regulations and instructions to define the most meticulous details of their manufacture: the length and width of materials, the number of threads in a warp, methods of dyeing, etc. During the first years of Colbert's administration some 150 regulations were issued laying down rules for the manufacture and dyeing of woven goods; one such instruction, dated 1671, contained no less than 317 articles relating to the 'decoration of woolen fabrics of all colours and to the elements and drugs thereby employed.' Special works inspectors were appointed to see that these rules were adhered to; they examined commodities both in the workshop and on the market, interfered in every detail of production, carried out searches, and so on. Commodities that had been manufactured in violation of the rules were seized and put on public display, together with the name of the industrialist or merchant concerned. For the violators there were fines and confiscations. This strict regulation over industry introduced by Colbert became even more petty and constraining under his successors.

At first the mercantilist policy of Colbert and of those who followed him was crowned with brilliant success; France held a place in the front rank of Europe's trading and industrial nations. But the successes were fragile, as became obvious almost immediately following Colbert's death, and even more so in the mid-18th century. French industry, it is true, was unrivalled in the production of luxuries for the needs of the court and the aristocracy; many of these luxuries even earned the title of 'French commodities.' The court at Versailles eclipsed and outshone all the other courts of Europe, and Paris became the acknowledged pace-setter in fashion and taste. Yet these outward achievements rested upon a frail base. In a country where the population was overwhelmingly made up of peasants ruined by the exactions of the gentry and the tax officers, capitalist industry had little scope for advancement. Instead of providing the state with a source of income, the new 'manufactories' demanded their usual privileges and subsidies, absorbing part of the state's resources. The number of centralized workshops remained insignificant, being in their majority simply distribution offices which farmed out work to cottage workers. France's dreams about her industry conquering *vast foreign markets and colonies went unrealized.* Her battle with England in the mid-18th century for domination over the world market ended with England's victory, as the latter took control of France's American colonies and consolidated its own position in India. In cloth, which was the most important branch of industry, England held first place. The petty regulation of industry, in which Colbert had invested such great hopes for bettering the quality with which commodities were manufactured, in reality became an obstacle to the introduction of technical improvements, inhibited the diversification of production, and prevented industrialists from rapidly adapting to the demands of the market. Bacalan, intendant[2] of manufactories, noted that regulations placed constraints upon the entrepreneurial activity of manufacturers, put a check on competition, and stood in the way of inventiveness. 'Freedom is preferable to regulation.' he wrote in 1761. 'At least it occasions no harm, while rules are always dangerous, and a good many of them are absurd.' In the middle of the 18th century not just entrepreneurs but even state officials were increasingly and persistently voicing the demand for *the abolition of the constricting regulation of industry* characteristic of mercantilist policy.

Certainly, the strongest brake upon the growth of France's capitalist industry was not the constraining influence of mercantilist policy in and of itself, but the fact that its authors pursued it in a country of impoverished peasants while simultaneously preserving a seigniorial system and absolute monarchy. Had France had a developed agriculture, industry could have reckoned upon an extensive internal market, especially when account is taken of the country's large population (at the beginning of the 18th century there were around 18 million people in France, compared to a population in England and Wales numbering no more than five or six million). France's backward and decimated agriculture proved too narrow a base, however, for the growth of capitalist industry. *The purchasing power of a half-starved peasantry*, compelled to hand over the better part of its meagre harvest to the gentry and the state, was inconsequential. Without the resources to purchase industrial articles the peasantry had cut its living requirements to a minimum. According to Young, who visited France before the revolution, the peasantry wore neither shoes nor stockings, and sometimes went even without clogs.[3] In Brittany the peasants were dressed from head to foot in the same coarse cloth usually used to fashion sacks. In the end, Colbert's mercantilist policy of building up a brilliant manufacturing industry upon the backs of an unclothed and unshod peasantry was bound to fail. In France in the mid-18th century, the conviction was becoming increasingly widespread that the primary condition for the durable growth of a capitalist economy was the advance of agriculture and the abolition of feudal survivals in the countryside.

The reality of the situation was that French *agriculture* during the 18th century was in deep decline and utter devastation. It is true that with all but the most insignificant exceptions serfdom had already died out: the individual peasant was a free man. But his land was still encumbered with innumerable *feudal payments and obligations*. Only a small percentage of the peasantry possessed land that was legally fully their own private property (*alodium*). The majority owned land upon which they paid a so-called *cens*. The peasant paying a *cens*, too, was seemingly in possession of his own land: he could sell it and transfer it as a legacy. But the extent of his ownership was limited by the feudal rights of the seigneur. Every parish had its supreme master, or seigneur. The latter would occasionally own a small piece of land or have a castle in the parish, though sometimes not even that; he hardly ever appeared in the parish to look around. Nevertheless, every peasant in the parish was obliged to pay the seigneur an annual *cens*, the size of which was determined by custom and never changed. On some lands the *cens* was replaced by payment in kind, where the peasant handed over to the seigneur one tenth, one eighth, or sometimes even one quarter of his harvest (the so-called *champart*). Over and above this when the land was sold by a peasant or transferred to an heir after his death, the seigneur was paid a certain sum of money by the new owner.

Even worse off were those peasants with only small plots or those with no land at all. Some of them were employed as handicraft workers or seasonal labourers, or hired themselves out as farm hands; others rented a piece of land from the seigneur or from another owner, paying for it in kind with half of their harvest. Having no resources for equipment, these share-croppers or *métayers* (so called because they gave up half of their harvest to the landowner) often received seed, livestock, or simple agricultural implements from the landlord. If lack of means meant that the *cens*-paying peasants worked the land by primitive methods, cultivation was even worse on lands worked by the *métayers*. Only a small portion of the land belonging to the gentry, the clergy, the crown, or wealthy members of the bourgeoisie was leased in large plots to better off peasants or tenant farmers who could invest a reasonable capital in their holdings and cultivate them along more rational principles. Unlike in England, the extensive spread of tenant farming in the 18th century, which went hand in hand with the improvement and rationalization of agriculture, was rarely to be found in France. In the French countryside of the 18th century the role played by *bourgeois forms* of landed property and rent was still insignificant compared to *ownership where a* cens *was paid* or to *sharecropping by métayers*, both of which were enmeshed in a vast number of survivals from the feudal system.

No less burdensome than the seigneur's exactions were the *state taxes* shouldered by the peasant economy. The absolute monarchy required vast sums to maintain its centralized bureaucracy and its army. The rush for colonies and foreign markets that made up mercantilist policy led to endless wars of devastation. Equally, support for the new manufactories absorbed substantial resources. The other side of this was that after its protracted struggle to deprive the members of the landed aristocracy of their political rights the crown attempted to recompense them by putting into place a resplendent court, by establishing for the gentry a multitude of court and other official positions, by reinforcing their seigniorial rights on the land, exempting them from the payment of tax, etc. The gentry were completely freed from paying the most important direct tax, the *taille*, out of which the clergy received an annual cash payment of a fixed amount. As the towns people were also able to evade both the *taille* and other direct taxes, the latter's entire weight fell on those at the bottom of the agricultural population—the peasantry. The latter suffered also from indirect taxes, especially that on salt. Both the size of the taxes and the way in which they would be levied often changed, so that the peasantry never knew in advance how much was going to be demanded of it. What usually happened was that the collection of taxes was entrusted to wealthy tax farmers, (*fermiers générals*)[4] who used this privilege to build up fortunes; sometimes the treasury would receive only a small share of the total taxes that had been exacted. These taxes to the state (to which church tithes must also be added) exhausted the peasant economy. Not long before the revolution, the Duke of Liancourt pointed out that a fiscal policy based upon 'the custom of constantly demanding money from the cultivator without giving anything back in exchange' would severely retard agricultural progress. Another brake on this progress was *corn-pricing policy*. Ever since Colbert, the French government had pursued with increasing diligence a mercantilist policy of *bringing down the price of corn*: its aim was first of all to cheapen the raw materials and the hands needed by industry, and second, to ensure that the urban population—that of Paris in particular—would be provided for. The export of corn abroad was forbidden; its import was permitted. Within the country the corn trade was subjected to extremely tight regulation: the sale of corn was prohibited other than at the markets, and it could not be sent out of a city; because of fears about speculation and rising prices, the activities of corn merchants were greatly restricted, and there was no free movement of corn between individual provinces. The result was that the high price of corn in some localities was accompanied by its undervaluation in others, and prices fluctuated sharply from year to year. Agriculture suffered at one and the same time from low corn prices and from the uncertainty engendered by their constant fluctuation.

Ruined by payments to the seigneur and the state, and suffering under the policy on the price of grain, the peasant economy was unable to accumulate the means for making improvements in *agricultural technique*. The three field system of cultivation predominated, although in many localities even the two field system still operated. The sowing of forage had been introduced only in certain northern provinces. While the patchwork of fields and compulsory crop rotation kept industrial crops from being widely cultivated, the raising of livestock was in a pitiable state and the fields went virtually unfertilized. A lean cow, a wooden plough, and a harrow made up the inventory of a French peasant—at the same time as the English farmer was in the main already practising crop rotation, had a flourishing animal husbandry, and was using iron agricultural tools. It was no wonder that the French harvest lagged far behind that in England (usually not exceeding a fifth of the latter), and that, from the beginning of the 18th century up to the revolution, France experienced no less than thirty famine years.

The poor productivity of agriculture, together with *the low price of corn* which prevailed until the middle of the 18th century, reduced the income side of the peasant's budget at the very time when his *payments to the state and to the seigneur* were stretching his outgoings to the absolute limit. As Taine aptly put it, the French peasant of the pre-revolutionary period resembled a man plunged up to his neck in water, who ran the risk of drowning with the slightest wave. Except for small groups of well-off peasants and farmers, the overwhelming mass of peasants lived a life of perpetual and brutal want never having enough to eat and never able to make ends meet. Bishop Massillon wrote in 1740: 'Our rural people lives in horrible destitution, without beds and without furniture, the majority even feeding themselves for half the year with bread made of barley or oats, this constituting their only food though they be compelled to snatch it out of their own mouths and those of their children just to pay taxes.' Moreau-de-Jeunesse, the famous statistician, characterized the pre-revolution state of the Breton peasant in the following terms: 'Out of the four sheafs which he brings in from the fields, one belongs to the seigneur, another is owing to the priest or to the priory of the neighboring monastery, a third goes entirely to the payment of taxes, and the fourth to cover his costs of production.' If this is a somewhat exaggerated calculation, it was in any event not at all unusual for half the gross harvest to go in payment to the state and the seigneur, so that with deduction for seed (and given the low yields, this could come to 1/5 of the harvest) scarcely any grain remained to provide for the farmer. There were times when payments owing to the state and the seigneur would be levied on the basis of the gross, and not the net yield (i.e., with no allowance being made for the deduction of seed); when this happened the peasants, especially the *métayers* would sometimes not even have enough left with which to keep themselves alive.

It was not simply that these conditions made agricultural improvement and expansion impossible; they profoundly disrupted the process of *simple reproduction of agriculture at its old level.* Many rural districts were actually being depopulated; in others, those cultivating the land either went off to do seasonal work or swelled the near countless ranks of beggars. The Marquis de Turbilly noted in 1760 that half the cultivable land lay empty, and at every step one could see fields abandoned by their tillers. Arthur Young's travel notes give a striking picture of how the decline of agriculture was affecting the majority of French districts, excepting but a handful of lucky provinces. In one province he describes how a third of the land was totally uncultivated while the remaining two thirds showed evident signs of devastation; in another he encounters nothing but 'poverty and poor crops'; and in a third, 'the poor people who cultivate the soil here, are *métayers*...who hire the land without ability to stock it...a miserable system that perpetuates poverty and excludes instruction.'[5]

The degradation of France's agriculture during the 18th century was a clear sign that there was now a glaring contradiction between the need to develop the productive forces and the antiquated socio-political regime. For France to develop her capitalist economy there would have to be an advance in agriculture—but the precondition for this was the replacement of the seigniorial system with bourgeois forms of land tenure. In the middle of the 18th century this had already become obvious but there were *two contradictory paths* by which this change could take place, depending upon whether the land, which under the feudal system was jointly owned by lord and peasant, became the private property of the former or the latter. If it went to the lord, this would mean that large-scale landowners would gradually drive the peasants paying the *cens* and the *métayers* off the land and would begin to lease it out in large tracts to well-to-do farmers. This process, whereby priority was given to the leasing of large plots on a capitalist basis while the majority of peasants was made landless, was what took place in England, which thus became a country of *large-scale land holdings*. The second path implied that the peasants would be freed of all seigniorial payments and obligations and that they would separately become the sole proprietors of the land they worked. This was the path followed by the Great French Revolution, out of which France emerged as a country of *small-scale peasant holdings*.

In the middle of the 18th century, however, a revolutionary solution to the agrarian question still seemed excluded. The only way by which an impoverished agriculture could be rationalized and given a boost appeared at that time to be along the English model of large-scale tenant farming, the spread of which had already played no small role in promoting the success of English agriculture. Above all, this type of *capitalist agrarian reform* lay in the interests of the farmers and the well-off layers of the peasantry, i.e., the *rural bourgeoisie*, or the so-called '*rural third estate.*' A reform along these lines could to some extent work to the advantage of the landowners, who would retain their right of ownership over the land and would receive rent. In the mid-18th century this type of agrarian reform found its proponents in the Physiocrats, who proposed to resolve the historical task of advancing agriculture by replacing the seigniorial system with capitalist tenant farming.

In their programme, the Physiocrats were concerned to create favourable conditions for the development of capitalist agriculture. We have seen that up to the mid-18th century, French agriculture suffered first, from low productivity of the land and poor harvests, second, from low corn prices, and third, from heavy seigniorial obligations and state taxes. The first two of these acted to reduce the income of the cultivator, while the third placed the most severe strain upon the budgeting of his expenses. In their programme the Physiocrats demanded that all of these unfavourable conditions be done away with. First, they championed the *rationalization of agriculture* along the lines of English farming. Second, they launched a furious attack against the mercantilist policy of *reducing corn prices*, and demanded both freedom of trade and the *free export* of corn. Third, their programme called for the complete relief of the farming class from taxation and for *all taxes* to be shifted *onto the rent paid to the landowners*.

The Physiocrats did not, however, confine themselves simply to preaching reforms in the interests of bolstering agriculture and making the rural bourgeoisie more wealthy. They tried to give their practical programme a theoretical foundation. They argued that only with the implementation of their reforms could it be assured that the process of social reproduction would proceed normally and provide a substantial net income (or in their terminology, 'net product'). The leader of the Physiocrats, Quesnay, with his theory of *social reproduction* and the theory of *net product* (or surplus value), made what was the first attempt to analyze the *capitalist economy taken as a whole*. And though the Physiocrats' practical programme came to nothing, their theoretical ideas, once freed from their onesidedness and errors, were taken up and developed by later currents within economics (the Classical school and Marx), and thus earned for Quesnay immortal fame as one of the founders of modern political economy.

1 Quotations in Part Two not otherwise referenced in the notes have been translated from the Russian.

2 The intendant was a representative of the crown in a particular province who was responsible for the inspection of various public services. Bacalan was in fact intendant of commerce; in 1764 he published his *Paradoxes philosophiques sur la liberté du commerce entre les nations*.

3 Arthur Young, *Travels in France During the Years 1787, 1788, & 1789*, edited by Constantia Maxwell, Cambridge University Press, 1929. 'All the country girls and women are without shoes or stockings; and the ploughmen at their work have neither sabots nor stockings to their feet. This is a poverty that strikes at the root of national prosperity; a large consumption among the poor being of more consequence than among the rich. The wealth of a nation lies in its circulation and consumption; and the case of the poor people abstaining from the use of manufactures of leather and wool ought to be considered as an evil of the first

magnitude. It reminded me of the misery of Ireland.' (*Travels*, pp. 23-24.)

4 The tax farmer (*otkupshchik*) was one who paid the government a fee for the right to collect taxes (*sdavat' vzimanie nalogov na otkup*, or literally, 'to farm out the collection of taxes').

5 Young, *Travels in France*, p. 16.

CHAPTER TEN

The History of the Physiocratic School

The Physiocratic theory was developed in France during the middle of the 18th century. The entire first half of that century, however, can be looked upon as the age of the *physiocrats' forerunners.*

The ruin of the peasantry and the decline of agriculture had attracted attention even at the close of the 17th century. Labrouier was already painting a sombre picture of the peasants' poverty, while Fénelon was writing that the people were starving and abondoning the cultivation of the land, and that 'France was being turned into a desolate and starving poor house.'[1] *Boisguillebert* (1646-1714), the economist who called himself the barrister of agriculture, came out against the Colbertist policy of lowering corn prices and demanded the free export of grain. He maintained that 'never is a people so unfortunate as when the price of corn is cheap.' He also stood opposed to the mercantilists' exaggerated assessment of the role of money. Money in his view ought only to be ascribed a modest and secondary role as a means of facilitating exchange. A contemporary of Boisguillebert, the famous Marshall *Vauban*, demanded an easing of the ruinous tax burden on the peasantry. Boisguillebert's writings brought him into disgrace, while Vauban died on the very day when his book was being ceremonially burned at the hands of the executioner.

These same ideas were subsequently developed further by the Marquis *d'Argenson* (1694-1757). The struggle against mercantilist protectionism had led him to make a principled defence of complete freedom of trade. 'Do not interfere (*laissez faire*)—such must be the motto of every public authority.' It is in the writings of Argenson that we first meet—and with some frequency too—with that famous formula of the free traders, '*laissez faire*' (later supplemented, probably by Gournay, with the words, '*et laissez passer*').

Thus, by the mid-18th century, there were to be found certain thinkers whose individual ideas and practical demands were to become part of the Physiocratic system. It was with *the middle* of the 18th century, however, that these ideas and demands became the subject of lively debate amongst broad sections of society. The degradation of agriculture and the stagnation of industry, the impoverishment of the peasantry and the incessant state deficits meant that for a wide public there was an obvious lack of credibility in the ***ancien régime***. Henceforward there began in France a pre-revolutionary epoch of dissatisfaction and ferment, of projects for reform and quests for new social and philosophical formulae. With the beginning of the 1750's, economic questions likewise began to be debated in books and periodicals, within the salons of high society, and in government commissions. The failures of the state's corn policy helped generalize the conviction amongst broad sections of society that the old prohibitions and restrictions on the corn trade would have to be revoked. It was partly under the impact of public opinion that in 1754 the state permitted the free transport of corn between individual provinces, although the ban on exporting it out of the country remained in force.

It was during these years of heightened social interest in economic issues that two groups of economists appeared on the scene: one around *Gournay*, the other around *Quesnay*. Both grew up in opposition to the prohibitions, monopolies, and regulations of mercantilist policy. But while Quesnay rejected this policy in the name of the interests of ***agriculture and the rural bourgeoisie***, Gournay's demand was mostly for the removal of those restrictions which were putting a brake upon the free development of ***urban industry and trade*** (the guilds, industrial regulations, and domestic tariffs). Unlike Quesnay's school, Gournay and his followers were primarily interested in practical matters, and left behind no works of theoretical value.

Francois Quesnay (1694-1774), born into a family of small-scale, semipeasant landowners, had made his way in life entirely unaided. A medical doctor by training, he earned a reputation as an eminent physician, publishing a number of scientific works on medicine and biology. In 1749 he was invited to the court as attending physician to the famous Madame de Pompadour, the favourite of Louis XV, and for three years was also the appointed physician to the king himself.

During this time Quesnay, who was already 55, abandoned his scientific medical pursuits and devoted his energy to working on the economic problems that had been stirring the public opinion of his day. In his first articles, published in 1756-1757 in the famous Encyclopedia, Quesnay attributed the decline of agriculture to ***heavy taxes*** and to the artificially *low price of corn* produced by the ban on its export abroad. Already in these articles Quesnay was depicting the superiority of *large-scale farming*, and advising that prosperous farmers, who could invest substantial capital in agriculture, be attracted to the countryside.

It was in Quesnay's later works that he laid down the theoretical foundations to his ideas. In 1758 he produced his famous *Tableau Economique*, later supplemented by the writing of his *Economie Générale et Politique de l'Agriculture*.[2] These two works set down the basic propositions of Quesnay's economic theory and policy. Quesnay gave a statement of the philosophical basis to his theory in his *Le Droit Naturel* of 1765.*

If the development of Physiocratic theory was the work of Quesnay alone, it found its popularizers and propagandists in a talented group of followers who gathered around him and who formed the tightly-knit Physiocratic school, or 'sect' as its opponents termed it.** Most active amongst this group were the marquis *Mirabeau the Elder* and *Dupont de Nemours* (with lesser roles being played by *Mercier de la Rivière, Le Trosne,* and *Badeau*). The only one of the Physiocrats' followers who can properly be considered an original and independent thinker is *Turgot*, who never in fact belonged to the 'sect' in the strict sense of the word.

The Physiocrats propagandized their ideas in books and journals, salons, and government commisions. At one point they even gained control over a semi-official journal put out by the government, but being soon ousted from it, they acquired their own journal, *Ephémérides*. In 1767 the Physiocrats began regular weekly meetings in the Salon Mirabeau in Paris, which for ten years served as a rallying point for these kindred spirits and the recruitment of new members. These meetings played no small part in fostering an 'agrarian vogue' through wide circles of society. The Physiocrats' ideas drew universal attention, attracting the interest both of Voltaire and Rousseau and of royal personages of such eminence as Catherine II.

As Quesnay's ideas gained currency it became increasingly clear that as a social current Physiocracy, though securing definite reforms, was doomed to failure in the conditions of pre-revolutionary France. France was moving inexorably towards a revolution, in which the

*The most notable of Quesnay's other works are his *Analyse* of the *Tableau Economique*, his *Despotisme de la Chine*, and his *Dialogues sur le commerce et sur les travaux des artisans*.

**Quesnay's followers called themselves 'economists,' but became generally known by the sobriquet 'Physiocrats' after Dupont had published Quesnay's works under the celebrated title *Physiocracy, or the Arrangement of the State that is Most Profitable for the Human Race*. Physiocracy means 'the rule of nature.'

broad popular (including peasant) masses led by the urban bourgeoisie were to assert themselves against the crown and the privileges of the aristocracy. The Physiocrats' endeavours to avoid the agrarian revolution by means of agrarian reform—to be carried out in the interests of a weak rural bourgeoisie and with the assistance of the crown and certain sections of the gentry—had no chance of being realized. By adhering to the absolute monarchy, the Physiocrats isolated themselves from that current of social thought which dominated their age, the Encyclopedists, who were the ideologues of the progressive urban bourgeoisie. At the same time they provoked attack upon themselves from those who directly defended the economic interests of the commercial-industrial bourgeoisie (including the followers of Gournay): from them came a bitter assault against the Physiocratic doctrine of the 'sterility' of the commercial-industrial class and the need for the free export of corn. In 1770, the famous economist *Galiani** counterposed to the Physiocratic ideal of an agricultural country exporting its corn and importing back from abroad cheap finished manufactures, the idea of a developed industrial nation which could consume all its own corn domestically, and even make additional imports. On the issue of corn exports the interests of agriculture and industry sharply conflicted, and the liberal law of 1764, which allowed for the free exportation of corn abroad, was abrogated in 1770.

Hopes that the Physiocratic programme could be carried out were given new life during the ministry of *Turgot* (1774-1776).** After his appointment as minister of finance, Turgot attempted to carry out a number of important reforms. He restored the freedom of the domestic corn trade, issued a law abolishing the guilds and establishing freedom of occupation, and replaced the obligation—so burdensome to the peasantry—of doing labour service upon the roads (the *corvée*) with a monetary payment falling on all landowners, including the gentry. But Turgot's reforms provoked intense dissatisfaction amongst the reactionary sections of society (the court aristocracy, the gentry, and the tax farmers), which led the minister-reformer to resign. The hopes of the Physiocrats notwithstanding, the absolute monarchy and the landowning class proved incapable of carrying out any reform of society, and France rapidly proceeded towards the formidable events of the Great Revolution.

*See Chapter Seven, above.

**Turgot's (1727-1781) main work is his *Réflexions sur la formation et la distribution des richesses*, written in 1766 and published in 1769-70.

Turgot's fall was the final blow completing the collapse of the Physiocrats as a defined social current. At first this collapse of their practical programme proved fatal even to their theoretical ideas, which for years, even decades, were either buried in oblivion or made the object of the cruelest derision. Marx, in the middle of the 19th century, was one of the first to point out the immense scientific service of the Physiocrats concealed beneath the fantastic form or the mistakes of their theory. At the end of the 19th and beginning of the 20th centuries the Physiocrats enjoyed a temporary rehabilitation, and Marx's high assessment was fully confirmed by the thorough study of their theoretical ideas. At present Francois Quesnay contends with Adam Smith for the honour of founder of political economy.

1 Cited in Gaetano Salvemini, *The French Revolution, 1788-1792* (London, Jonathan Cape, 1954), p. 33.

2 This is in fact the subtitle of the *Philosophie Rurale*, written by Mirabeau with Quesnay's collaboration, and published in 1763. Certain extracts appear in Ronald L. Meek's *The Economics of Physiocracy* (London, George Allen & Unwin, 1962).

CHAPTER ELEVEN

The Social Philosophy Of the Physiocrats

The Physiocrats, as we know, considered it essential to replace the small-scale peasant holding with large-scale farming, wishing to guarantee these farmers the free export of corn and freedom from taxes. By what means did they hope to carry out their programme? In their answer to this question the Physiocrats differed sharply from the members of the Englightenment, the ideological vanguard of the urban bougeoisie. The latter were harshly critical of the absolute monarchy, and counterposed to it the political ideal of either a constitutional monarchy with separation of powers (Montesquieu) or a democratic state based on the idea of popular sovereignty (Rousseau). In this manner the Enlightenment, despite its failure to work out its ideas to completion, put before the bourgeoisie the task of the revolutionary conquest of political power. The Physiocrats had a different solution to this political question. They were *defenders of an enlightened absolutism*, of an absolute monarchy which, in Quesnay's words, was 'the only power standing above all of the different exclusive interests, which it must restrain'. The Physiocrats expected such an enlightened monarch to put through the economic reforms that they were recommending.

There have been many authors who have pointed out the logical contradiction between the Physiocrats' monarchist views and their economic demands for maximum individual freedom. Their adherence to the monarchy can, however, be explained from their general social and class position. The Physiocrats did not so much try to rely upon an already existing rural bourgeoisie, which in any case was numerically insignificant and without influence, as to create conditions that would favour this class's economic development. Under these conditions it was clearly hopeless to dream of the rural bourgeoisie conquering political power. If the monarchy were overthrown power promised to pass into the hands either of the gentry, which was still holding onto its political privileges, or of the youthful and wealthy urban bourgeoisie. Both of these prospects threatened the Physiocratic programme with collapse. If the gentry came to power, any tax reform in the Physiocratic spirit would be precluded, and the burden of taxation would be hoisted upon the class of farmers. Seizure of power by the urban bourgeoisie, on the other hand, might further reinforce (as the Physiocrats feared) the hated mercantilist policy of promoting trade and industry at the expense of agriculture. With the overthrow or undermining of the monarchy (coupled with a stronger political role for the gentry or the urban bourgeoisie) threatening the Physiocratic programme with total failure, there was nothing left for the Physiocrats to do but to invest all their aspirations in the absolute monarchy and declare themselves its upholders.

When the Physiocrats expressed their support for keeping the absolute monarchy in power, this was certainly not so that the latter should continue its ruinous policy of sustaining feudal and mercantilist privileges. In the Physiocrats' scheme of things these privileges went against reason and '*natural right*', i.e., those eternal and immutable laws, preordained for all time by the Creator, and obligatory both for individuals and for the state power. The crown must not issue laws simply at will, since these might turn out to be in conflict with natural right and bear incalculable harm. It is this lack of knowledge of the eternal laws of natural right that explains the multitude of harmful '*positive*' laws which the state issues. If the crown is to avoid confusion and disorder in social life it must see to it that all its laws adhere strictly to the prescriptions of natural right. For the Physiocrats' political ideal was a 'legal despotism', i.e., a monarchy which carried out the dictates of natural right or (as we shall see) encourages the development of the bourgeois economy.

The Physiocrats thereby placed a limit upon the crown's arbitrary legislative authority, in the form of eternal and inviolable natural laws, which stand above the 'positive' laws of the state. The Physiocrats were here following the *doctrine of 'natural right'* developed by earlier bourgeois thinkers of the 17th century (Grotius, Hobbes, and Locke). As part of its struggle against the antiquated feudal regime the bourgeoisie was raising the demand for a new social order, a 'natural order', which, in its eyes, was one of justice and reason. As a counterweight to the privileges that were sanctified by the authority of the crown and its laws, the bourgeoisie claimed sanctity for its demands in the authority of supreme, eternal natural laws, before which both crown and the positive laws of the state were obliged to defer. Just what made up this ideal 'natural order' that the thinkers of the 17th and 18th centuries were demanding be put in place?

In essence what they meant was the *bourgeois social order*, freed of feudal survivals and granting to its individual members the possibility of an unfettered pursuit of profit based on free competition with other members of society. *The right of the individual* to satisfy his natural wants and to acquire the things necessary to do this, the right of *personal freedom* (i.e., the freedom of the individual from serfdom), freedom of *private property* (i.e. property free of feudal obligations and restrictions), *freedom of individual competition* (i.e., the abolition of feudal and guild restrictions over economic activity)—such were the most essential 'natural rights' of the individual which the bourgeois ideologists demanded be established.

The doctrine of natural right played a crucial *revolutionary* role, as a battering ram which burst through the strongholds of the feudal regime and the absolute monarchy. The more radical thinkers of the 18th century had demanded that personal 'natural right' be assured not only within the economy but in politics as well, i.e., they called for a democratization of the governmental system (Rousseau's doctrine of the social contract and popular sovereignty). In keeping with their conservative political inclinations, the Physiocrats attempted to blunt the revolutionary edge of the theory of natural right, and refrained from using it to draw political conclusions. On the other hand, it was to their credit to have applied the ideas of natural right most consistently *to the realm of economic life*. For the Physiocrats, the 'natural order' was the totality of economic conditions that were necessary for the unhampered development of the bourgeois-capitalist economy (above all, within agriculture). The laws of bourgeois economy were declared by the Physiocrats to be natural laws, and no legislator had the right to violate them. In this way the Physiocrats helped to liberate economic life from state interference, while they, on the other hand, began to arrive at a conception of an internal, 'natural', law-determined regularity to economic life, existing independently of the arbitrariness or intervention of any legislator.

From the very outset natural right acquires with the Physiocrats an *economic* colouring, being defined by Quesnay as 'the right which man has to things suitable for his use.' Now while this formulaion prompted more radical thinkers to criticize the unequal distribution of goods between the rich and the propertyless, Quesnay is quick to limit it: man's natural right is reduced simply 'to a right to the things whose use he can obtain.' But to ensure that things are actually acquired, men 'must possess ... bodily and mental faculties, together with ... means and instruments.' The inequality that exists in the faculties and 'resources' (i.e., wealth) that people possess will create tremendous inequality in the use that they can make of their natural right. While to communist theorists, like Mably, and even to the more radical petty-bourgeois ideologists, like Rousseau, this inequality appeared as a violation of natural right and impelled them towards a critique of private property, Quesnay easily accepted it as resulting inevitably 'from the arrangement of the laws of nature.' Inequality of property is an unavoidable but minor evil which, in view of the immense benefit that private property itself brings in encouraging personal diligence, must necessarily be accommodated. Natural right, therefore, reduces itself to man's right to freely *apply his labour* and to the *right of private property*. 'Personal freedom and property are guaranteed to people from without, by natural laws, upon which reposes the basic harmony of well-ordered societies.'[1]

This formula can be simplified even further: for personal freedom is nothing but the basic form of property, i.e. '*private property*', or the individual's right to freely apply his labour. From private property derives '*movable property*', or man's right to the things which he creates with his labour. Finally, man, who with the help of his labour and movable things has made virgin land suitable for agriculture, acquires in perpetuity '*landed property*'. In essence, the natural right of man comes down to these three forms of property, all closely associated with one another. In their theory of property the Physiocrats were repeating the ideas of Locke, making, however, one characteristic departure: while Locke was by no means sure that man had a right to those fruits of the earth that could not be cultivated by his own efforts, the Physiocrats justified large-scale landed property on two grounds: firstly, in the course of making the land suitable for cultivation landowners (or their forefathers) had made certain outlays (labour and movable things); secondly, these landowners would only invest large capitals in the land if their ownership over it was firmly guaranteed, and such investment was a necessary condition of agricultural prosperity. The Physiocrats were agreed that the owners of large estates should be left their land, provided they leased it to capitalist farmers.

As we see, the Physiocratic doctrine of natural right bears the visible marks of the ambivalence of their moderate-bourgeoise social and economic programme. In so far as they leaned towards political compromise with the monarchy and economic compromise with the large landowners, they declared themselves defenders of the first and offered a rationale for the second. On the other hand, in so far as they hoped that the result of this type of social and political compromise would be the free development of capitalism within the economy proper, they proclaimed as their 'natural order' the system of bourgeois economy freed of feudal survivals and based on private property. The guarantee of 'private property' meant that the producer would be liberated from serfdom and from the fetters of feudalism and the guilds. The guarantee of 'movable property' meant the assertion of the power of capital and victory for free competition within commodity exchange. Finally, 'landed property' for the Physiocrats meant the bourgeois form of land tenure, disencumbered from seigniorial customs and premised upon the capitalist form of rent.

To establish this type of 'natural order' within the economy it would be necessary *to do away with the constraints of state tutelage*. Let the state allow broad scope to the workings of natural laws and the free play of individual interests, and concern itself simply with eliminating the artificial barriers that put a brake upon the action of these laws. 'What is demanded to make a nation prosperous? To cultivate the land with the greatest possible success and to safeguard society from thieves and beggars. The realization of the first of these demands is left to each person's own individual *interest*, that of the second is entrusted to the *state*'. Woe to that country whose government keeps not to the modest task of protecting society from the danger of insidious elements, but begins to interfere into the economic activity of individuals. The Physiocrats saw the mercantilist policy of strictly *regulating individual economic activity* as a source of constant disorder and enfringment of wise and natural laws. For them, *the free and uninhibited activity of individuals* was the best assurance that 'natural order' would be established within the economy. The Physiocrats were fervent advocates of *economic individualism*, as was typical of the ideologists of the nascent bourgeoisie.

1 The bulk of the discussion in this paragraph appears in *Le Droit naturel*, translated in Meek, *op cit*, pp. 43-56. All quotations but the last are taken from the passages on pp. 43-47.

CHAPTER TWELVE

Large-Scale and Small-Scale Husbandry*

Physiocratic economic theory, which we shall now begin to analyze, had as its task the investigation into, and discovery of the natural laws of the economy. The Physiocrats were convinced that they could find eternal and immutable economic laws which would accord with the laws of nature and bear maximum advantage to mankind. Though they themselves were not aware of it, what they meant by natural economic laws were the laws of bourgeois economy. The Physiocrats chose as the object of their theoretical study and as the ideal of their economic policy the *large-scale capitalist farm*: This predilection expressed both their social and class sympathies and their over-riding interest in how to maximize the growth of the productive forces associated with agriculture. We have already seen just how far the degradation and devastation of French agriculture had gone by the mid-18th century. To bring France's economic life and state finances back to health there would have to be an advance in agricultural productivity, which to the Physiocrats was conceivable only within the context of capitalist farmsteads.

The Physiocrats looked to the England of their day as an example of the rapid spread of large-scale farms and concurrent agricultural rationalization. The contrast between the backward three field system used by the small-scale French peasantry and the improved system of crop rotation employed by the English farmer leaped to the eye. The Physiocrats became zealous defenders of *new agricultural methods*. If the productivity of agriculture was to be raised it would be necessary, in their view, to introduce the rotation of crops, raise greater numbers of livestock, begin feeding cattle in stalls, make extensive use of fertilizers, and expand the sowing of industrial crops. But a holding run along such rational lines demands the investment of substantial capital, and can be tended only by large-scale farmers.

*By husbandry [*zemledelie*, the word used by Rubin in the title to this chapter—*Trans.*] we mean here, as throughout our discussion, agriculture in the general sense.

And so the Physiocrats became the defenders of 'large-scale, prosperous and scientific cultivation,' which they counterposed to the backward, 'small-scale cultivation' of the peasantry.

The Physiocrats never tired of emphasizing the *low productivity* of the holdings worked by the peasants paying a *cens* and the *métayers*. Small-scale peasants work with only the most primitive implements, have insufficient cattle, and use practically no fertilizer on their fields. As a result, the amount of produce they get from the land is insignificant, barely able to satisfy their most pressing needs. 'Husbandmen who make a wretched living from a thankless type of cultivation serve only to maintain profitlessly the population of a poor nation.'[1] Small-scale agriculture yields practically no 'net product,' or net income over and above the worker's necessary means of subsistence. It therefore follows that to make agriculture more productive small-scale peasant holdings will have to be replaced by large farms. 'The land employed in the cultivation of corn should be brought together, as far as possible, into large farms worked by rich husbandmen; for in large agricultural enterprises there is less expenditure required for the upkeep and repair of buildings, and proportionately much less cost and much more net product, than in small ones. A multiplicity of small farmers is detrimental to the population.'[2] Turgot, too, agreeing with these words of Quesnay, shows decided preference for tenant farming over holdings of *cens*-paying peasants and *métayers*: 'This method [i.e. of tenant farming, Ed.] of putting out land to lease is the most advantageous of all to the Proprietors and to the Cultivators;* it becomes established in all places where there are wealthy Cultivators in a position to make the advances involved in cultivation; and as wealthy Cultivators are in a position to provide the land with much more labour and manure, there results from it a huge increase in the product and the revenue of landed property.'[3]

The Physiocrats, therefore, proposed an agrarian reform directed at breaking those feudal-seigniorial bonds that had tied the aristocratic landowner to the *cens*-paying peasants and the *métayers*. The land ought gradually to be cleansed of the latter and leased out in large tracts and as unrestricted private property to large farmers, whether these be the more prosperous peasants or independent tenants resettling from the towns. For the poorest strata of the peasantry there would be no option but to become wage labourers for the new farmers. Seigniorial obligations would be replaced by a voluntary contract between landowner and tenant; the small-scale, semi-feudal

*By Proprietors is meant landowners, while Cultivators are the tenants.

peasant holding by the large capitalist tenant. This type of agrarian reform would represent a sort of compromise between the rural bourgeoisie, who would profit from it considerably, and the large landowners, who would retain their right of ownership over the land and would receive rent. The seigniorial order in the countryside, which had so retarded the development of the productive forces, would be supplanted by a more progressive capitalist agriculture. On the whole, however, this reform was to be carried out at the expense of the broad mass of peasants, who, as in the case of England, would be made landless and proletarianized.

The Physiocrats, however, were by no means horrified by such a prospect. On the contrary, it appeared to them as the only way out of the agricultural crisis, and they depicted it in glowing terms: 'Farmers... would increase in numbers; small-scale cultivation would disappear in one case after the other; and the revenue of the proprietors and the taxes would be proportionately increased owing to the increase in the produce of the landed property cultivated by rich husbandmen.'[4] The Physiocrats demanded of the state that it even take active measures to encourage farm holdings at the expense of small peasant holdings (e.g., by freeing farmers, but not the small peasants, from militia service, relieving them of the *corvée*, etc.).

So that this type of farming should gain extensive implantation, as great a number of tenant-capitalists as possible would have to be attracted to the countryside. This was the basic task that the Physiocrats set themselves—*to attract capitals into agriculture*. 'The government ought to be more concerned with attracting wealth to the countryside than with attracting men. Men will not be lacking if there is wealth there; but without wealth there is a general decline, the land becomes valueless, and the kingdom is left without resources and power.'[5]The whole of Physiocratic economic policy was designed to assist in drawing capitals out of the towns and into the countryside, out of trade and industry and into agriculture. To this end the price of corn should be set high, as this would render agriculture an especially profitable undertaking. Also to this end the farmer must be assured of the inviolability of the capital that he has sunk into the land, and be relieved of personal obligations and land taxes, which ought to fall *in toto* on the landowners. 'Thus there must be complete security for the ready employment of wealth in the cultivation of the land, and full freedom of trade in produce.' Otherwise 'wealthy inhabitants in essential occupations... [would] carry off into the towns the wealth which they employ in agriculture, in order to enjoy there the privileges which an unenlightened government, in its partiality towards town-dwelling hirelings, would grant them.'[6] The greatest harm done by mercantilist measures is that because they artificially stimulate trade and industry and because of the system of state loans and tax farming, they 'separate finance from agriculture, and deprive the countryside of the wealth necessary for the improvement of landed property and for the operations involved in the cultivation of the land.'[7]

The Physiocratic ideal, then, was not the natural agriculture of a patriarchy, but commodity agriculture, producing for the market and organized by capitalist farmers.They consciously grasped that only the application of capital to agriculture would raise the latter's productivity and make possible the extractionof a 'net product' (net income). The small-scale peasant holding delivers no net product. The greater the capital invested in agriculture, the larger the yields, the lower the costs per unit product, and the higher the net agricultural income. The capital advanced by cultivators, says Quesnay, must be of sufficient size, 'for if the advances are not sufficient, the expenses of cultivation are proportionately higher and yield a smaller net product.'[8] 'Thus the more insufficient the advances are, the less profitable the men and the land are to the state.'[9] In other words, the smaller the total invested capital, the greater the unit costs and the lower the productivity of agriculture. *The investment of large capitals is a necessary condition for raising the productivity of agriculture.*

It follows that when the Physiocrats talk about agriculture as the sole source of wealth it is not agriculture in general that they have in mind, but capitalist agriculture. Likewise, when they describe a net product coming from the land they mean land that has been fructified by capital. When Quesnay says that 'it is the *land* and the *advances** of *the entrepreneurs* of cultivation which are the unique source of the revenue of agricultural nations',[10] we should see in these words the correct formulation of Physiocratic theory. Land in and of itself, without the application of capital, is possessed of no miraculous ability to yield a net product. In view of this fact, 'the most fertile land would be worth nothing without the wealth [capital—I.R.] necessary to provide for the expenses involved in cultivation'.[11]

Thus, capital's function within production is to increase vastly the productivity of agriculture; the source of net income is simply land plus the capital applied to it, i.e., capitalist agriculture.

Just what is this capital that carries out such crucial functions in

*Quesnay uses the term 'advances' in the sense of capital that is advanced for production.

production? As the Physiocrats are concerned primarily with the effect of capital on rises in agricultural productivity, it is natural that they should view it from its *material-technical* aspect, as the totality of *the means of production* in the broad sense of the term. As against the mercantilist confusion of capital with money, the Physiocrats stressed persistently that it was not money in and of itself that constitutes capital, but the means of production that money purchases and which contribute to an increased productivity of labour. 'Look over the farms and workshops,' says Quesnay,' 'and you will see just what constitutes the fund of these so precious advances. You will find there buildings, livestock, seed, raw materials, movables and implements of all kinds. All of this, without doubt, is worth money, but none of it is money.' Turgot describes capital in very similar terms: 'The more that cultivation is perfected and the more energetic it becomes, the larger are these advances. There is need of livestock, implements of husbandry and buildings in which to keep the livestock and to store the produce; it is necessary to pay a number of people proportionate to the extent of the undertaking and enable them to subsist until the harvest.'[12] Hence the Physiocrats were the forerunners of the so-called 'national economic' concept of capital (as the totality of produced means of production) still common in bourgeois economic science. Despite the fact that this concept of capital suffers from its disregard for capital's social aspect, it nevertheless represents real progress over mercantilist doctrine, in that it shifts the focus of analysis away from the realm of exchange and into that of production.

As is obvious from the preceding quotations, the Physiocrats analyzed capital in its different *material elements*. They included amongst its constituents livestock and agricultural implements, seed, means of subsistence for workers, fodder for livestock, etc. Beyond this analysis of capital's material components, the Physiocrats were also the first to make distinctions within capital according to the *velocity with which it circulates*: as we will see in the next chapter, they differentiated *fixed* capital from *circulating* capital.

1 Quesnay, 'The General Maxims,' translated in Meek, *op cit*, p. 243.
2 *Ibid*, p. 235.
3 Turgot, *Reflections on the Formation and the Distribution of Wealth* in *Turgot on Progress, Sociology and Economics*, translated and edited by Ronald L. Meek (Cambridge University Press, 1973), p. 133.
4 Quesnay, 'Maxims,' p. 242.
5 *Ibid*, p. 254.
6 *Ibid*, pp. 254-55.
7 *Ibid*, p. 238.

8 *Ibid*, p. 233.
9 *Ibid*, p. 242.
10 *Ibid*, p. 238, Rubin's italics.
11 *Ibid*, p. 242.
12 Turgot, *Reflections*, Meek edition, p. 147.

CHAPTER THIRTEEN

Social Classes

For the Physiocrats, as we have seen, the *leitmotif* behind both their practical programme and their theoretcial argumentation was large-scale capitalist agriculture, which presupposed the separation of the class of *landed proprietors* from the class of *capitalist farmers*, the organizers of production. Obviously, besides these two classes there also existed a class of direct producers, i.e., *agricultural wage labourers*. To the Physiocrats the presence of this class was inevitable; nevertheless, it was upon the contradiction between the first two of these classes— the landowners (whom Quesnay calls 'proprietors') and the farmers (called by Quesnay the cultivators, or the 'productive class')—that their entire attention was usually focused. Just as the 'third estate' (the bourgeoisie) of 18th century France included among its numbers the wage labourers who had not yet managed to crystallize themselves into a distinct social class, so, too, in Quesnay's scheme did the agricultural workers form but a background to the productive class of cultivators, without being distinguished separately from them. This is not surprising if we remember that the class contradictions between capital and labour were only weakly developed at this time. As the working class did not yet play an independent role within social life, the relations between the farmers and their workers held little place in Quesnay's thinking. His attention was drawn first, to the conflict of interests between *the town* (i.e., industry and trade) and the *countryside* (the cultivators), and second, within the sphere of agriculture itself, to the conflict of interests between the *landowners* and the *farmers*. Quesnay therefore sees a tripartite class division of society: within agriculture he distinguishes the class of *landowners* from the *'productive'* class (the farmers), while to both of these he counterposes the urban *commercial-industrial* population, which he designates as the *'sterile'*[1] class (including members of the liberal professions, servants, etc.).

Each of these latter two classes—the productive and the sterile—can in fact be broken down into two different classes: *entrepreneurs* and *wage labourers.* The great service of Turgot was to have emphasized this class distinction with great precision: 'Thus the whole Class which is engaged in meeting the different needs of Society with the vast variety of industrial products finds itself, so to speak, subdivided into two orders: that of the Entrepreneurs, Manufacturers, and Masters who are all possessors of large capitals which they turn to account by setting to work, through the medium of their advances, the second order, which consists of ordinary Artisans who possess no property but their own hands, who advance nothing but their daily labour, and who receive no profit but their wages.'[2] This same class distinction exists within the 'productive' cultivator class: 'The Class of Cultivators, like that of Manufacturers, is divided into two orders of men, that of the Entrepreneurs or Capitalists, who make all the advances, and that of the ordinary Workmen on wages.'[3] Thus the *three* classes of Quesnay are converted by Turgot into *five*. For the sake of clarity, the different class distinctions of Quesnay and Turgot can be presented as follows:

The Division of Classes According to Quesnay	*The Division of Classes According to Turgot*
1. The Class of Proprietors	1. The Class of Proprietors
2. Productive Class (Cultivators)	2. Capitalist Farmers
3. The Sterile Class (Commercial and Industrial)	3. Agricultural Workmen
	4. Industrial Capitalists
	5. Industrial Workmen

Although Quesnay's analysis designates only three classes, his actual argument, as we have already noted, equally presupposes the presence of wage labourers. It thus seems permissible for us to take Turgot's schema as a clearer and more consistent formulation of the views of Quesnay himself.

Let us now look at agriculture and examine the characteristics of the classes involved.

The Proprietors, according to Quesnay's theory, acquired their land either as an inheritance or by purchasing it from the people who had been its original owners. The latter, by means of their labour and movable property took the virgin land and made it fit for cultivation: they uprooted trees, drained the land or watered it, enclosed it, provided roads, etc. By making these basic advances, the so-called '*avances foncières*,[4] the landowners consolidated their perpetual right to ownership of the land. As its owners, they would now receive from the tenant farmers *a rental payment* (or rent), equal to the whole of the net income, or 'net product' remaining after the farmer has deducted his costs of production from his gross income.

The farmers lease the land from the proprietors on a more or less long-term basis, and husband their holdings with their own capital. Here the farmer must invest two types of capital: first, he must immediately lay out large sums of money to purchase his agricultural stock, both dead and living (agricultural implements, livestock, etc.), which depreciates slowly and can remain in service over several years, for example, ten years; second, the farmer must lay out annually a fixed sum for running expenses, and these he receives back *in toto* over the course of the year out of the returns from the sale of the harvest—in this category fall outlays on seed, fodder, and workers' wages (or what is the same thing, means of subsistence for the workers). Thus the farmer invests in his business first, a *fixed capital* (or to use Quesnay's expression, *avances primitives*) and second, a *circulating capital* (or *avances annuelles*). Quesnay accords special significance to increases in fixed capital: the greater they are the more productive the holding. Quesnay assumes that the size of the fixed capital is to be five times that of the circulating capital; for example, the class of farmers taken as a whole invests in agriculture a fixed capital of 10 billion livres, and a circulating capital of two billion, making twelve billion in all.*

What are the farmers' earnings from their holdings? After gathering in the harvest and selling it, they must use their receipts above all to cover their total costs of production, that is, all of their circulating capital and that portion of the fixed capital that has depreciated over the year. Taken all together, the class of farmers has first of all to recoup a circulating capital of two billion livres which was spent: 1) on raw materials (seed, and so on), and 2) on means of subsistence for all those taking part in production, i.e., not just the wage workers, but also for the farmers themselves and their families. As we see, Quesnay lumps together the means of subsistence *for the workers* (that is, their wages) and the means of subsistence consumed by *the farmers themselves* (purchased in fact out of their profits). Thus the farmers' outlays on their own consumption are treated not as profit, but as necessary costs of production: it is as if the farmers were paying themselves a wage (even though a high one), which, just like the workers' wages, represents one portion of the advanced circulating capital.

*The numerical examples used here and in the ensuing discussion are taken from Quesnay's *Analyse* of his *Tableau Economique*; livres have been changed to rubles. [Translated in R. L. Meek's *Economics of Physiocracy*, pp. 150-167. We have restored the term livres—*Trans.*].

Over and above the replacement of the two billions' worth of circulating capital, the farmers receive, at the end of the year (on Quesnay's assumption), a further sum, equal to 10% of their total invested fixed capital, i.e., another one billion livres. This sum, however, is not in fact a profit on capital. Quesnay assumes that it goes merely to replace the part of the fixed capital that has worn out through the year, plus any losses that might be incurred through accidental misfortune (crop failures, floods, hail, and the like)—in short, it is an amortization and insurance fund. If, as we have assumed, the value of the fixed capital (implements, livestock, etc.) is ten billion livres, and it has a lifetime of 10 years, it is clear that every year one tenth of it will wear out; hence, to keep the fixed capital constantly in good condition a billion livres will have to be spent annually on its repair and renewal (we are leaving any insurance funds out of account here).

Thus the farmers, who have invested a capital of twelve billion livres in their holdings, receive three billions in the course of each and every year: two billion as replacement for their circulating capital, and one billion as compensation for the worn-out portion of fixed capital. If the farmers sell the entire harvest for five billion livres, the two billions that stand as a surplus over and above production costs makes up the net income, or 'net product', and is paid to the landowners as rent. The farmers themselves receive no net income, but merely indemnification for the capital that they have spent. The only personal gain that they derive from the process of production is the receipt of the necessary means of subsistence for them and their families (even if greater in quantity and of better quality than those going to the workers). Consequently, even though the farmers are capitalists, they are in receipt of no profit on their capital, but obtain simply their necessary means of subsistence, or a kind of wage, albeit a more sizable one.*

*It is only in Turgot that we find any clear indication that the farmer (like the industrial capitalist) receives 'a profit sufficient to compensate him for what his money would have been worth to him if he had employed it in the acquisition of an estate...without any exertion,' and which stands over and above a replacement for his expended capital and the wage for his personal labour within the enterprise. Turgot is one of the first writers to attempt to provide a theory of profit and to try to determine its magnitude. In his view the profit on capital is equal to the total rent that the owner of a money capital would receive if he had used it to purchase a tract of land; if a parcel of land purchasable for 1,000 livres were to yield a net income (rent) of 50 livres, then a capital of 1,000 livres must also yield a 50 livre profit—in other words, the rate of profit will be established at 5%. Turgot's mistake is to derive the size of profit from the price of land, when in fact it is the other way around: changes in the price of land depend upon

This failure to grasp the social nature of the farmers' income and to *give due regard to the category of profit* is one of the Physiocrats' most serious errors. In the Physiocratic scheme of things the farmer figures simultaneously as capitalist, investing a sizable capital in his business, and as worker, drawing a mere wage. And while it is true that there is a further income of 10% charged on fixed capital, this represents a replacement of capital, rather than profit. Quesnay senses that the farmer derives some sort of income in proportion to the size of his invested capital, but he does not wish to present this as a net income (profit) left over after production costs have been covered. Quesnay, as a defender of the farming class, wants to 'reserve' a minimal income (profit) for the latter which would be secure from the claims of rapacious landowners and an extravagant government. The only way that he can do this is to depict the farmers' entire income as a compensation for their capital and the means necessary for their subsistence. To render the farmers' income secure Quesnay has transferred it out from under the heading of net income and under that of capital replacement or costs of production, leaving the rent paid to the landlords as the only item under net income. Quesnay, to protect the farmers' profit, has dressed them up as workers or peasants, who receive nothing but their necessary means of subsistence.

One other reason for this disregard of farmers' profit lies in the backward economic conditions of 18th century France, where farmers were few in number and lost in a sea of peasants and *métayers*. At that time in France the tenant farmer was not always clearly distinguished from that other tenant, the *métayer*, even though the latter (like the peasant) really did derive only his necessary means of subsistence from his holding. In addition, the farmer often worked his holding himself alongside his workers, and seemed to merge with them socially. The nature of the farmer as a capitalist had not yet crystallized itself with sufficient clarity; the ties between the farmer, peasant, *métayer*, and agricultural labourer made the transition from the one to the other barely perceptible.

---

fluctuations in the rate of profit (or interest). With a rate of interest of 5%, a tract of land yielding a net income of 50 livres will sell for 1,000 livres; if the rate of interest is 10% the price of this same plot of land will not exceed 500 livres. Turgot's example shows that even the finest intellect amongst the Physiocrats continued to seek explanations for the laws of the capitalist economy (in this case, the rate of profit) uniquely within the sphere of agriculture (here, the price of land). This testifies to the backwardness of French economic conditions and the continued predominance of agriculture. Turgot's explanation of the profit level has close similarities to Petty's explanation of the level of interest (see Chapter Seven, above).

We have already had to make passing reference to the third class of people employed in agriculture (and whom Quesnay fails to distinguish as a specific group), that is, the *agricultural wage labourers*.

These agricultural workers sell their 'labour,' or labour power to the farmers, and receive back from them a wage. What is the size of this wage? According to Physiocratic theory, the level of wages does not exceed the minimum necessary to sustain the workers' existence. In Quesnay's words, 'the level of wages, and consequently the enjoyments which the wage-earners can obtain for themselves, are fixed and reduced to a minimum by the extreme competition which exists between them.'[5] The wage depends upon the price of the workers' food, above all corn. 'The daily wage of a labourer is fixed more or less naturally on the basis of the price of corn.'[6] This so-called '*iron law of wages*,' which during the 17th and 18th centuries had many defenders amongst the mercantilists, was formulated even more precisely by Turgot (who is thus often considered to have been its author): 'Since [the hirer] has a choice between a great number of Workmen he prefers the one who works most cheaply. Thus the Workmen are obliged to vie with one another and lower their price. In every kind of work it is bound to be the case, and in actual fact is the case, that the wage of the Workman is limited to what is necessary in order to enable him to procure his subsistence.'[7]

If, as we have seen, the Physiocrats confuse the capitalist farmer with the peasant and the agricultural worker, their error is repeated even more crudely when applied to industry. The farmer, as depicted by Quesnay, though receiving no profit, is nevertheless a capitalist, in that he advances substantial sums for fixed capital and for hiring workers. According to Quesnay's picture, the industrialist figures as an artisan who makes no advances for fixed capital and hires no labourers. These artisans (members of the 'sterile class') expend nothing but raw materials and their personal labour and, upon selling the products that they have fashioned, receive back merely a compensation for their *raw materials* plus the value of the *means of subsistence* needed for themselves and their families. As with the farmers, the profit of the industrialists is ignored, being seen by the Physiocrats as either the artisan's 'subsistence' or the worker's 'wages'. Also, like the farmers, industrialists receive only a replacement for their capital or costs of production, that is, their outlays for raw materials and for supporting themselves and their families while working. This theoretical *confusion of industrial capitalists with artisans* could have been facilitated by the fact that there were very few large-scale capitalists in

France during the 18th century, and handicrafts continued to predominate.

1 Rubin throughout his discussion of the Physiocrats uses the term 'unproductive' (*neproizvoditel'nyi*), instead of 'sterile' (*stérile*), the term actually used by the Physiocrats.
2 Turgot, *Reflections*, Meek edition, p. 153.
3 *Ibid*, p. 155.
4 Literally, 'advances on the land', translated by Meek as 'ground advances'.
5 Quesnay, 'The Second Economic Problem', in Meek, *op cit*, p. 194.
6 Quesnay, 'Maxims', p. 258.
7 Turgot, *Reflections*, p. 122.

CHAPTER FOURTEEN

The Net Product

The analysis of the class division of society leads us to the central point of Physiocratic doctrine, its theory of *the exclusive productivity of agriculture*. According to Physiocratic theory, agriculture is a 'productive employment' because the product of cultivation does not simply replace the farmer's overall costs of production, but earns over and above this a certain surplus, a 'net product', or 'revenue', which is paid to the landowner as rent. Industry constitutes a '*sterile*' employment, since the value of industrial products does not exceed production costs. [As for trade, which is also a 'sterile' pursuit, see below—Ed.] It is only in agriculture that wealth actually grows or that new wealth is created.

Below we will see that there is a distinctive *duality* to this doctrine of the exclusive productivity of agriculture. On some occasions the Physiocrats talk of agriculture yielding a 'revenue', that is, a surplus of *exchange value* over and above the value of costs of production; on others they talk in terms of agriculture yielding a 'net product', i.e., a surplus of *articles of consumption* beyond those necessary for the subsistence of the actual cultivator. In other words, the Physiocrats understand the exclusive productivity of agriculture sometimes as agriculture's ability to yield a surplus *quantum of value*, at other times as its ability to produce a surplus *quantity of material products*. The fact that agriculture is *productive of value* is confused with the *physical productivity* of the land, a duality which renders Physiocratic theory prone to confusion and contradiction.

What was it that prompted the Physiocrats to seek an explanation of 'revenue'? It was the fact that the value of industrial products contains their costs of production (plus profit), whereas the value of agricultural produce includes, besides these elements, also a rent paid to the owner of the land. The Physiocrats were thus in essence confronted with *the problem of rent*: how were they to account for this manifestly greater value of agricultural produce which yields a surplus quantum of value, ground rent?

With the Physiocrats the problem of rent takes on a particular form owing to their failure (as we have already seen) to note the existence of profit, and their inclusion of the farmers' income (as well as the industrialists') as part of necessary costs of production. Now, if profit is included in production costs the entire problem appears as follows: why is it that the value of industrial products replaces only the cost of production, or capital, while the value of the produce of cultivation yields an excess value, a net income beyond that necessary to replace production costs? Rent is here converted from a surplus over *the costs of production plus profit* into a surplus over *costs of production*—that is, into surplus value. Rent, which in reality is part of surplus value, as is profit, is taken to be the sole form of surplus value, the sole net income. The problem of *rent* is thus turned into the problem of *net income*, or *surplus value*.

Having posed the problem of surplus value, however, the Physiocrats could find no way to solve it, for a correct solution is possible only with a correct theory of value. In so far as the Physiocrats have a *theory of value* it is ill-formed and unable to explain the origins of surplus value. According to Physiocratic doctrine, a product's value is equal to *its costs of production*; in consequence, when a product is sold at its value there can accrue no net income (or surplus value). The Physiocrats distinguished: 1) the *'fundamental price'* of a product (i.e. its cost price, or costs of production) and 2) the *price upon its sale at first hand* (that is, the price at which the product is sold by the direct producer). With respect to industrial products the Physiocrats maintained that totally free competition between industrialists (artisans) would cause the selling price of these products to tend to fall to the level of their costs of production (which would include the industrialist's own necessary means of subsistence). The 'price upon sale at first hand' will not exceed the product's 'fundamental price' (its cost price), and industry will provide no 'revenue' over the compensation for production cost.

In their theory of value the Physiocrats adhere, therefore, to a *theory of 'production costs,'* which then, with complete consistency, leads them to deny the possibility that industry can receive a *net income*, or *surplus* value. As soon as the Physiocrats move onto the realm of agriculture, however, their theory of value comes up against the fact that a net income, or rent exists. Where does this rent, which figures as an excess in the value of the product over its costs of production, derive from? The 'price upon sale at first hand' of agricultural produce clearly exceeds its 'fundamental price' by the full amount of the rent. In turn, this means that the law of production costs fails to apply to the products of agriculture; the latter are subordinate to a completely different law of value than the products of industry.

What is the law of value that governs agricultural produce? Quesnay at one point attempted to argue that because of rapid population growth, demand for these products is constantly in excess of supply, and they are thus sold at a price that *exceeds their costs of production*: the margin between the former and the latter is what makes up net income (rent). In essence, however, such an assertion, that the price of agricultural produce is forever greater than its value, is tantamount to a complete repudiation of a theory of value.

Quesnay's attempt to explain the origin of net income from a rise in the *value* of agricultural produce proved bankrupt. Being shut off from the only methodologically correct means of explaining surplus value—i.e., on the basis of the theory of value—the Physiocrats had no choice but to fall back on another and principally false approach. Once it becomes impossible to derive net income from a rise in the value of agricultural produce does not its origin then have to be explained completely independently of that produce's exchange value? If there is no means of demonstrating that agriculture has the power to produce an increase in value, do we not then have to try to derive net income directly from the land's greater physical productivity? And so it was that Quesnay arrived at the central idea of Physiocratic doctrine, that the source of net income ought to be sought in the *physical productivity of the land*.

First the problem of rent was turned into a problem of net income. Now the latter is transformed into a problem of 'net product': the fact that in agriculture an excess appears in the *value* of the product over the value of the costs of production is accounted for by the physical productivity of the land, which yields an *in natura surplus of produce* over and above the quantity of *products* laid out as production costs. An enquiry into the relation between the *value* of the product and its costs of production is cast aside and replaced by an enquiry into the relation between different quantities of *in natura produce*—that spent on production, on the one hand, against that got from the harvest, on the other. To be able to make this comparison between a harvest and production costs on an *in natura* basis, the Physiocrats resort to two simplifications: first, they ignore those costs of production made up of fixed capital (ploughs, implements, etc.) and assume that in agriculture the only *costs of production* are themselves *agricultural products or corn* (seed, fodder, and the cultivators' means of subsistence); second, in calculating production costs the Physiocrats give greatest weight to the *means of subsistence going to the cultivators*. Now, once production costs *become equated with the cultivators' means of subsistence*, the question of the *in natura surplus of the harvest* over and above the *costs of production* is transformed into the following: whence derives *the surplus in means of subsistence that the harvest yields over and above those means of subsistence required to maintain the cultivators during the period of their labours*?

For the Physiocrats this surplus is to be explained by the physical productivity of the land and its ability to create new material substance. Following Cantillon, the English economist, the Physiocrats maintain that in agriculture there is a process of generation that creates *new material substance* over and beyond that which existed beforehand, a quantitative increase that cannot take place in industry since the latter is confined simply to imparting to this substance a *different form*. In Quesnay's words, the work of the shoemaker 'consists merely in giving the raw materials a definite form'; this is 'simply production of forms, and not a real production of wealth'. In agriculture there is a 'generation or creation of wealth', a real increase of substance. In industry there is only a 'combining' of raw materials with expenditures on the artisans' means of subsistence; the finished product is simply the result of combining these raw materials and means of subsistence, both of which were already in existence prior to industrial production and were obtained from agriculture.[1] In agriculture wealth is 'multiplied'; in industry it is merely 'composed'. The Italian Physiocrat, Paoletti, expressed this idea quite clearly: 'Give the cook a measure of peas, with which he is to prepare your dinner; he will put them on the table for you well cooked and well dished up, but in the same quantity as he was given, but on the other hand give the same quantity to the gardener for him to put into the ground; he will return to you, when the right time has come, at least fourfold the quantity that he had been given. This is the true and only production'[2] Only agriculture gives birth to new matter in exchange for that which is consumed and destroyed by man. Industry can create no such new substance, but only transform or modify its shape.

Agriculture generates new material substance for human society. Since the better part of this consists of the means of human subsistence, agriculture is not simply *the source of new material substance, but also the sole source of these means of human subsistence*. This in turn means that agriculture yields the means of subsistence not simply for the actual cultivators, but for other classes of society as well. 'It is the cultivator's labour which regenerates not only the subsistence goods which he himself has destroyed but also those destroyed by all the other consumers.' It is this that gives supreme social superiority to the class of cultivators, which 'can always subsist on its own, off the fruits of its own labours. The other if left to itself could obtain no subsistence by its own sterile labours in and of themselves', unless it could receive means of subsistence from agriculture.[3]

But we also know that these necessary means of subsistence constitute the wages of both agricultural and industrial workers. It therefore follows that agriculture is *the source of wages* for both the *agricultural* and *industrial* populations. 'Whatever [the Husbandman's] labour causes the land to produce over and above his personal needs is the unique fund from which are paid the wages which all the other members of society receive in exchange for their labour.'[4] By giving over part of their means of subsistence to the industrial class in exchange for the latter's manufactures the cultivators seemingly pay them their maintenance, or wages. The cultivators are the class which pays the labour of the industrial population; the latter is 'salaried' by the agricultural class.

The Physiocrats' train of thought here can be summarized as a set of propositions, each of which helps detail the general features of agriculture:

1) agriculture is the source of *rent* (which is the margin between the value of the product and the farmer's costs of production plus profit);

2) agriculture is the source of *net income* (which is the margin between the value of the product and its costs of production, the latter also containing in concealed form the farmer's profit);

3) agriculture is the source of *new material substance*, which it puts at the disposal of society for satisfying the needs of its members;

4) agriculture is the source of *the net product* (which is the surplus of the produce of agriculture over the products that are spent on the process of production);

5) agriculture is the source of the *surplus of means of subsistence* over and above those means of subsistence necessary for the actual cultivators;

6) agriculture is the source of *means of subsistence* for both the agricultural and *industrial* populations;

7) agriculture is the source of the *wages* that pay the labour of the industrial population.

The starting point for the Physiocrats is the greater value produced by agriculture as the source of rent, or net income. To explain this phenomenon they look to agriculture's physical productivity as the source of new material substance, and to the *in natura* form of its produce as means of subsistence. The Physiocrats then shift back again from this physical 'primacy' to agriculture's *social* primacy as the sole source of the wages that nourish and 'maintain' the industrial population.

In this manner the Physiocrats' entire theory of net income is infused with a fundamental dualism between two points of view: the *value* and the *physical*. They commit two basic errors. First, the basic *physical difference* that they discern between agriculture and industry does not exist. Agriculture, no matter what the Physiocrats think, produces no new substance, but simply converts the diffuse material substance of soil, air, and water into grain, in other words, it endows matter with a form that is suitable for the satisfaction of human needs. Yet industry does exactly the same thing. Similarly, there is no basis for attributing to agriculture any special superiority in permitting a collaboration between human labour and the forces of nature, since this same collaboration with the forces of nature (steam, electricity) takes place within the industrial labour process.

The Physiocrats' second and in principal more important mistake is to take the peculiar *physical* productivity of agriculture (even if it were to exist) and deduce from this that agricultural produce has a higher *value*. 'Their error,' wrote Marx, 'was that they confused the *increase of material substance*, which because of the natural processes of vegetation and generation distinguishes agriculture and stock-raising from manufacture, with the *increase of exchange value*.'[5] The Physiocrats did not suspect that the inability of industrial labour to create new material substance in no way precludes it from being a source of surplus value. Had the Physiocrats not artificially included capitalist profit in production costs, they would have been forced to conclude that industry, too yields a profit, or net income beyond the mere restoration of its costs of production. On the other hand, the Physiocrats failed to grasp that this increase in the material quantity of agricultural products (an increase which they attributed to the land's greater physical productivity) still does not signify any growth in their *mass of exchange value*. The Physiocrats confused the production of *in natura* products (use values) with the production of exchange value. This confusion merely reflects the backward state of French agriculture in the 18th century, as it was going through a transitional stage from a natural to an exchange economy.

Despite the depth of these errors, the Physiocrats' theory of net income contained fertile ideas for future development.

The Physiocrats saw that the decisive feature of economic prosperity was the growth of net income or surplus value, and the main aim of the productive process was to increase this. They were mistaken to attribute the ability to yield a net income only to agriculture, but having done that the Physiocrats were perfectly consistent in drawing the conclusion that agriculture alone constitutes 'productive' employment. Their erroneous doctrine of the exclusive productivity of agriculture was, therefore, premised upon a correct idea, namely that from the point of view of the capitalist economy only *labour which yields surplus value* can be deemed productive.

The Physiocrats performed a second and still greater service, however, in that they took the question of the origin of surplus value out of *the sphere of exchange* and into production. The mercantilists had known surplus value primarily as profit upon trade, in which they saw nothing more than the mark up which the merchant adds to the price of the commodity. In their view profit has its source within exchange, especially within foreign trade, and it was this that they affirmed to be the most profitable occupation. The mercantilist doctrine that trade is the source of net income (or profit) was sharply refuted by the Physiocrats. For them trade brings no *new* wealth into a country, since free competition, and the abolition of all exclusive monopolies and commercial restrictions reduce it to an exchange of one material product for another of equal value. 'For my part, I can never see anything in this trade but the exchange of value of equal value, without any production, even though circumstances render this exchange profitable to one or other of the contracting parties, or even to both. In fact, it must always be assumed that it is profitable to both; for both sides procure for themselves the enjoyment of wealth which they can obtain only through exchange. But there is never anything here but an exchange of an item of wealth of one value for another item of wealth of equal value, and consequently no real increase in wealth at all.'[6] For all its advantages and its necessity, trade cannot be considered a 'productive' occupation. The source of new wealth (net income) must be sought within production itself (agriculture) and not in exchange.

The Physiocratic theory that *exchange is an exchange of equivalents* presupposes that products have a value determined even before they enter into the process of exchange. 'The formation of prices always precedes purchases and sales'. 'The real price of produce is established prior to its sale.' Quesnay expressed here a theoretical proposition of extreme importance, to be developed subsequently by Marx: the *value of the product is established in the process of production*, even before it enters the process of exchange.

The basis on which the mercantilists recognized foreign trade as most profitable was that it would allow a country to receive greater value for less, and exchange an *in natura* product for money or precious metals. In their doctrine on the equivalence in value of exchangeable products, the Physiocrats refuted the first of these mercantilist prejudices; in their *theory of money* they took up arms against the second. According to the Physiocrats what should be strived for was to produce as many products *in natura* as possible; selling them or transforming them into money neither presents any special difficulty nor yields any particular advantage. 'Is there really a greater need for buyers than for sellers? Is it really more profitable to sell than to buy? Is money really to be preferred to the good things of life? Certainly it is these things that are the true object of all commerce. Money merely facilitates the mutual exchange of this ordinary wealth by its circulation and is acquired by this party or that in the process.'

Money, in other words, is not true wealth, but only a *means for the more convenient mutual exchange* of the use values that *genuinely comprise wealth*. 'Thus money does not constitute the true wealth of a nation, the wealth which is consumed and regenerated continually, for money does not breed money.'[7] Therefore, 'it is in this renascent wealth, and not... in the nation's money stock, that the prosperity and power of a state consist.'[8] Money plays only the role of 'a token intermediating between sales and purchases.'[9] Coined money 'has no other use than that of facilitating the exchange of produce, by serving as an intermediary token between sales and purchases.'[10] 'Thus it is not money which we ought to think about, but rather the exchanges of the things which are to be sold and those which are to be bought, for it is in these exchanges themselves that the advantage which the contracting parties wish to procure for themselves resides.'[11] The mercantilist policy of attracting money into the country via a favourable balance of trade is mistaken. Concern ought to be with multiplying the produce of cultivation, rather than with increasing the country's stock of money; if produce is abundant and its price advantageous, there will be no shortage of ready money. A nation assures itself the largest possible net product or net income not by increasing its quantity of money through trade but by enlarging its volume of produce through production (agriculture).

1 'We have to distinguish an *adding together* of items of wealth which are combined with one another, from a *production* of wealth. That is, we have to distinguish an increase brought about *by combining* raw materials with expenditure on the consumption of things which were in existence prior to this kind of increase, from a *generation* or creation of wealth, which constitutes a renewal and *real* increase of renascent wealth.' (Quesnay, *Dialogue sur les travaux des artisans* ('Dialogue on the Work of Artisans'), translated in Meek, *op cit*, p. 207.)

2 Cited in Marx, *Theories of Surplus Value*, Part I (Moscow, Progress Publishers English edition, 1969), p. 60.

3 Cited in Georges Weulersse, *Le mouvement physiocratique en France (de 1756 à 1770)*, Volume I (The Hague, Editions Mouton, 1968, photographic reprint of the 1910 edition), p. 256.

4 Turgot, *Reflections*, Meek edition, p. 122. 'The Husbandman, generally speaking, can get on without the labour of the other Workmen, but no Workman can labour if the Husbandman does not support him. In this circulation, which by means of the reciprocal exchange of needs renders men necessary to one another and constitutes the bond of society, it is therefore the labour of the Husbandman which is the prime mover. Whatever his labour causes the land to produce over and above his personal needs is the unique fund from which are paid the wages which all the other members of society receive in exchange for their labour. The latter, in making use of the consideration which they receive in this exchange to purchase in their turn the produce of the Husbandman, do no more than return to him exactly what they have received from him. Here we have a very basic difference between these two kinds of labour...' This passage shows, beyond the point that Rubin is trying to make here, Turgot's genuine insight into the nature of wages as being *advanced* by the capitalist to the worker and necessarily returning to him, i.e., wages as being part of the capitalist's *circulating capital*. Marx was to demonstrate this quite clearly throughout his discussions of capitalist circulation in Volume II of *Capital*, especially in his schemes of simple reproduction.

5 Marx, *Theories of Surplus Value*, Part I (Progress Publishers English edition), pp. 62-63 (Marx's italics).

6 Quesnay, 'Dialogue on the Work of Artisans', in Meek, p. 214.

7 Quesnay, 'Maxims', p. 252.

8 *Ibid*, p. 251.

9 *Ibid*, p. 251.

10 Quesnay, 'Dialogue on the Work of Artisans', p. 218.

11 *Ibid*, p. 219.

CHAPTER FIFTEEN

Quesnay's *Tableau Economique*

Having dealt with Physiocratic doctrine on different social classes and branches of production, we can now move on to analyze Quesnay's famous *Tableau Economique*, the short lines of which sketch out the picture of *reproduction and distribution of the entire social product* between the various classes and branches of production.

Quesnay first wrote the *Tableau Economique* in 1758, and a small number of copies was run off in the court printshop. This initial text of the *Tableau* vanished and was only discovered in 1894 by a scholar working on the papers of Mirabeau. Complaints about its lack of clarity and incomprehensibility led Quesnay in 1766 to publish the *Analyse du Tableau Economique*, an explication of which we give below. The Physiocrats hailed the *Tableau* as a scientific discovery of momentous import; Mirabeau compared it with the invention of paper and money. Its opponents poured ridicule upon this 'hardly intelligible work', and it remained, in Engel's expression, 'an insoluble riddle of the sphinx',[1] undeciphered and unavailed of by scientific thought until the mid-19th century. Marx was one of the first to demonstrate the *Tableau's* immense scientific significance, a judgement now acknowledged by all researchers.

Now to the *Tableau* itself.[2] As we know, Quesnay divides society into three basic classes: 1) the class of '*proprietors*' (the landowners, including the crown and the clergy); 2) the '*productive*' class (the farmers, who represent the entire agricultural population); and 3) the '*sterile*' class (the commercial-industrial population, professional people, etc.). How, then, is the total social product that is created in the course of the year distributed between these three classes?

Let us take the point at which *the production year ends* and the new year is just beginning, that is, the autumn, when the productive class (whom we shall henceforth term the farmers) have already gathered in the harvest, the value of which we assume amounts to five billion livres.[3] To obtain this harvest the farmers have laid out the following payments over the year just ended: 1) a circulating capital of two billion livres (subsistence for all those engaged in cultivation, fodder, seed, etc.) and 2) one billion livres for repair and renewal of fixed capital (implements, livestock), that is, 10% of ten billion livres, which is the value of the total fixed capital stock. The farmers, then, have spent a total of three billion livres, and have received a harvest worth five billions. The surplus of two billions is the net product, or net income that the cultivation has yielded, and it goes to the landlords (whom we shall henceforth refer to as the 'proprietors') as rent. The farmers had already paid these two billions rent to the landlords in cash at the start of the year, and it is as cash that the latter are currently holding it. Finally, the 'sterile' class (whom we will call the 'industrialists') starts off the new production year with a stock of industrial goods worth two billions, which they would have manufactured during the year just ended. Thus, at the start of the new production year our three classes are holding the following, either as produce or as cash:

1) the *farmers* have a stock of agricultural produce worth five billion livres (of these four billions' worth are in foodstuffs and one billion's worth in raw materials for working up by industry);

2) the *proprietors* have two billion livres in cash;*

3) the *industialists* have in stock two billion livres worth of industrial manufactures.

There now begins a process of exchange or circulation between these three classes, which consists of a series of acts of purchase and sale between them. To give clarity to our exposition we present two schemes: the first depicts the transfer of products between the different classes, the second the transfer of money.**

As is obvious from Scheme I, the *first* act of circulation is for the proprietors to purchase one billion's worth of foodstuffs from the farmers for their own maintenance over the coming year. These one billion livres in foodstuffs pass in this initial act of circulation from *F* to *P*, while the same amount of cash moves in the reverse direction from *P* to *F* (Scheme II). As a result of this first act of the circulation we

*The entire stock of ready money in society is limited to these two billion livres which start out the year in the hands of the landowners. Quesnay himself had posited this stock to equal three billion livres, but this does not affect the problem.

**In the schemes each line indicates an act of circulation entailing one billion livres. The direction of the arrows shows from and towards which social class the products and money are being transferred (in each act the money moves in the opposite direction to the products). The figures indicate the sequence of the individual acts of circulation. The circle with the letter *F* represents the class of farmers, that with the letter *I*, the industrialists, and that with the letter *P*, the proprietors, or landlords.

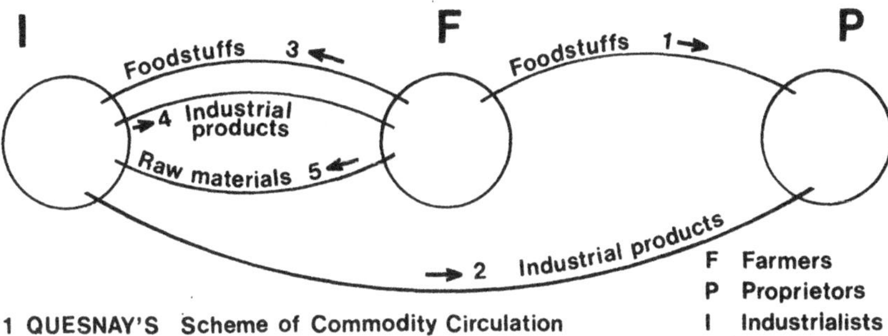

1 QUESNAY'S Scheme of Commodity Circulation

get the following distribution of produce and money: the farmers have agricultural produce worth four billion livres (three billions' worth of foodstuffs and one billion's worth of raw materials) plus a billion in cash; the proprietors have a billion's worth of foodstuffs and a billion in money; the industrialists have two billions' worth of manufactures.

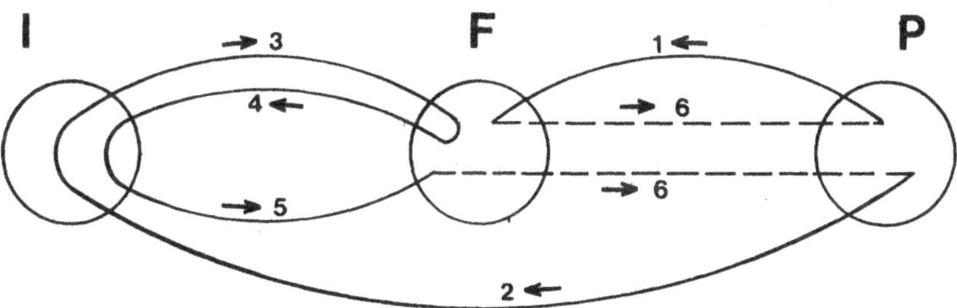

2 QUESNAY'S Scheme of Monetary Circulation

In the *second* act of circulation, the proprietors take their remaining billion livres and purchase industrial products for their own consumption from the industrialists; these products move from *I* to *P*, while money moves from *P* to *I*. The result of this second act of circulation is that: *F* has four billions' worth of agricultural produce and one billion in cash; *P* has a billion's worth of foodstuffs and one billion livres' worth of industrial products; and *I* has a billion in industrial products and another billion livres in cash.

In the *third* act of circulation the industrialists, who have received one billion in cash from the proprietors,* take this money to buy from the farmers the foodstuffs necessary for their own upkeep over the coming year. The result of this third act of circulation is that: *F* has three billion livres' worth of agricultural produce (two billions' worth of foodstuffs and one billion in raw materials), plus two billions in cash; *P* has one billion livres' worth of foodstuffs and one billion's worth of industrial goods; *I* has industrial products worth one billion plus a billion's worth of foodstuffs.

In the *fourth* act of circulation, the farmers take the money just obtained from the industrialists and use it to purchase back from them one billion livres of industrial items, which we assume to be made up of tools and other implements needed to repair or restore their fixed capital. After this fourth step: *F* has three billions' worth of agricultural produce (two billions in foodstuffs and one billion in raw materials), industrial goods worth one billion livres, and a further billion livres in money; *P* has foodstuffs worth one billion livres plus a billion in industrial products; *I* has foodstuffs worth a billion livres and a further billion in cash.

Finally, in the *fifth* act of circulation the industrialists take the money that they have just received from the farmers to buy from the latter one billion livres' worth of raw materials, which will be worked up within industry. Following this fifth act of circulation: *F* has foodstuffs worth two billion livres (which they retain for their own subsistence), one billion in industrial products and two billions in money; *P* has foodstuffs worth one billion livres and one billion's worth of industrial products; *I* has a billion in foodstuffs and one billion worth of raw materials.

For all its simplicity, Quesnay's scheme was the first ingenious attempt to depict as a unified whole the entire *process of the reproduction, circulation, distribution, and consumption of a society's product*. Quesnay's wish is to trace out the path of social reproduction, that is to reveal those conditions that make possible the uninterrupted, *periodic repetition* of the production process. Quesnay starts his *Tableau* from the point where the harvest has been collected, when the entire annual social product—which he takes as a single entity—has been produced. Upon completion of *production* this product

*Since the industrialists in the third act of circulation give to the farmers the same money that they had themselves received from the proprietors in the second act, line no. 2 in the scheme of monetary circulation runs directly into line no. 3 (as does the latter into line no. 4, and line no. 4 into line no. 5). Using an unbroken line shows that it is physically one and the same coinage that is changing hands here. In the scheme for commodity circulation, the lines are not connected up with each other, but are broken, since in the *Tableau* each product is transferred only once, from producer to consumer.

enters the *process of circulation*, made up of a series of acts of purchase and sale. In the *Tableau* the whole circulation process is reduced to five acts of purchase and sale between different classes. In fact, of course, each of the acts of circulation listed in the *Tableau* consists of a multitude of discreet transactions between separate individuals. The first act, for example embraces many thousands of separate purchases on the part of the proprietors from the farmers, but which, being of a similar nature, the *Tableau* combines together into one. It is the social and class aspects of these acts of circulation that interest Quesnay: i.e., how they promote the transfer of the social product among the social classes. For this reason Quesnay leaves out of his scheme of circulation the transactions carried out between members of the same class (e.g., what the farmers buy from or sell to each other).

As Quesnay has it, the circulation process embraces not simply the movement of *products in natura*, but equally the movement of *money* back in a direction opposite to the flow of products. Quesnay's scheme shows clearly that the movement of money is secondary and subordinate—merely to service the movement of products. The circulation of five billion livres' worth of products is serviced by a total cash sum of two billions. Of the latter, one half goes simply to service the circulation of one billion in produce (the first act of circulation); the other billion by being passed from hand to hand services all the remaining four stages in the circulation process. The end result is that the entire two billion in cash, which started out in the hands of class *P*, finishes up in the possession of class *F*. What, then, does the latter do with it! As soon as the circulation process is complete it passes it back to class *P* as the upcoming year's rent. This one-way transfer of two billion in money from *F* to *P* has been indicated in our second scheme by two broken lines (step number 6), each of which represents a transfer of one billion livres.* In sum, the second scheme clearly shows the unceasing circular movement of money; it passes from one hand to the next and eventually returns to its point of departure. One billion livres goes from *P* to *F*, and then back to *P*; the other billion livres is transferred from *P* to *I*, then from *I* to *F*, from *F* back to *I*, and then back again from *I* to *F*, at which point it passes as rent to *P*.[4]

As a result of this process of circulation the entire social product winds up distributed between the different social classes *in such a manner as to permit the process of production to be renewed*

*We have indicated the movement of money (Scheme II) during the two-sided acts of purchase and sale (i.e., within the sphere of commodity exchange) with solid lines (nos. 1, 2, 3, 4, and 5).

at its former level. The farmers have two billions' worth of foodstuffs (as well as seed, fodder, etc.) with which to maintain both themselves and their workers for the whole year; in addition they have a billion in industrial goods for renewing the depreciated portion of their fixed capital. They have thus been compensated for the whole of their circulating capital plus their worn out fixed capital and can begin the production process over again at its previous volume, obtaining at the end of the year a harvest worth five billion livres. The industrial class has its necessary means of subsistence (one billion livres' worth) plus raw materials (also worth a billion) which when worked up over the coming year will result in the manufacture of finished goods once more worth two billion livres. As the value of the industrial goods equals the value of the raw materials plus the value of the means of subsistence consumed by the industrialists, it is obvious that industry creates no net income. The farmers and industrialists have, therefore, sufficient stock of products both for their *personal consumption* and *for repeating the process of production*. Finally, the landowners have those foodstuffs and industrial wares that they will require for the year's consumption.

The case Quesnay is examining in his *Tableau* is one of *simple reproduction* where reproduction is on the same scale as before. He was, nevertheless, totally aware that there are two other forms of reproduction: reproduction on an *extended* scale, and reproduction on a *declining* scale. The difference between them consists in the differing magnitudes of net product that are produced, or—since this in turn depends upon the volume of capital invested in agriculture—in differences in the amount of such capital. If there is a rise in total outlays on agricultural production (at the expense either of the net income going to the proprietors or of the outlays on industry whose own level of reproduction is assumed not to change) the net product will grow, and with it the whole of the reproduced social product. If the fund for agricultural advances remains at the old level 'unconditionally needed to maintain cultivation *in status quo* or to restore the expenses of cultivation', reproduction will proceed on the same level as before. Finally, 'if the cultivators cannot be assured of receiving back all of their productive expenditure', outlays on cultivation will decline, and so, too, will the scale of social reproduction; in that case 'advances, wealth, useful enterprises, necessary employments, produce, revenue, population—all this will decline under a *force majeur*. This constitutes a physical law, established by nature making it possible to judge the past, present, and future fortunes of governments by the manner in which they have been or are presently conducting themselves.' This 'physical law established by nature' is the basic law of social reproduction: *whether an economy prospers, remains stationary, or declines depends upon the expansion, stagnation, or reduction of the basic outlays on agriculture*—or, in other words, upon the capital at the disposal of the farming class. If a state is to prosper it has no means of doing so other than increasing the capital invested in agriculture; nor will it have any means by which to forestall its decline if it violates the necessary laws of reproduction, i.e. if through taxes or excessively high rents it squanders or eats away at the capital of the farmers. Hence ensue the two basic principles of the Physiocrats' *economic policy*: first, the necessity to introduce free trade and to raise the price of corn, so as to increase the flow of capital into agriculture; second, the need to protect this agricultural capital from the excessive claims of the landlords and the state.

1 Engels, Preface to the Third German Edition of *Anti-Dühring* (Moscow, Progress Publishers English edition, 1969) p. 20.

2 Rubin's discussion is based on the *Analyse*, which is translated in Meek, *op cit*, pp. 150-67.

3 Rubin uses rubles; we have changed this to livres throughout, to correspond with the original French text.

4 This is one of the basic laws of simple reproduction established by Marx in Volume II of *Capital*, namely that the money advanced to initiate the process of circulation must return to its original holder; should it not, the circulation of the annual product will be broken and reproduction cannot take place. See *Capital*, Vol. II, (English edition, Moscow, Progress Publishers, 1967), Chapter XX, Section III.

CHAPTER SIXTEEN

Economic Policy

The Physiocrats were fervent supporters of the *freedom of trade and industry* from state interference. They demanded that mercantilism's strict and petty regulation over economic life be removed. The Physiocrats were the ideologues of *free trade*, and in this sense were the forerunners of the Classical school. But there was a fundamental difference between the kind of free trade advocated by the Physiocrats and that proposed by the classical economists. This difference emerges from their different social and class positions. Both the Physiocrats and the Classical economists protested against mercantilist policy, which had brought wealth to certain privileged sections of the commercial bourgeoisie; they both demanded that the interests of merchant capital be subordinated to those of productive capital. Yet while the Classical economists understood productive capital to mean primarily industrial capital and wanted to pave the way for industry's powerful advance, it was the interests of productive agricultural capital that, as we have seen, the Physiocrats pushed to the fore. The Classical economists took up the cause of the industrial bourgeoisie; what they expected from free trade was the import of cheap foreign corn into England. The Physiocrats, being defenders of the rural bourgeoisie, saw free trade and the free export of corn as a means of raising the price of grain. The Classical economists were the spokesmen for *industrial free-trade*; the free trade defended by the Physiocrats was *agrarian*.

The reason why the Physiocrats attacked mercantilism so furiously was that the latter had, in their view, created a sharp *divergence* between the prices of industrial and agricultural products: while the monopolies enjoyed by industrialists, traders, and the guilds had made industrial products excessively dear, prices on corn were being depressed artificially by the prohibition on the latter's export. The Physiocrats wanted this divergence (known in our own day as '*the scissors*') to be eliminated, and sought to have *corn prices raised and industrial prices lowered*.

The Physiocrats attempted to give their practical demands a theoretical foundation; they wanted to demonstrate theoretically that high corn prices were advantageous. To do this they made use of their *theory of reproduction*, which occupies a central place in the Physiocratic system. By *reproduction* the Physiocrats mean the renewal of the capital advanced (or production costs) together with the production of a net product (surplus value). Taken in *this* sense reproduction for them occurs only in agriculture. Obviously, then, any transfer of capitals out of *agriculture and into industry* would be accompanied by a *curtailment* in the overall reproduction process (since industrial capitals are renewed, or circulate without any 'increase'), while a flow of capitals *from industry into agriculture* would bring with it an *expanded* process of reproduction and an *increase in net income*. It therefore follows that to permit the flow of capitals from agriculture into industry (or trade) is incompatible with the aims of normal continuity or possible expansion of the reproductive process; to the contrary, it is the reverse flow that ought to be encouraged. To achieve this goal corn prices should be high, as they make agriculture a more profitable employment and attract new capitals to it; because of this the 'fund of advances on cultivation' grows—as does the net product (net income)—the process of reproduction takes place on an extended scale, and the entire economy receives a powerful stimulus towards prosperity and expansion.

Out of this doctrine of the benificial effect of high corn prices arises the basic Physiocratic maxim of economic policy: *'That the prices of produce and commodities in the kingdom should never be made to fall.'* 'Only high prices can guarantee and maintain the well-being of a people and the state through the successes of cultivation. This is the alpha and omega of economic science.' By high price, or *bon prix*, Quesnay does not mean the excessively high price of corn that comes with harvest failure (in France such years had alternated with years of cheap corn and had made for terrible economic uncertainty); what he wants is for the price of corn to reach the *high* and *stable* level that prevails 'amongst trading nations', i.e., on the world market, and which exceeds its price in agricultural countries such as France. For corn prices within France to rise to their world market level, French corn would have to gain free and open access to the world market—hence the Physiocrats' persistant struggle against mercantilist *prohibitions on the export of corn*. Originally the Physiocrats had taken 'free trade' to mean above all the free exportation of corn; it was Quesnay's view that its free importation could only be allowed in years when the harvest had been bad. Quesnay, therefore, was advocating free trade mainly to the extent that the interests of agriculture demanded it.

It was Quesnay's students who gave the slogan 'free trade' a broader and more absolute character, and it was only with them that the famous formula of the free traders, *laissez faire, laissez passer*, began to be more and more frequently repeated.

The Physiocrats did not seek freedom of trade simply as a way of raising prices on agricultural produce; it was equally a means by which *prices on industrial products could be lowered.* The free importation of cheap manufactures from England or other industrial nations would undermine the monopoly position of the local manufactories and the guild master craftsmen, whose inflated prices on finished goods worked to the detriment of their agricultural consumers. Let no one complain about the foreigners flooding France with cheap manufactures and destroying its local industrialists. The country will only gain if these French industrialists find it unprofitable to continue producing and put their capital to more profitable employment in agriculture: every livre invested in agriculture yields a net product, while in industry it circulates without any 'increase'. 'An agricultural nation should facilitate an active external trade in raw produce, by means of a passive external trade in manufactured commodities which it can profitably buy from abroad.' Thus the ideal of Physiocratic foreign commercial policy—an ideal dictated by the interests of agriculture and the farming class—is to sell *corn* abroad at high prices and *buy cheap foreign industrial manufactures in return.*

Thus, the first benefit of free trade is that it guarantees a country 'an advantageous price in its sales and purchases' (i.e., a high price on agricultural produce and a low price on industrial goods). The second benefit of free trade is that mutual competition between merchants compels them to accept a lower rumuneration and reduce their *profit on trade to the level of their necessary means of subsistence.* Free competition alone can compel industrialists and merchants to give up their excess monopoly profits, the entire burden of which falls upon the class of cultivators. Thus we have Quesnay's famous VIIIth Maxim: 'That the government's economic policy should be concerned only with encouraging productive expenditure and trade in raw produce [i.e., production and circulation of agricultural products—*I. R.*], and that it should refrain from interfering with sterile expenditure [i.e., industry and commerce—*I.R.*].'[2] If the class of cultivators is to reduce its burden of 'maintaining' industry and commerce, the latter must be freed of state interference and turned into an arena for the unbridled competition between industrialists and merchants (both native and foreign).

For the Physiocrats free trade appeared as a means of making the 'scissors' go back in the opposite direction, where prices on industrial products would fall to the level of necessary costs of production and prices on agricultural produce would rise to the level of the world market. The farming class, however, had to defend itself against more than the mercantilist policy of onesidedly encouraging industry and trade at the expense of agriculture. Its interests also had to be protected against the overbearing claims of the *landowners* and the *government*. We saw in chapter nine that once the cultivator had taken out his rent and taxes there was often barely enough grain remaining for his meagre subsistence. It was understandable, then, that under these conditions those owning capital should evince little desire to lease land. To attract capitals into agriculture farmers would have to be guaranteed that the combined total of their rent and taxes (along with the church tithe) would not exceed the total 'revenue' left over after their capital and profit on farming had been covered. The Physiocrats' *doctrine on taxes* was a demand for just such a guarantee.

The Physiocrats demanded that all forms of direct and indirect taxes be replaced by a single direct land tax falling on 'revenue'. The tax must be proportional to net income, raisable only in step with this income's growth. But since net income goes to the landowners as rent, *the tax must fall exclusively on the landowners* and comprise a certain proportion of the rent they receive.* This plan of a *single tax on the landowners' rent*, was later advanced by a number of bourgeois-radical reformers (including Henry George). For 18th century France it was a daring project, tantamount, in Marx's words, to 'the partial confiscation of landed property';[3] but it would also mean the abolition of the aristocracy's exemption from taxes, since a single tax would have to be levied upon all landowners, including the gentry.

As with their slogan of free trade, the Physiocrats tried to give a *theoretical* underpinning to their demand for tax reform. This was their doctrine on *net income and reproduction*. As we know, the value of the total annual product is divided into two parts: one replaces the advanced capital (costs of production, within which is concealed the farmer's profit); the excess makes up the net income. It is obvious that the first portion is a 'fixed property', with a clearly defined function—to be reinvested in production. It is only net income, that is 'transferrable property', which can be 'disposed of as is seen fit': to be

*Quesnay, in his *Analyse*, assumes that out of the total net income four parts in seven are retained by the landowners, two parts go to the state in the form of taxes, and one part to the church as tithes.

spent on the needs of the landlords, the state, and the church (as well as on further improving the land). Should any tax fall not on net income, but on farming capital, this would lower outlays on cultivation, make it impossible for reproduction to take place on its former scale, and lead to a reduction of the net income and to the ruin of landlord and state alike. 'Taxes should ... not be taken from the wealth of the farmers of landed property: for *the advances of a kingdom's agriculture ought to be regarded as if they were fixed property requiring to be preserved with great care in order to ensure the production of taxes, revenue, and subsistence for all classes of citizens.* Otherwise taxation degenerates into spoliation, and brings about a state of decline which very soon ruins the state.'[4] The Physiocrats' basic demand vis-à-vis taxes was that, in the interests of keeping the process of reproduction on its proper path, *farming capital be treated as inviolable.*

If farming capital must remain inviolate, is it then not possible to put the burden of taxation onto *workers'* wages or onto the *commercial-industrial* class? The Physiocrats rejected both of these schemes. Any tax on workers would necessitate a rise in their wages, since they receive only their necessary means of subsistence; and this would inevitably be paid by those 'people who hire the workers', i.e., by the very same capitalist farmers. As for the merchants and industrialists, under free trade they receive, as we know, only their capital (costs of production) and their necessary means of subsistence. A tax on commercial or industrial turnover would inevitably push up expenses on industry and trade, which would be paid, in the last instance, by the agricultural population. Since neither industry nor commerce create any new wealth (net income), any tax on them—as with the working class—will ultimately fall upon agriculture, to be levied either out of farming capital or out of net income. The first case, as already noted, entails violating the entire process of reproduction and bringing the country to ruin. In the second case, if the tax is eventually to be transferred to net income in any case, would it not simply be better to place it upon this sole reserve of 'transferrable' resources right away? Not only is it cheaper to tax net income (i.e., the landowners' rent) directly, but it makes it possible to keep the amount of tax exactly proportional to the size of net income.

These basic principles of Physiocratic economic and taxation policy were closely tied both to their *social and class position* and to their *theoretical outlook*. The introduction of a free trade and the single land-tax would inevitably clear a path for the growth of capitalist agriculture. On the one hand, the corn trade would be freed from arbitrary administrative regulation and subjected to the vicissitudes of the world market, the state of which was strong and profitable; on the other hand, farming capital would be protected from the claims of the landowners and the treasury and the latter's appetite restricted to the realm of net income; both of these conditions would of necessity promote the flow of capitals into agriculture, which would be reorganized along capitalist lines leading to the enrichment of the farming class. What is more, these principles of economic policy followed logically from the theoretical *laws of reproduction* discovered by Quesnay. For the process of reproduction to proceed normally farming capital has to be safeguarded, first from being reduced during the process of circulation and exchange between agriculture and industry—which requires in turn *freedom of trade*, with corn prices high and industrial prices cheap—and second, from any reduction owing to deductions being made to meet the needs of the landowners and the state—which requires that rent and taxes be limited by the size of net income, i.e., the introduction of a single *tax on rent*. Just as the Physiocrats' economic theory aimed to discover the laws of capitalist reproduction, so their *economic policy* had to assure that this *process of reproduction* proceeded normally. Yet, as we saw in our chapter on natural right, the Physiocrats took the laws of capitalist reproduction that they had discovered to be eternal and immutable 'natural' laws. It is therefore understandable that they passed off their principles of economic policy as being commanded by natural law. They declared free trade to be a 'sacred freedom, which can be looked upon as a summary of all the rights of man', in exactly the same way as 'taxation is subordinated by the Creator of nature to a definite order', prescribed by natural laws and coinciding with the taxation programme of the Physiocrats. All these parts of the Physiocratic system—the philosophical conception of natural laws, the theoretical laws of reproduction and the principles of economic policy—were inextricably bound together by the unity of their social and class position, itself exemplified by their system.

1 Quesnay, 'Maxims', in Meek, p. 235 (Quesnay's emphasis).
2 *Ibid*, p. 233.
3 Marx, *Theories of Surplus Value*, Part I (Progress Publishers English edition), p. 52.
4 Quesnay, 'Maxims', p. 232 (Quesnay's italics).

CHAPTER SEVENTEEN

The Theoretical Legacy of the Physiocrats

The main theoretical service of the Physiocrats is that they attempted to lay bare the *mechanism of the capitalist economy as a whole*. The mercantilists had been taken up with analyzing individual economic phenomena, primarily those which had immediate practical interest. In the best of cases they confined themselves to a study of the causal connection between a number of separate phenomena; their theory of the balance of trade, which elucidated the connection between the movement of commodity imports and exports and fluctuations in a currency's rate of exchange represents the highest level of generalization that mercantilist thought was able to attain. What characterized Physiocratic theory is that it makes broad generalizations and seeks to uncover the connection between all the basic phenomena of the capitalist economy. This is why their theory of social reproduction as a unified process, embracing all aspects of economic life, lies at the centre of the Physiocratic system.

The theory of social reproduction as set out in Quesnay's *Tableau Economique* represents the Physiocrats' most valuable theoretical legacy. For in it economic thought displayed a capacity for generalization that few other examples can equal. Casting aside all particulars and details, Quesnay, in a few bold and ingeniously simple strokes, depicts the entire process of capitalist reproduction as it embraces the production, circulation, distribution, and consumption of products. Here Quesnay's thought attains the highest levels of generalization: the entire economy is conceived of as an exchange of material objects between *agriculture and industry*—society is explained as a totality composed of specific *social classes*; the products that are produced and dispersed throughout the entire country are aggregated into a single *social* product, and this—by means of a few essential *acts of circulation* (each of which is itself a generalization of an infinite multitude of specific acts of purchase and sale)—is then *distributed* amongst the main social classes. The concept of the economy as a periodically repeating process of reproduction; the idea that a nation's wealth is the outcome of a process of production that renews itself with every year; the idea that the national product is distributed amongst individual social classes—every one of these fundamental ideas of Classical political economy, which were to be further developed by Smith and Ricardo, belongs to Quesnay.

The individual errors in, and clumsiness of the *Tableau Economique* notwithstanding, the theory of social reproduction that Quesnay created all by himself can on the whole be said to be the most mature and thoroughly thought out of his creations. Its basic ideas became part of the reserve fund of economic science, where they remain to this day. Just how far this theory of social reproduction transcended its own age can be seen from the fact the Classics not only failed to develop its ideas any further, but in this area at least, actually trailed behind Quesnay. This is even more true of the epigones of the Classical school, who failed to make any scientific use whatsoever of the *Tableau Economique's* seminal ideas. Whilst Quesnay's work in other areas (the problem of surplus value, capital, wages, and money) was improved upon by Smith or Ricardo, it was more than a century before anyone was found to carry on the work of developing the theory of social reproduction. Only Marx, in the second volume of *Capital*, took up the thread of Quesnay's initial investigation, improved upon the theory of social reproduction contained in the *Tableau*, and brought it to completion.

This theory takes us directly onto the problem of *capital and surplus value*, and it is the Physiocrats' development of this problem that constitutes their second great scientific service. The Physiocrats understood reproduction as the production of a product which both replaces its own value (the capital advanced) and yields over and above this a certain excess, or net income (surplus value). The reproduction process thus encompasses the replacement of capital and the production of surplus value. By sharply counterposing the costs of production (capital) to net income (surplus value) the Physiocrats incisively characterized the capitalist economy as an economy whose aim is the production of surplus value. By making this distinction they brought greater clarity both to the problem of capital and to the problem of surplus value.

Unlike the mercantilists, whose attention was focussed upon the *money* form of capital, the Physiocrats advanced a concept of productive capital as the totality of the *means of production*. They made the first and what for their day was the best analysis of capital, both from the point of view of its *material elements* and from that of its *rate of circulation*. By their use of the terms *'avances primitives'* and *'avances annuelles'*, they made a seminal distinction between *fixed* and *circulating* capital, a distinction which Smith took over *in toto* and which prevails in economic science even to this day.[1] What is inadequate about the Physiocratic theory of capital (as with the theory of the Classics) is that it ignores capital's social form and concentrates on the technical functions of those means of production which function as capital. Yet this failing—shared by both the Physiocrats and the Classical school—is inherent in any scientific tendency which, bounded by a bourgeois horizon, takes the bourgeois form of economy as the eternal and 'natural' form of economy in general. It is just this conception which infused the progressive ideologists of the bourgeoisie during the period when the latter was still playing a revolutionary role in its struggle against the remnants of the feudal order.

The same basic failure emerges, but with even greater force, in their doctrine of net income (*surplus value*). Because the Physiocrats took no notice of profit, surplus value was known to them solely in the form of *ground rent*, and so they sought its source in the specific properties of agriculture. The problem of the inter-relation between different social classes (the problem of surplus value) was confused with the problem of the inter-relation between different branches of production. Once the Physiocrats had failed in their attempt to explain surplus value (rent) on the basis of the greater value of agricultural produce, they had no other recourse but to look for its source in the physical productivity of nature. The Physiocrats confounded a surplus quantum of value with a surplus product *in natura*, the production of value with the production of material substance, and the ability of agriculture to produce value with the physical productivity of the land. Thus what the Physiocrats arrived at was a *physical-naturalistic* solution to the problem of surplus value: their doctrine of nature as the source of value, and a theory of the exclusive productivity of agriculture. What this reflects is the limitation imposed upon Physiocratic thought not simply by the horizons of bourgeois economy but by the even narrower perspective of its most backward sector, the semi-natural agrarian economy. The narrowness of this outlook left its mark upon the whole of Physiocratic theory, leading it to form an incorrect understanding of the role of industry and to ignore industrial profit:* once the production of surplus value is confused with the

*Turgot alone stands out as having had a broader outlook and having been more

production of material substance industry becomes a 'sterile' occupation incapable of yielding any 'revenue'; and once industry yields no net income industrial profit becomes simply compensation for the industrial capitalist's necessary means of subsistence. These closely interconnected errors—the *physical-naturalistic* solution to the problem of surplus value, the doctrine of the unproductiveness of *industry,* and disregard of the most basic category of capitalist economy, *profit*—constitute between them the main defects of Physiocratic theory and were what most often gave the Physiocrats' opponents grounds for reproof and derision. The other form of income inherent in capitalist economy—wages—fared better amongst the Physiocrats than did profit. For their time, Quesnay and Turgot gave one of the best formulations of the *iron law of wages*, a formulation which Ricardo was to develop and which still has its scientific partisans.

However mistaken their solution to the problem of surplus value may have been, the Physiocrats nevertheless performed a great service by having posed it in a clear cut fashion and by having taken it out of *the sphere of exchange* and into that of *production.* The mercantilists knew net income solely as commercial profit, as 'profit upon alienation', the source of which lay in a non-equivalent exchange of products, which in turn meant that one of the contracting parties in the exchange gained at the expense of the other. The Physiocrats were the first to pose the question of absolute rather than relative income, and of the possibility therefore of there being an increase of wealth (value) even where there was an exchange of equivalents. It is obvious that were this to be the case, an increase in value takes place not in the process of exchange, but in the process of production preceding it. The idea that value *is created within the production process* and determined *prior to the product entering the process of circulation* belongs to the Physiocrats, and forms the necessary basis for the theory of surplus value. If the mercantilists (and especially Petty) gave one of the earliest formulations of the *labour theory of value*, the merit of posing the problem of *surplus value* goes to the Physiocrats (even though their lack of a correct theory of value kept them from solving this problem correctly). Subsequent scientific progress consisted of an

---

inclined to show concern for the interests of industry and the commercial-industrial bourgeoisie—in keeping with which he displayed a greater theoretical interest in the problem of profit (see above, Chapter Thirteen). [Turgot's conception of the nature of profit and surplus value is discussed by Marx in Part I of *Theories of Surplus Value* (Progress Publishers English edition), pp. 54-59.—*Trans.*]

attempt to create a ***synthesis between the theory of value and the theory of surplus value*** (Smith and Ricardo), a synthesis which only Marx was able to carry out with success.

1 It is interesting to contrast this statement by Rubin with the view of Marx, who argues in Volume II of *Capital* (English edition, Progress Publishers, 1967), Chapter X, that Smith's only advance upon the Physiocrats' correct distinction between fixed and circulating capital was his ability to generalize it to all spheres of capitalist production and not confine it simply to agriculture. In all other respects, however, Marx held Smith's discussion and understanding to be a genuine step backward from the Physiocrats. For Smith's adoption of what was correct in Physiocratic doctrine co-exists with his appropriation of some of their basic errors (most importantly, the confusion of circulating *capital*, which is a value relation, with the *physical* means of subsistence of the labourers), errors which had a logical basis in the Physiocrats' system but which for Smith only obscured the more essential relations between constant and variable capital. For a fuller discussion see Editor's Note 4 to Chapter 24.

Part Three

Adam Smith

CHAPTER EIGHTEEN

Industrial Capitalism in England During the Mid-18th Century

In France, mercantilism, which reflected the interests of commercial capital, had provoked the opposition of the Physiocrats, who were defenders of the rural bourgeoisie. This opposition came to nothing in practice, however, as the Physiocrats' programme was not carried out. The only forces that could crush mercantilism were those of the urban industrial bourgeoisie. It fell to the Classical school, founded by Adam Smith, to complete the conquest of mercantilism, in practice as well as in theory. If the Physiocrats dreamt of rapid successes for productive *agricultural capital*, the Classical school struggled against mercantilism in the name of the free development of *industrial capitalism*. To best understand Smith's doctrine we must first know something about the state of industrial capitalism in England round about the middle of the 18th century, on the eve of the industrial revolution.

The 18th century was a transitional period in the history of English industry, and was characterized by the coexistence of different forms of industrial organization: first there were *independent handicrafts*, which still existed as a relic of the past; second, there was a widely-diffused system of cottage, or *domestic* large-scale industry; and third, there had by now appeared large, centralized capitalist enterprises, or *manufactories*.

At the beginning of the 18th century there were still large numbers of independent *craftsmen* in England. Defoe has left an interesting portrait of the life of the independent master cloth-makers who lived near Halifax: 'at almost every House there was a *Tenter*, and almost on every Tenter a Piece of *Cloth*...or *Shalloon*...' '...every Clothier must keep a Horse, perhaps two, to fetch and carry for the use of his Manufacture ... so every Manufacturer generally keeps a Cow or two, or more, for his family ... ' '... a House [is] full of lusty fellows, some at Dye-fat, some dressing the Cloths, some in the Loom ... ' '... Women and Children ... are always busy Carding, Spinning, &c.

so that no Hands being unemploy'd, all can gain their Bread, even from the youngest to the antient; hardly any thing above four Years old, but its Hands are sufficient to it self.'[1] The craftsmen preserved their independence thanks to the fact that it was they themselves who were carting their commodities to nearby markets for sale.

However, once at this market the craftsmen usually had to sell their commodities not directly to the consumer, but to a *middleman*. The cloth makers who lived near Leeds brought their cloth into Leeds twice a week, where trade was first carried out on a bridge and later on in two covered markets. Each cloth maker had his own stall to which he brought his cloth. At six or seven in the morning, at the peal of the bells, the merchants and middlemen would appear and start bargaining with the cloth makers, concluding all their business in about an hour. By around nine o'clock the benches had been cleared and the market was deserted. Under this set-up the masters, though still maintaining their independence, were already selling their commodities to the merchant, rather than to the consumer.

This need to sell to the merchants was caused in most cases by the *specialization* of the crafts, by the fact that each was concentrated in a specific region, and by the *expansion of the market*. If the cloth makers living around Leeds, for example, specialized in the manufacture of a particular type of cloth, its consumption was obviously not limited to the Leeds area alone; it would be exported to other English towns or even abroad. As the master could not deliver his cloth himself to such far-flung markets, he would sell it to merchants whose loaded caravans used to take the goods to the various fairs and trading towns of England.

Also, the *remoteness of raw materials markets*, for example, the impossibility of going to the large trading centres to buy wool, led to the same result: the raw materials were purchased by merchants, who distributed them to the masters for working up. Thus, in Lancashire, weavers used to supply themselves with warps and wefts, work them up, and transport the finished products to market. Gradually, however, it became more difficult to acquire thread, at which point the Manchester merchants began to distribute warps and cotton to the weavers, and the weavers became dependent upon them.

In other situations the dependence of the craftsmen upon the merchants was brought about by the need to buy new *means of production*. Advances in weaving technology demanded that each master have a greater number of looms. Lacking the means for this, it was the buyers up who ordered the additional looms and passed them out to the masters.

Thus the changing conditions of producing and selling commodities (the specialization of crafts, the wider market over which these commodities were sold, the remoteness of markets for the purchase of raw materials, the need to expand the means of production) caused *the master craftsman to be gradually subordinated to the buyer up*. In Leeds the master still brought his own commodities to the merchant in town. Gradually, however, the merchant began to come to the master for them. The London merchants themselves travelled to the masters, bought up their commodities, and paid them in ready cash. In Birmingham the buyers up went around the master lock-smiths on pack-horses buying up their commodities. Cut off from the market, the craftsman became dependent on merchant capital.

So long as the craftsman could sell his commodities to a number of merchants he could still retain a degree of independence. But little by little he would become increasingly dependent upon one merchant in particular, who would buy up his entire output, place advance orders for his wares, extend him advances, and, finally, begin to supply him with raw materials (and, less frequently, with implements of production). From this moment on, the product belonged no longer to the craftsman (who was now receiving simply a recompense for his labours), but to the buyer up. He, in his turn, became a putter out, with many small-scale master craftsmen—craftsmen who had become dependent cottage labourers. Independent handicrafts gave way to the *cottage, or domestic system of large-scale industry*, the spread of which signified the penetration of *commercial capital into industry*, and paved the way for the complete reorganization of industry on a capitalist basis.

During the 17th and 18th centuries, concurrently with the spread of the domestic, or decentralized system of large-scale industry, *manufactories* made their appearance. These were more or less large-scale, centralized capitalist enterprises. The manufactory differed from the domestic system in that the workers worked not alone at home, but on a single premises, which had been set up by the entrepreneur. It was distinguished from the later factory by the predominance of manual labour and the absence of any application of machinery.

The manufactories came about sometimes *independently* of the domestic system and sometimes directly *out of it*. They arose independently wherever it was a case of a new, previously unknown branch of production being implanted in a given country: either foreign entrepreneurs, together with their hired personnel, or individual masters, who would subsequently join together into a single 'manufactory', would be sent for to come from abroad. It was in this manner that many manufactories arose in France—with the active participation of the state. In other cases they grew directly out of the domestic system: the buyer up, who had previously put out raw materials for individual cottage workers to work up at home, would gather these workers together onto one premises where they would have to work under his direct supervision. The dependent cottage labourer was being converted into a hired worker (a proletarian) receiving a wage. The buyer up-putter out was becoming the direct organizer of production, an industrial capitalist. If the spread of the domestic system was a sign of commercial capital's penetration into industry, the setting up of manufactories signified the completion of this process and the coming into being of *industrial capitalism* in the strict sense of the word.

By bringing the workers together under one roof the entrepreneur rid himself of the unnecessary expense involved in distributing the materials to the individual cottage labourers and in transferring the output of some workers to others for further processing; at the same time he gained better control over the raw materials, since under the domestic system the putters out were always complaining that the cottage workers were keeping back part of the raw materials for themselves. On the other hand, the domestic system did relieve the entrepreneur-buyer up of all fixed-capital costs (buildings, implements of production), while it made it possible for the cottage workers to work at home and combine their activity with subsidiary occupations (agriculture, growing fruit and vegetables, etc.). It was because of these advantages that the domestic system proved able to compete with the manufactories, all the more so since the latter held no special advantages in terms of technology. The manufactories were, therefore, unable to oust and replace the domestic system on any significant scale—this was a task that only the factories, with their extensive application of machinery after the industrial revolution of the end of the 18th century, had it within their power to accomplish. Independent handicrafts and the domestic system existed side by side with the newly established manufactories which did not so much replace them as wrest from them individual processes of production which, because of the complexity of their production process, the high quality of the raw materials involved and so on, demanded special supervision over the workers. Often only the very first and last production processes would be carried out within the manufactory, with intermediate processes being done at home by cottage labourers. Hence we very often see the *combination of the manufactory with the domestic system*: a few dozen workers (in rare cases a few hundred) would be labouring in the manufactory, while its owner would at the same time be distributing a substantial amount of work for cottage workers to work up at home.

Although the manufactory did not become as widespread during the 17th and 18th centuries as did the domestic system or the 19th century factory, it nevertheless played an important role in the history of economic development. It signified the appearance of *industrial capitalism*, with its own characteristic *social and technological* features: 1) the division of society into a class of *industrial capitalists* and a class of *hired labourers*; and 2) the domination of *large-scale production* based on the *division of labour* (although without the application of machinery).

In the age that preceded the appearance of the manufactories the money capitalist (the usurer and financier), the merchant capitalist (the merchant), and the buyer up-putter out were familiar figures. The latter represented a hybrid between the merchant and the entrepreneur. His primary line of business was still trade, and he undertook the organization of cottage industry only insofar as this was necessary for the more successful vending of commodities. His income was equally hybrid in character, being made up partly of commercial profit ('profit upon alienation') earned by selling commodities where there was a favourable market, and partly from the exploitation of the cottage worker-producer. With the appearance of the manufactories the *industrial capitalist* in the narrow sense of the word gradually emerged with his own characteristic form of income—*industrial profit*. The owner of the manufactory saw his main job as organizing the process of production. He gave up his commercial role, usually selling his commodities to merchants, who received the profit from trade.

At the same time, it was in the manufactory that the process of forming an *industrial proletariat* was being completed. Of course, the socio-economic processes that created the preconditions for the proletariat's appearance had been going on long before the spread of manufactories, proceeding with especial intensity in the 17th and 18th centuries (the creation of a landless peasantry, the impoverishment of the craftsmen, the exclusiveness of the guilds and the difficulty of becoming a master, the separation of the journeymen from the masters). The industrial proletarians had their forerunners in the *journeymen* and *cottage labourers*. The journeymen, however, never gave up hope of acquiring simple instruments and becoming master craftsmen; the cottage labourers, recruited from the semi-proletarianized craftsmen and peasants, preserved an illusory independence thanks to the fact that they worked at home, had their own implements of labour, and drew subsidiary earnings from agriculture. The journeymen and cottage workers represented an intermediary type, between the independent producer (the craftsman and peasant) and the wage labourer. The workers in the manufactories were proletarians in the exact sense of the term: the large-scale nature of production left most of them with no hope of joining the ranks of the entrepreneurs. Deprived of all implements of production, they received their income strictly from the sale of their labour power, i.e., quite precisely, a *wage*. And although there were still innumerable threads tying the manufactory workers to craft production and cottage industry (they had often been craftsmen and cottage labourers before, had hopes of going back to their previous illusory independence, sometimes drew an auxiliary income from a plot of land or a vegetable patch, and in a few cases even retained their own simple instruments which they carried with them to work in the enterprise), their work in the manufactory put them in the social position of hired proletarians and gave their income the social character of a wage.

Moving from industrial capitalism's social characteristics to its technological ones, one can say that in terms of its *implements of labour* the manufactory still preserved a continuity with handicrafts, while in terms of its *organization of labour* it paved the way for the factory. The extensive application of machinery, which was to ensure the factory production of the 19th century its rapid development, was still unknown in the manufactory. The basic form of the capitalist organization of labour had, however, already been created: large-scale production based upon the *division of labour*. Alongside the previously existing *social* division of labour between individual enterprises appeared a manufacturing, or *technical* division of labour within the enterprise itself.

The break down of the production process into separate stages had also existed within guild handicrafts. There, however, it occurred simply as a social division of labour between individual craft enterprises: the carders worked up the wool, after which they passed it onto the master spinner who prepared the yarn; the weaver wove the material, the dyer dyed it, and so on. Within each workshop the division of labour was practically non-existant. The transition from handicrafts to the manufactory was a twofold process: in the first place previously independent crafts or processes of production were grouped together in a single manufactory (for example, a manufactory making cloth would join together carders, spinners, etc.); in the second place, each individual process of production (e.g., carding or spinning) would be further broken down into a series of even more detailed operations. By *breaking down the process of production* and then *combining* them according to a single plan, the manufactory acquired the features of a complex, differentiated organism, in which individual jobs and workers formed a necessary complement to one another.

Hand in hand with this break down of the production process went the *specialization of the workers*. A specific worker was assigned to each detailed operation, to be occupied with this and this alone. The master craftsman possessing more or less universal technical knowledge (within his own profession, of course) was replaced by a worker concerned with only a detail or *part* of the process, and who, by constantly repeating one and the same simple, monotonous operation became capable of performing it with great perfection, speed, and dexterity. Although the majority of operations was still performed by workers who were trained craftsmen, the more simple jobs were already beginning to be carried out by workers who were untrained—a group completely unknown in the period of the guilds. On the other hand, the need to co-ordinate the joint work of many individuals within a single enterprise led to a division within the leading organizing personnel: besides the entrepreneur, who was the ultimate organizer of the enterprise, there appeared foremen, overseers, checkers, etc. With the manufactory, workers began to be broken down into horizontal groups: although trained craftsmen or *skilled* workers still formed the basic nucleus, they now had *untrained* workers underneath them and *managerial personnel* above them.

Finally, parallel with this specialization of the workers came the specialization, or *differentiation of the implements of labour*. A particular tool would be modified depending on the nature of the operation it was meant to perform. Hence appeared different types of hammers, cutting tools, etc., each of which was adapted as best as possible to a given detailed operation. Tools, however, continued to be manually operated, with their action dependent on the strength and dexterity of the hands that guided them. They were little more than a supplement to the living workers, who still occupied the primary place within the production process. The manufactory relied on *manual technology*, the *high level of productivity* of which was owing to the *break down* of the process of production, the specialization of the *workers*, and the differentiation of the *implements of labour*.

Thus in 18th century England new, capitalist relations were developing within industry alongside the previously-existing guild handicrafts: the *domestic system* had become widespread; less so the *manufactory*. In the course of its growth capitalist industry came up against obstacles created by outmoded yet extant legislation: in particular the *guild system*, which in its day had been set up to protect the interests of the crafts, and the *policy of mercantilism*.

The guild regulations extended the right to engage independently in industry only to those persons who had taken a seven-year course of study and had become members of a guild (this was Elizabeth I's law on apprenticeship, issued in 1562 and still in force in the 18th century). These same regulations forbade the sale of commodities to any buyer up who was not in a guild. The prohibition on taking in more than a certain number of journeymen and apprentices held back the construction of manufactories. Strict compliance with guild regulations would have made it impossible for the domestic system and the manufactories to spread. But the demands of economic development proved stronger than outdated legislation. The guilds themselves were gradually compelled to allow work to be done for buyers up, since craftsmen were now producing for far away markets and could not have managed without their assistance. Already in 16th century Strasbourg, for example, weavers unable to find a market for their goods were beseeching merchants in every way possible to buy up their wares. The guilds were more stubborn in their struggle against the manufactories, but they still could not halt their development. To escape the guild restrictions the putters out and entrepreneurs transferred their activities to *rural areas*, or to new towns which were not subject to the guild regime. Yet even in towns where the guild system was in force, regulations were completely *by-passed* in the interests of the capitalist-entrepreneurs—*new branches of production*, non-existent when the guild laws had been issued (e.g., cotton textiles), were exempted from their application. The law providing for Justices of the Peace to set compulsory wage levels also gradually fell into disuse: as late as the mid-18th century, Parliament reaffirmed the legal force of this law in the interests of the small-scale master cloth-makers, but was soon compelled to repeal it under pressure from the capitalists engaged in cloth making.

Mercantilist policy, which in its day had served to implant the capitalist economy, over the course of time turned into a brake on its further development. The zealous patronage afforded to favoured branches of native industry harmed the growth of industrial capitalism *in other sectors*. For many years, for instance, the English government, acting in the interests of the cloth industry, had forbidden, or put all kinds of constraints upon the development of the cotton textile industry that was later to assure England her dominant position in the world market. *The monopolies of the privileged trading companies* were hampering the initiative of individual private traders and industrialists. The system of rigid *protectionism*, which it is true still found support from some industrialists, was already becoming superfluous and even harmful to the most important sectors of English industry—textiles and metallurgy—which were in no way threatened by foreign competition and had everything to gain from doing away with the obstacles standing between them and the world market.

To ensure the powerful development of industrial capitalism and to turn England into the world's factory required that trade and industry be freed from the restrictions of the guilds and mercantilism. *The ideas of free trade* that North had expounded and Hume had developed (as did the Physiocrats in France) had gained wide currency by the second half of the 18th century. Adam Smith owed his book's brilliant success above all to its eloquent sermons on behalf of the freedom of trade and industry.

Adam Smith can be called the *economist of the manufactory period* of capitalist economy. Only an economist who had observed the growth of industrial capitalism through large-scale manufactory enterprises could present a general picture of the capitalist economy and analyze its separate elements in a way so markedly different from the Physiocrats. Smith for the most part portrays the capitalist economy as a manufactory with a complex division of labour: hence his theory of *the division of labour*. Smith opposes the Physiocrats' false ideas about the class division of society, by consistently and correctly *dividing society into the classes of capitalists, wage labourers, and landowners*. He clearly differentiates the *forms of income* appropriate to each of these classes and isolates the category of *industrial profit*—an enormous advance over the Physiocrats' naive notions of profit. Once profit is identified as a specific category one does away both with the identification of rent with surplus value and with the theory that the origin of surplus value lies in the physical productivity of the land. Smith seeks the source of value and surplus value in *labour*—not simply agricultural labour, but industrial labour as well. Despite falling into some fatal errors in formulating this theory of value and in attempting to deduce from it the phenomena of distribution, Smith is nonetheless the first to make the ***labour theory of value*** the touchstone of his entire economic theory. Smith's theory of capital marks a tremendous step forward. The technical features of industrial capitalism characteristic of the manufactory period find their theoretical reflection in Smith's doctrine on the division of labour; its social characteristics are reflected in his theory of social classes and forms of income (especially his theory of industrial profit), in his labour theory of value, and in his theory of capital.

1 Daniel Defoe, *A Tour Thro' the Whole Island of Great Britain*, Vol. II (London, Peter Davies, 1928), pp. 601-02. A tenter is a rack used for stretching cloth; a shalloon is a thin piece of cloth used for coat linings. Although Rubin presents this in his Russian text as one continuous passage, he has in fact strung together individual sentences taken from separate paragraphs in Defoe's narrative. We have broken up the sentences as they appear in Defoe's original.

CHAPTER NINETEEN

Adam Smith, the Man

On the surface Smith's life is very straightforward. He was born in 1723 into the family of a customs official, in the small Scottish town of Kirkcaldy. Displaying exceptional abilities at an early age, he devoted himself primarily—and assiduously—to the study of philosophy. Beginning in 1751 Smith spent 13 years as a professor at Glasgow University, where he taught a highly successful course in *'moral philosophy'*. Following the spirit of the 18th-century Encyclopedists, the course was not confined simply to ethics, but covered theology, ethics, natural right, and, finally, a section which would now be most accurately called economic policy. Smith's economic theory grew out of the last of these. At that time Glasgow University had no separate chair of political economy, which is not surprising since political economy had not yet formed into an independent science: mercantilist writings were largely practical in character, while for those thinkers disposed towards theory, political economy still remained a subordinate part of philosophy and natural right. At first economic questions had this same subordinate status in Smith's thinking. He devoted his main efforts to his work on ethics, and in 1759 he published *The Theory of Moral Sentiments*, which earned him great renown.

When Smith incorporated economic problems into his course on moral philosophy he was possibly following the example of his predecessor in the department, the famous philosopher, Hutchison. However, whereas Hutchison used to deal with economic questions only in passing, Smith gradually made them the focus of his scientific activity. Smith moved from philosophy to political economy, just as Quesnay had followed the same path from philosophy and medicine. In neither case can this transition be seen as purely accidental: if Quesnay's evolution could be explained by his growing concern with the economic problems of mid-18th century France, what influenced Smith was firstly, the great *changes* taking place at the time *in English economic life*, and secondly, the influence of his elder contemporaries, Hume and Quesnay.

England was in transition from the age of commercial capital to that of *industrial capitalism*, and the changes in economic life were so considerable that they could not fail to attract the attention and interest of anyone living at the time. Nor should it be thought that these changes went unfelt in far away Scotland. The implantation of industrial capitalism was proceeding there with especial success and rapidity. During the first half of the 18th century the number of large-scale manufactories was actually greater in Scotland than in England; *joint stock companies* had been set up in the cloth and linen industries. In the Scottish mountains the *metallurgical* industry had made great headway: it was there, in the celebrated factories of Corran that the famous Watt, the future inventor of the steam engine, built his first improved machine in 1769—the pump. The years when Smith lived and taught in Glasgow saw an unusually rapid development of trade and industry in the city—large-scale manufactories were established, banks were set up, and port and shipping facilities were improved.

Scotland's rapid economic development in the 18th century explains why it was that commercial-industrial and intellectual circles in Glasgow displayed what for their day was a lively interest in economic questions. A political-economy club had already been formed in Glasgow in the 1740's, which, given the date it was founded, would obviously make it the first in the world. Smith was an habitué of this club and met there weekly with his friends. Both the conversations inside, and local events going on outside the club's walls gave economists food for thought. Watt, whom we have already mentioned, had his workshop in Glasgow, where he carried out experiments on a new type of machine. When the local guild corporation forbade him in 1757 from conducting any further experiments Smith earnestly took up his case, and Watt was soon allowed to continue his work in the University workshop.

Besides his observations on what was actually going on around him, Smith's thinking was also nurtured by literary influences. Hume (a close friend of Smith) had published his economic works at the beginning of the 1750's. A few years later appeared Quesnay's first articles and his *Tableau Economique*. Both Hume and the Physiocrats (whom Smith got to know personally later on in Paris) exercised a strong influence on him.

Smith later recalled his thirteen years as a professor as the most useful and happiest period of his life. He closed these years as the celebrated author of *The Theory of Moral Sentiments*, and with a plan for a general economic work. In 1764 he gave up his professorship at his own request in order to travel to France as the preceptor of a young lord. In all, Smith spent more than two and a half years in France, including nine months in Paris where he met with eminent philosophers and teachers including Quesnay and his followers. In Paris Smith was already known as a philosopher but had still not proved himself as an economist; in the words of the Physiocrat, Dupont, 'he has still not shown the stuff that he is made of'.

At the time of his Paris visit Smith was already telling his friends that he was contemplating a substantial work on economic questions. Upon his return to England at the end of 1766 he decided to devote all his efforts to carrying out this plan. Rather than returning to university life, he settled in his native Kirkcaldy, that small town where for seven years he led a secluded existence working on his opus. None of his friends' efforts to induce him to give up his isolation met with any success. 'I want to know', wrote Hume to him, 'what you have been doing, and propose to exact a rigorous account of the method by which you have employed yourself during your retreat. I am positive you are in the wrong in many of your speculations, especially where you have the misfortune to differ from me.'[1] Hume again writes, a few years later, 'I shall not take any excuse from your state of health, which I suppose only a subterfuge invented by indolence and love of solitude. Indeed, my dear Smith, if you continue to hearken to complaints of this nature, you will cut yourself out entirely from human society, to the great loss of both parties.'[2]

The years in isolation had not been in vain. In 1776 Smith's great work, *An Inquiry into the Nature and Causes of the Wealth of Nations* was presented to the world; it earned him universal acclaim and opened up *a new era in the history of economic thought*. From this moment onwards, political economy ceased to be either an aggregation of separate discourses or an appendage of philosophy and natural right: it emerged as a systematically and coherently expounded independent theoretical science. Even before Smith the need had been felt for such a scientific synthesis. It was no accident that, just as they were about to pass from the scene both the economic schools that preceded Smith had, as it were, wished to present the world with a synthetic exposition of their knowledge and ideas: approximately 10 years prior to the appearance of Smith's work the world had received a general statement of the mercantilist position in James Steuart's *An Inquiry into the Principles of Political Oeconomy*, while Turgot had generalized the work of the Physiocrats in his *Réflexions sur la formation et la distribution des richesses*. Neither of these books, however, was capable of opening a new scientific age: the first because its underlying theoretical ideas were either not worked out or mistakenly presented, the second because the Physiocrats' horizon never looked beyond the sphere of agriculture. It fell to Smith to give a theoretical formulation of the phenomena of rising industrial capitalism.

Smith's book owed its immense success on the one hand to its quality of *theoretical generalization*, and on the other to the eloquence with which it *propounded the ideas of free trade*. The struggle for and against mercantilist policy was still being carried out at too topical a level to afford Smith the luxury of a purely theoretical investigation. Of the five books of *The Wealth of Nations* only the first two are dedicated to theoretical questions while descriptive material and problems of economic policy predominate in the other three, with special consideration being given to the polemic against mercantilism. Today these sections of Smith's work hold merely an historical interest; the first two books, on the other hand, were to form the basis for theoretical economy's future development.

Smith lived for a further fourteen years after the publication of *The Wealth of Nations*. The growing pressures of his work on the Board of Customs and the infirmities of old age left him little time and energy for scientific labours. It is true that right up to his death he continued to entertain his life-long dream of rounding off his scientific-philosophical system by writing those parts that were still missing. He gathered together materials for works on law and the history of literature, but not long before his death in 1790, he burned his manuscripts.

1 Hume's letter to Smith of 20 August 1769, in *The Correspondence of Adam Smith*, edited by Ernest Campbell Mossner and Ian Simpson Ross (Oxford, Oxford University Press, 1977), p. 155.

2 Hume, letter to Smith of 28 January 1772, *ibid*, p. 160.

CHAPTER TWENTY

Smith's Social Philosophy

Smith's economic system, like that of the Physiocrats, was intimately linked with his doctrine of *natural right*. In 18th-century England, as in the France of the same period, the bourgeoisie, as we have seen, had still not managed to completely emancipate the capitalist economy from the bonds of antiquated legislation; it is therefore understandable that it sought to sanctify its class demands (which coincided in this period with the interests of overall national economic development) with the authority of an eternal, rational, 'natural' right. But it is noticeable that Smith's views on natural right depart significantly from those of Quesnay. The idea of natural right was central to Quesnay's system. In his view, any positive legislation contradicting natural right would bring ruin to the country and the degradation of its economy: *economic progress or regression* depends upon whether the dictates of *natural right* are carried out or violated.

Smith ascribed to legislation a more modest impact upon economic life. 'Mr. Quesnai', he wrote, 'seems to have... imagined that [the political body] would thrive and prosper only under *a certain regimen, the exact regimen of perfect liberty* and perfect justice. He seems not to have considered that in the political body, the natural effort which every man is continually making to better his own condition is a principle of preservation capable of preventing and correcting, in many respects, the bad effects of a political oeconomy, in some degree, both partial and oppressive. Such a political oeconomy, though it no doubt retards more or less, is not always capable of stopping altogether the natural progress of a nation towards wealth and prosperity, and still less of making it go backwards.'[1] *Economic progress forces a way for itself*, whatever the retarding influence of poor legislation that violates the principles of natural right.

The explanation for this marked difference in the views of Quesnay and Smith lies in the *differing economic conditions* of France and England in the 18th century. In France, capitalist agriculture was not so much an actually-existing phenomenon as a Physiocratic slogan that had still to be put into practice. Given France's feudal survivals and absolute monarchy, the extensive development of capitalism was genuinely impossible without a fundamental social and political revolution and the implementation of the 'natural law' of bourgeois society. This accounts for the extreme importance of natural right in Quesnay's system. England in the 18th century found itself in a different situation. Despite the continuing political domination of the landowning oligarchy, the basic social preconditions for the development of capitalism were already present. The capitalist economy was developing rapidly, either breaking or bypassing the separate guild or mercantilist restrictions which, despite slowing down the former's growth, could not halt it—hence Smith's view that economic progress is continuous, even where legislation is bad and contradicts the principles of natural right.

Thus for Smith *economic forces prove stronger* than legal and political obstacles. There follows from this an important methodological principal: it is possible to study the action of economic forces independently of the legal and political environment within which this activity takes place. Smith, in this way, cautiously cuts the umbilical cord binding political economy to natural right—a cord which for Quesnay had formed an unbreakable thread. Political economy becomes an *independent science*, and this is one of the great achievements of the Classical school. On the other hand, the ground is being prepared for counterposing *eternal and immutable economic laws* to historically transient and alterable socio-political conditions, and this is one of the Classical school's flaws. In their view, the nature of economic forces does not alter, even though they may be compelled to operate in different social surroundings. In Smith's eyes economic life is a combination between economic forces, the nature of which does not alter, and historical conditions, which do; the latter accelerate or slow down the movement of the former, but do not change their nature. Although an interest in changes in historical conditions is not foreign to Smith, he sees the economist's main task as studying the activity of economic forces which by nature are immutable.

What do these economic forces consist of? As is clear from the passage quoted above, Smith has in mind 'the natural effort which every man is continually making to better his own conditon.'[2] These *'natural efforts' of each individual* are a perpetual *stimulus to economic progress*. The constancy and immutability of their action flows from the *constancy of human nature*. Man, who by virture of his egoistical nature strives constantly to improve his own condition, is 'far more interested in that which directly concerns himself than he is in that which concerns others'.[3] Within the complex and changing web of economic phenomena we will find one *constantly acting force*: 'the uniform, constant, and uninterrupted effort of every man to better his condition, the principle from which publick and national, as well as private opulence is originally derived.'[4] For Quesnay the necessary condition of economic progress is the implementation of an *immutable system of natural right*; for Smith it is the activity of the *immutable nature of 'economic man'*. The type of 'economic man' at the centre of the Classical school's constructs, in independent pursuit of his own personal interests through free competition with others, is none other than an idealization of the independent commodity producer tied to other members of society by relations of exchange and competition. The Classical economists took the socially-conditioned and historically changing nature of the commodity producer and elevated it to being the naturally-conditioned and immutable nature of man in general.

Once the aspiration of the individual to better his situation is made to flow from the constancy of human nature, it is obvious that it will be operative in *all historical epochs* and under *any social conditions*. Smith challenges the view (which he attributes to Quesnay) that the individual exhibits this striving only under conditions of complete freedom. Smith's view is that it has been operating many hundreds of years before complete freedom (i.e., the bourgeois order) was ever realized, gaining victory over bad administration and legislation. Unfavourable social conditions are certainly able to retard the activity of these economic forces. Under slavery, for instance, the workers had no personal interest in the progress of production, whereas 'on the contrary, when they are secure of enjoying the fruits of their industry, they naturally exert it to better their condition'.[5] Invariable human nature manifests itself most forcefully under definite social conditions, namely those of the bourgeois order based on *private property* and unrestricted *competition*. Instead of explaining the nature of man-as-commodity-producer by the conditions of this social system, however, Smith sees the latter simply as an additional condition for the full outpouring of the individual forces located within man's permanent nature. The victory of one social system over another (the bourgeois order over the feudal) appears to Smith (as to other members of the 18th-century Enlightenment) as a victory of man's 'natural', immutable nature over the 'artificial' social institutions of the past. And as the new bourgeois social institutions are a necessary condition for the complete manifestation of the invariable nature of the individual, they thereby take on the character of eternal, 'natural' forms of economy.

Thus the starting point of Smith's investigation, his abstract economic man, is studied, so to speak, within a bourgeois encirclement, i.e., the commodity-capitalist economy. This abstraction from social factors, for all the errors it produced in evaluating such factors through the prism of human 'nature,' proved to be the saviour of Classical theory. For it allowed it to become a *theory of commodity-capitalist economy*.

How does Smith bridge the gap from his abstract individual to commodity-capitalist society? True to his original individualistic principles, Smith moves from the individual to society. Society is composed of separate, independent individuals: the social phenomenon is the result of these different individuals interacting with one another; social unity (insofar as we are talking about the economic side of society) is fashioned out of, and held together by these individuals' personal interests. So far as their economic contacts are concerned each individual enters into intercourse with others only insofar as this is dictated by his own *personal interests* and promises him some form of gain. The form of this intercourse is *exchange*. 'The propensity to truck, barter, and exchange one thing for another' is an essential principle of human nature. This permanent characteristic causes individuals to join together into an exchange society.

Society looked at as an economic unit, is an *exchange society* which separate persons enter into out of their *personal interests*. Already in Smith's early work, *The Theory of Moral Sentiments*, we find this extremely revealing passage: 'Society may subsist among different men, as among *different merchants*, from a sense of its utility, without any mutual love or affection; and though no man in it should owe any obligation, or be bound in gratitude to any other, it may still be upheld by a *mercenary exchange* of good offices according to an agreed valuation.'[6] Smith conceives of economic intercourse between people as a form of exchange, in other words, as *economic intercourse between the owners of commodities*. Smith develops this idea further in the second chapter of Book I of *The Wealth of Nations:* 'But man has almost constant occasion for the help of his brethren, and it is in vain for him to expect it from their benevolence only. He will be more likely to prevail if he can interest their *self-love* in his favour, and shew them that it is for *their own advantage* to do for him what he requires of them. Whoever offers to another a bargain of any kind, proposes to do this. *Give me that which I want, and you shall have this which you want*, is the meaning of every such offer; and it is in this manner that we obtain from one another the far greater part of those good offices which we stand in need of. It is not from the benevolence of the butcher, the brewer, or the baker, that we expect our dinner, but from their regard to *their own interest*.' [7] An individual's personal interest prompts him to enter into exchange with other people; and the aspiration to exchange, as we shall see, calls forth in turn the division of labour between people.

The argument just presented brilliantly characterizes Smith's *individualist and rationalist* method. Smith explains the origin of the most important social *institutions* (in this instance, exchange and the division of labour) by the undeviating nature of the *abstract individual* —his personal interest and conscious striving for the greatest gain. He thereby attributes to abstract man *motives and aspirations* (here, the striving to barter or exchange) that are in fact the *result* of the influence exercised on the individual by these same *social institutions* (the division of labour and exchange) over long periods of time—influences which he then adduces as a means of explaining these institutions. Smith deduces the basic socio-economic institutions that characterize the commodity-capitalist economy from the nature of man; what he takes as human nature, however, is the determinate nature of man as it takes shape under the influence of the commodity-capitalist economy.

Smith applies this same method of moving from the individual to society when explaining other socio-economic institutions. He explains the appearance of money by the simple fact that, owing to the inconvenience of *in natura* exchange, '*every prudent man in every* period of society, after the first establishment of the division of labour, must *naturally* have endeavoured to manage his affairs in such a manner, as to have at all times by him, besides the peculiar produce of his own industry, a certain quantity of some one commodity or other, such as he imagined few people would be likely to refuse in exchange for the produce of their industry.' [8] The words that we have italicised are those which especially characterize Smith's method. We should look for explanations of social institutions in the nature of '*every man*', that is, in the personal interests of each individual; hence we call Smith's method *individualist*. We call it *rationalist* because, in talking about the '*prudent*' man who consciously weighs up his advantages, Smith takes the rational calculation of the benefits and losses inherent in distinct economic activities—a calculation which only develops within the soil of highly developed commodity and capitalist economy—to be a property of human nature in general. Moreover these actions of the individual take place 'in *every* period of society' (once the division of labour has been established); this assertion reveals the *anti-historical* nature of Smith's method. Finally, Smith takes these activities of the individual as *'natural'*; here Smith grounds himself on the theory of natural right, introducing, however, important improvements that we will need to dwell on further.

According to Smith's basic *sociological conception*, socio-economic phenomena result from the actions of individuals as dictated by personal interest; it follows from this—and this conclusion is extremely important—that *economic phenomena are 'natural' in character*. The concept 'natural' is being used here in two different senses, one theoretical, the other practical. The basic proposition of Smith's *theoretical* system states that *economic phenomena possess an inherent, 'natural', law-determined regularity,* which exists independently of the will of the state and is based on the immutable 'natural' inclinations of the individual. The basic proposition of Smith's *economic policy* states that only when economic phenomena proceed 'naturally', unconstrained by the state, do they *bring maximum benefit both to the individual and to society as a whole*. The first of these propositions made Smith one of the founders of *theoretical economics*; the second made him the town crier of *economic liberalism*.

Let us begin with the second proposition. Once the individual's personal interest is seen as the stimulus of economic progress and the source of all economic institutions, the individual must be given the possibility to develop his economic powers freely, without any obstacles. The main precept of economic policy is *freedom of individual economic activity* and the elimination of *state interference*. There is no danger that in struggling for his own personal interest the individual will violate the interests of society as a whole. The interests of the individual and those of society are in *complete harmony*. Out of this mutual interaction of individuals—each of whom pursues only his correctly-understood personal interests—arise the most valuable social institutions, which in turn foster a tremendous rise in the productivity of labour: the division of labour, exchange, money, the accumulation of capitals, and their proper distribution between the different branches of production. A man 'by pursuing his own interest... frequently promotes that of the society more effectually than when he really intends to promote it.'[9] Thus 'every man, as long as he does not violate the laws of justice, is left perfectly free to pursue his own interest his own way, and to bring both his industry and capital into competition with those of any other man, or order of men. The sovereign is completely discharged from a duty, in the attempting to perform which he must always be exposed to innumerable delusions, and for the proper performance of which no human wisdom or knowledge could ever be sufficient; the duty of superintending the industry of private people, and of directing it towards the employments most suitable to the interest of the society.'[10] The government refrains from interfering in economic life, and preserves for itself only the modest functions of defending the country's external security, protecting individual persons from oppression by other members of society, and concerning itself with certain social undertakings. Economic life is given over wholly to the free play of individual interests. Smith, like the Physiocrats, expected that the realization of this 'obvious and simple *system of natural liberty*' [11] would result in maximum benefit both for society as a whole and for the separate classes of the population.

Smith's optimistic views—which for all the reservations that he placed upon them made him the founder of *economic liberalism*—could make their appearance only in an epoch when the industrial bourgeoisie still played a progressive role and its interests coincided with the needs of the overall economic development of society. Smith's aim had never been to defend the narrow interests of merchants and industrialists, towards whom he evinced no particular sympathy. He spoke about the condition of the workers, often with ardent feeling, and he wanted to improve it. But he was deeply convinced that only with complete freedom of competition and the powerful development of the capitalist economy would it be possible to expect any improvement in the position of the lower classes. He believed that the working class would receive an ever-increasing share in the growing mass of wealth of capitalist society. Capitalism's future development was to prove Smith's optimistic expectations wrong and lay bare the irreconcilable contradictions between the interests of the bourgeoisie, on the one hand, and those of the working class and the economic development of society as a whole, on the other. In its day optimistic liberalism played a positive role as a tool for freeing the productive forces of capitalist economy from the fetters of the old regime and of mercantilism, but later on, in the hands of Say, and especially of Bastiat, it was turned into an instrument for defending capitalism against the attacks of the socialists.

Smith, therefore, considered the economic phenomena of bourgeois society to be 'natural,' in the sense that they had been arranged in the best possible fashion and required no conscious intervention by any agencies of the state or of society. In this sense, to identify a phenomenon as 'natural' is the same as judging it as something positive. Here, to be *'natural'* means that it corresponds to the principles of *natural right*. In addition to using the term 'natural' in an evaluative sense, however, Smith also employs it when making purely theoretical judgements, where his task is to investigate a phenomenon as it exists, independently of any positive or negative assessment. Here to identify a phenomenon as 'natural' has a *purely theoretical* meaning, indicating, as we have already noted, that economic phenomena possess an inherent, *'natural' law-determined regularity* independently of any interference from the state. When Smith says that the 'natural price' (the value) of a commodity replaces its costs of production and earns an average profit, he means that where there is free competition and no intervention by the state the prices of commodities will have a tendency to establish themselves at the level indicated. This spontaneously established normal level for the price of the commodity in question, constitutes its 'natural' price. What is 'natural' in this instance is the result, reached regularly and spontaneously without the state placing any constraints upon the free competition of individuals. Hence the concept 'natural' embraces two characteristics: 1) *spontaneity*, and 2) *law-determined regularity*. As to the first, a price is only recognized as 'natural' when it is the *spontaneous result* of free competition and the conflict of individual personal interests; in this sense the *'natural'* (free) price is to be counterposed both to the *'legally set'*, fixed price established by the state or the guilds, and to a *'monopoly'* price. As to the second attribute, not every market price is identified as 'natural,' but only 'the central price, to which the prices of all commodities are *continually gravitating*,'[12] in other words, that level of prices which must be established under conditions of *market equilibrium*, where there is a balance between supply and demand. In this sense Smith differentiates *'natural'* price (value)—which expresses the law-determined regularity of market phenomena—from *'market'* prices, which constantly fluctuate depending upon fluctuations in supply and demand.

This second concept of 'natural' plays an extremely important part in Smith's theoretical system: he speaks of natural price, the natural level of wages, of profit, and of rent. Here the concept 'natural' means not that the precepts of natural right are being adhered to, but is a recognition of the *spontaneous law-determined regularity of market phenomena*. Although Smith from time to time uses the term in its first, evaluative sense, he most frequently employs it in its second, purely theoretical meaning; in any case, he does not confuse the practical and theoretical meanings of the term. Smith's transition from an evaluative to a theoretical understanding of the term 'natural' marked a great step forward for the *purely theoretical, scientific-causal study of economic phenomena.*

The economic investigations of the mercantilists were practical in character. Their works were overwhelmingly a collection of *practical prescriptions* recommended to the state for implementation. The embryos of a theoretical analysis that we find in Petty had little impact upon the general train of mercantilist thought. With the Physiocrats as well, attention was focused not so much upon investigating that which existed (i.e., the real phenomena of the capitalist economy) as upon elaborating that which ought to have existed (i.e., the conditions which had to be realized if the nation's economy was to flourish). They looked upon their economic laws and propositions as the *prescriptions of natural right*. It is only because they took capitalism as the ideal natural order that the Physiocrats' analysis contains theoretically valuable elements for an understanding of capitalist economy. If the mercantilist system was by nature *practical*, and that of the Physiocrats *teleological*, Smith consciously set himself the task of studying the capitalist economy *theoretically*. It is true that questions of economic policy are for Smith extremely important and are often interwoven with his theoretical analysis in the course of his exposition; nevertheless, in the main the latter is kept methodologically distinct and isolated from his considerations of practical issues. It is true that some of Smith's more serious errors can be explained by his confusion of theoretical and practical problems (see the chapter below on the theory of value), but in this there is no cause for surprise: because it had grown out of practical needs and had been dissolved into economic policy in its primitive stages, economic theory was not immediately capable of gaining a clear awareness of itself as a method of purely theoretical analysis. In any event, Smith's analysis represented a great and methodologically decisive service: he set political economy onto the path of *theoretically studying the real phenomena of capitalist economy*. Smith's reputation as the founder of political economy rests upon this.

1 Adam Smith, *An Inquiry into the Nature and Causes of the Wealth of Nations*, edited by R.H. Campbell, A. S. Skinner, and W. B. Todd (Oxford University Press, 1976), Book IV, Chapter 9, p. 674. Rubin's italics.

2 *Ibid*, Book IV, Chapter 9, p. 674.

3 Translated from the Russian.

4 *Wealth of Nations*, Book II, Ch. 3, p. 343.
5 *Ibid*, Book III, Ch. 3, p. 405.
6 Adam Smith, *The Theory of Moral Sentiments* (London, George Bell & Sons, 1875), Part II, Section II, Chapter 3, p. 124. Rubin's italics.
7 *Wealth of Nations*, Book I, Chapter 2, pp. 26-27. Rubin's italics.
8 *Ibid*, Book I, Ch. 4, pp. 37-38. Rubin's italics.
9 *Ibid*, Book IV, Ch. 2, p. 456. This is the passage where Smith articulates his famous concept of the 'invisible hand'. 'As every individual, therefore, endeavours as much as he can both to employ his capital in the support of domestic industry, and so to direct that industry that its produce may be of the greatest value, every individual necessarily labours to render the annual revenue of the society as great as he can. He generally, indeed, neither intends to promote the public interest, nor knows how much he is promoting it. By preferring the support of domestic to that of foreign industry, he intends only his own security; and by directing that industry in such a manner as its produce may be of the greatest value, only his own gain, and he is in this, as in many other cases, led by an invisible hand to promote an end which was no part of his intention. Nor is it always the worse for the society that it was no part of it. By pursuing his own interest he frequently promotes that of the society more effectually than when he really intends to promote it.'
10 *Ibid*, Book IV, Ch. 9, p. 687.
11 *Ibid*, Book IV, Ch. 9, p. 687. Rubin's italics.
12 *Ibid*, Book I, Chapter 7, p. 75. Rubin's italics.

CHAPTER TWENTY-ONE

The Division of Labour

Smith's very first lines show that he had clearly introduced something new into economic science. It is interesting to compare the beginning of Smith's work with that of Mun's 'mercantilist gospel'. 'The ordinary means therefore to increase our wealth and treasure is by *Forraign Trade*.'[1] That is how Mun—who sees commerce, or the sphere of circulation as the source of all wealth—begins his book. Smith, like the Physiocrats, shifts the focus of analysis onto production but in doing so avoids their onesidedness: it is *labour in general* that he proclaims the sole source of wealth, i.e., the entire labour of a nation as distributed over the different branches of production and divided up between society's individual members: 'The annual labour of every nation is the fund which originally supplies it with all the necessaries and conveniences of life which it annually consumes.'[2] The source of *wealth* is *labour*. Here 'labour' is to be understood as the total, aggregated labour of a nation having the form of a social division of labour, and 'wealth' as the totality of material products or articles of consumption.

If it is labour that creates wealth, then *increases* in the latter can take place under one of the following two conditions: 1) there is a rise in the individual worker's *productivity of labour*, or 2) *the number of productive workers* increases compared to other members of society. A rise in the productivity of labour, however, is a result of the *division of labour*, while an increase in the number of productive workers demands an increase and *accumulation of the capital* spent on maintaining them. Smith divides up the first two theoretically orientated books of *The Wealth of Nations* accordingly. Book One begins by describing the *division of labour*; from here Smith passes to the closely associated phenomena of exchange (*money, value*) and the distribution of what is produced (i.e., *wages, profit, and rent*). Book Two contains his theory of *capital* and his doctrine on the *accumulation of capital* and *productive labour*.

The first chapters of *The Wealth of Nations*, devoted to the division of labour, have always been considered among the most brilliant; it is they that have made the greatest impact by virtue of their sweep and eloquence of description. For all practical purposes Smith says little that is new compared to his predecessors (Petty, Furguson); yet what a happy intuition it was that led him to place his description of the division of labour at the very beginning of the book. Because of this, commodity society at once emerges as a society based on the one hand on the *division of labour* and on the other upon *exchange* between individual economic units—in other words, as a society based on *labour* and *exchange* (a 'commercial society,' to use Smith's term).

Smith begins with his well-known description of a *pin-making manufactory*, with its detailed division of labour between ten workers: one draws the wire, another straightens it, a third cuts it, etc. By breaking down the labour process into extremely simple operations, each of which is assigned to an individual labourer, the productivity of labour is raised 100 times: those ten workers produce 48,000 pins a day, whereas each of them working separately could barely produce twenty pins in a full day. Smith enumerates three reasons why the division of labour raises labour productivity: 1) each worker acquires greater *dexterity* by constantly repeating the same operations; 2) there is no *time* lost in switching from one operation to another; and 3) breaking the labour down into basic operations facilitates the invention of labour-saving *tools*.[3] The arguments used by Smith are characteristic of the manufactory period, which was itself characterized by the specialization of workers to a few partial operations and by the differentiation of tools. Smith's assertion that the *division of labour* is the main reason for the growth in labour productivity places him squarely in his context. His underestimation of the role played by the implements of labour, and by *machinery* in particular is quite understandable given that his was an age still prior to the onset of the industrial revolution and the manufactories' technical superiority relied on a minutely executed division of labour. Although at the beginning of his book Smith describes only the beneficial aspects of the division of labour inside the manufactory, in other passages he explains how humiliating the monotonous character of the work is to the individuality of the worker performing only partial operations and how it makes him 'stupid and ignorant'.[4]

From the pin-making manufactory Smith quickly moves on to other examples of the division of labour. Here he takes as his example not the division of labour *within a single* enterprise, but the division of labour between *different* enterprises belonging to different branches of production. Smith brilliantly depicts how cloth passes through a series of economic units, beginning with the sheep farmer, whose labours are devoted to obtaining the wool, and ending with the worker employed at dyeing and finishing the cloth. It is here, when describing this type of division of labour, that Smith is at his most eloquent. 'Observe the accommodation of the most common artificer or day-labourer in a civilized and thriving country, and you will perceive that the number of people of whose industry a part, though but a small part, has been employed in procuring him this accommodation, exceeds all computation. The woollen coat, for example, which covers the day-labourer, as coarse and rough as it may appear, is the produce of the joint labour of a great multitude of workmen. The shepherd, the sorter of the wool, the wool-comber or carder, the dyer, the scribbler, the spinner, the weaver, the fuller, the dresser, with many others, must all join their different arts in order to complete even this homely production.'[5] Over and above this were also employed merchants and carriers, shipbuilders, workers who fashioned the tools, etc. Here it is everywhere a question of a division of labour between different commodity producers or individual enterprises.

We see here that Smith confuses the *social* division of labour with the division of labour within the *manufactory*, which is *technical*. He fails to perceive the deep social distinction that exists between these two forms of the division of labour. The social division of labour between individual enterprises, being based on the exchange of their products, comprises the basic feature of any *commodity* economy and is already significantly developed under craft production; the technical division of labour within a single enterprise appeared only with the emergence of large-scale, *capitalist* enterprises, i.e., the manufactories. The first of these forms presupposes that the means of production are *broken up* between independent commodity producers; the latter presupposes the *concentration* of substantial means of production in the hands of a single capitalist. The separate, independent commodity-producers (handicraftsmen) are bound to one another only by exchanging their products on the market. In the manufactory the individual workers are bound to each other by the general direction of the capitalist. In the first instance the nature of the bond between people is *disorganized, spontaneous, and through the market*; in the second it is *organized and planned*.

Smith failed to take account of these distinctions because his attention—and this is generally speaking *one of the characteristic features of the Classical school*—was focussed not on the *social forms* of the division of labour but upon its *material and technical advantages* in raising the productivity of labour. From this standpoint, since both forms taken by the division of labour act to raise labour productivity, they can be treated as identical. The different social natures of the mutual relations between independent commodity producers, on the one hand, and the workers in a single manufactory, on the other, recede into the background, escaping the author's attention.

In his first chapters Smith's main task is to describe the *social division of labour* based on exchange and characteristic of any commodity economy. Greatly influenced, however, by the type of division of labour to be found within the manufactory, Smith also adduces examples from this sphere, and is in general inclined to depict the social division of labour as a form of the division of labour within the enterprise. To Smith, the whole of society appears as a gigantic manufactory, where the work is divided up between thousands of separate but mutually complementary enterprises. *The material connection and interdependence* between commodity producers is placed in the forefront. Each member of society is useful to all the others, and is compelled in turn to enlist their assistance. 'Without the assistance and cooperation of many thousands, the very meanest person in a civilized country could not be provided, even according to ... the easy and simple manner in which he is commonly accommodated.'[6] All people, though each of them be animated simply by the pursuit of personal gain, in reality work for one another: 'the most dissimilar geniuses are of use to one another';[7] *a complete harmony of interests* exists between society's individual members.

Here we *come accross a second feature of the Classical school*, closely tied to the first. Because Smith has directed his attention towards the material and technical interdependence between the individual members of society, he assumes that these individuals enjoy *a complete harmony of interests*. Through their labour the spinner and weaver mutually complement one another; the one could not exist without the other. Smith forgets, however, that both are commodity producers who sell their products on the market. The struggle over the price of the product (e.g., that of yarn) creates a deep antagonism between them; both branches of production, under the pressure of fluctuations in market prices and through the ruin of numerous producers, adapt to one another spontaneously. Smith's concern for the material and technical advantages of the division of labour, rather than for the social form that it assumes in a commodity-exchanging economy, leads him to overestimate the elements of *harmony* in such an economy and to ignore the *contradictions and antagonisms* that it produces.

Despite this, Smith did grasp the close connection between *the division of labour* and *exchange* and in fact lays great stress on it. A feature of the Classical school is not that it completely abstracts the material and technical side of production from its social from, but that it confuses the two. To the Classical School it was inconceivable that the process of production could have any social form other than a commodity capitalist one, which in their eyes is the rational and natural form of economy. Once it is assumed that the process of production always takes place within a specific social form, it becomes superfluous to carry out a special analysis of that form; rather it is enough simply to study the process of production in general. However, because the process of production in general is tied irrevocably to a given social form, the conclusions obtained from studying the former are fully applicable to the latter. Hence it happens that the Classical economists constantly *confuse* the *material-technical* and *social* points of view, an example of which is afforded by Smith's doctrine on the division of labour.

Smith cannot imagine any division of labour other than one based on exchange—for him a necessary property of human nature, one which distinguishes man from animals. This *propensity to exchange* called forth the *division of labour*. On this point Smith is mistaken, since the social division of labour has existed—albeit on a modest scale—even where a commodity economy had been absent, e.g., in the Indian commune. At another point Smith correctly notes that the development of exchange provides an impetus for the further division of labour: 'the extent of this division must always be limited by the extent of that power [the power of exchange—Ed.], or, in other words, by the extent of the market.'[8]Though he lays great stress upon the effect of exchange in bringing about and developing the division of labour, Smith nevertheless ignores the role of exchange as that *specific social form* that the division of labour assumes in commodity economy. He is constrained by his analysis of the division of labour in general and its material and technical advantages.

For all its inadequacies, Smith's *theory of the division of labour* did him a *great service*: by starting out from a conception of society as a gigantic workshop with a division of labour, Smith arrived at the extremely valuable idea of society as a society of people who *labour* and who simultaneously *exchange*. The division of labour makes all members of society *participants in a single process of production*. The products of labour of all members of society are 'brought, as it were, into a common stock, where every man may purchase whatever part of the produce of other men's talents he has occasion for.'[9] Each man becomes dependent on the labour of other people. 'But after the division of labour has once thoroughly taken place, it is but a very small part of these [the 'necessaries, conveniencies, and amusements of human life'—*Trans.*] with which a man's own labour can supply him. The far greater part of them he must derive from the labour of other people.'[10] Each man acquires the produce of other people's labour, and they are thus united together into a single *labouring society*. Smith conceives of his labouring society strictly as an *exchange society*: 'When the division of labour has been once thoroughly established, it is but a very small part of a man's wants which the produce of his own labour can supply. He supplies the far greater part of them by exchanging that surplus part of the produce of his own labour, which is over and above his own consumption, for such parts of the produce of other men's labour as he has occasion for. Every man thus lives by exchanging, or becomes in some measure a merchant, and the society itself grows to be what is properly a *commercial society*.'[11] The social division of labour appears to Smith only in the form of exchange, while, on the other hand, the *exchange of the produce of labour* is reduced, according to this view, to *an exchange of the labouring activities* of individual producers. Commodities 'contain the value of a certain quantity of labour which we exchange for what is supposed at the time to contain the value of an equal quantity.'[12] By acquiring the product of someone else's labour I thereby acquire the labour of its producer.

The Smithian conception of society as at one and the same time a labouring and an exchanging society can be expressed by the following two propositions: 1) what appears as a market exchange of commodities *for money* is in reality the mutual exchange of the *products of labour* of the different persons who, between them, perform the whole of social labour; 2) the exchange of the *products* of the different people's labour reduces itself to the mutual exchange of the producers' very *labour*. With the first proposition Smith took his distance *from the mercantilists*; with the second he differentiates himself from the *Physiocrats*.

The *mercantilists*, though focussing their attention upon exchange, were blinded by its market, *monetary* form: they saw only the exchange of an *in natura* product for money, i.e., for social wealth, and wanted to limit the entire exchange process to the sale, C-M, and then convert the money into treasure. Smith, following the example of the Physiocrats, saw exchange as a unity of the acts of sale (C-M) and purchase (M-C_1), in other words, as an exchange of one *in natura* product (C) for another (C_1) through the medium of money; the latter plays only a transitory role as *means of circulation*. Hence Smith's *assessment of the role of money* is the opposite to that of the mercantilists. Money does not constitute the wealth of society. 'The revenue of the society consists altogether in those goods, and not in the wheel which circulates them.'[13] Money is needed merely as an auxiliary for facilitating the circulation of products. 'The gold and silver money which circulates in any country may very properly be compared to a highway, which, while it circulates and carries to market all the grass and corn of the country, produces not a single pile of either'.[14] Money is simply 'dead' capital: an increase in the quantity of money in a country correspondingly lowers outlays on the material production of products and consequently reduces society's real income which consists in what it produces. Any savings on the outlays needed to maintain the monetary system (e.g., replacing gold with bank notes) are to society's overwhelming advantage.

Thus, *the exchange of a commodity for money* is in essence nothing but an *exchange of one product for another*. Thus far Smith is in agreement with Quesnay, whose *Tableau Economique* presented the first overall picture of the circulation of products.* Beyond this, however, they begin to diverge.

Although there were a number of particular questions where Smith was simply repeating the views of the Physiocrats,** in essence he overcame their onesidedness through his theory of the division of labour and value. The point of view from which Smith starts out is that labour creates wealth. The circulation of products is, in his view, not a movement of the *substance of nature*, but a circulation of the *products of labour*. Because for Smith society is a labouring society, he sees the exchange of the products of labour as an *exchange of the labouring activities* of society's individual members. Once the division and mutual exchange of labour are made the basis of commodity economy, it is evident that the different branches of production are bound to each other by relations of *mutual dependence*, rather than

*See above, Chapter Fifteen.

**Thus, for example, he considered agricultural labour to be more productive than industrial labour, asserted that in the 'natural' course of development capitals would first be invested in agriculture and only later on in industry, etc.

of *one-sided subordination*. Industry is not subordinated to agriculture but coordinated with it. Smith posits, in place of the *unidirectional flow of the substance of nature* from agriculture to industry,* a *two-directional transmission of the products of labour* originating from wherever it is that human labour is being applied: one flow of products passes from agriculture to industry, a counter flow moves from industry to agriculture. The two flows cross each other and are balanced out on the basis of an *exchange of equivalents*, which is the theory of value's object of study.

Smith could accord a central role to the *theory of value* (a theory that was virtually non-existant amongst the Physiocrats) precisely because he was able to identify the problem of how the different *branches of production* were economically *coordinated*, and to keep this question separate from the problem of the economic *subordination* of different *social classes*. He took up the latter in his theory of distribution; the first he dealt with in his theory of value. Although theoretically the two problems were closely interconnected, and the theory of distribution was built up on the basis of the theory of value, it was nevertheless necessary that they be studied separately; this in turn helped Smith to do away with the conceptual confusion that had kept the Physiocrats from correctly grasping both the class structure of society and the interdependence that exists between branches of production (agriculture and industry). Smith, too, continued to confuse these two problems, as we will see, and in so doing introduced contradictions into his theory of value. All the same, his merits were enormous: he identified the problem of coordination between *branches of production* of equal standing; he depicted the interrelation between them as a *mutual exchange of products of labour*; and he perceived that behind this exchange of products lies an *exchange of labour*. By doing this he assigned the *labour theory of value* that central place which it continues to occupy in economic science.

*In Quesnay's scheme industry simply returns to agriculture in another material form the substance of nature that it received from it.

---

1 Mun, *England's Treasure by Forraign Trade*, McCulloch edition, *op cit*, p. 125 (Mun's italics).

2 Smith, *The Wealth of Nations*, 'Introduction and Plan of the Work', p. 10.

3 *Ibid*, pp. 14-17.

4 'In the progress of the division of labour, the employment of the far greater part of those who live by labour, that is, of the great body of the people, comes to be confined to a few very simple operations; frequently to one or two. But

the understandings of the greater part of men are necessarily formed by their ordinary employments. The man whose whole life is spent in performing a few simple operations, of which the effects too are, perhaps, always the same, or very nearly the same, has no occasion to exert his understanding, or to exercise his invention in finding out expedients for removing difficulties which never occur. He naturally loses, therefore, the habit of such exertion, and generally becomes as stupid and ignorant as it is possible for a human creature to become. The torpor of his mind renders him, not only incapable of relishing or bearing a part in any rational conversation, but of conceiving any generous noble, or tender sentiment, and consequently of forming any just judgement concerning many even of the ordinary duties of private life ... The uniformity of his stationary life ...currupts even the activity of his body, and renders him incapable of exerting his strength with vigour and perseverance, in any other employment than that to which he has been bred. His dexterity at his own particular trade seems, in this manner, to be acquired at the expence of his intellectual, social, and martial virtues. But in every improved and civilized society this is the state into which the labouring poor, that is, the great body of the people, must necessarily fall, unless government takes some pains to prevent it.' *The Wealth of Nations*, Book V, Chapter 1, pp. 781-82.

5 *Ibid*, Book I, Ch. 1, p. 22.
6 *Ibid*, Book I, Ch. 2, p. 23.
7 *Ibid*, Book I, Ch. 2, p. 30.
8 *Ibid*, Book I, Ch. 3, p. 31.
9 *Ibid*, Book I, Ch. 2, p. 30.
10 *Ibid*, Book I, Ch. 5, p. 47.
11 *Ibid*, Book I, Ch. 4, p. 37; Rubin's italics.
12 *Ibid*, Book I, Ch. 5, pp. 47-48.
13 '... as the machines and instruments of trade, &c. which compose the fixed capital either of an individual or of a society, make no part either of the gross or of the neat revenue of either; so money, by means of which the whole revenue of the society is regularly distributed among all its different members, makes itself no part of that revenue. The great wheel of circulation is altogether different from the goods which are circulated by means of it. The revenue of the society consists altogether in those goods, and not in the wheel which circulates them. In computing either the gross or the neat revenue of any society, we must always, from their whole annual circulation of money and goods, deduct the whole value of the money, of which not a single farthing can ever make any part of either'. *Ibid*, Book II, Ch. 2, p. 289.
14 *Ibid*, Book II, Ch. 2, p. 321.

CHAPTER TWENTY TWO

The Theory of Value

In setting out to analyze the concept of value, Smith draws a primary distinction between *use value* and *exchange value*: the former he places outside the scope of his investigation and devotes his entire attention to the latter. In this way Smith grounds himself firmly in the study of commodity economy, where each product is designated for exchange rather than for the direct satisfaction of the needs of its producer. Smith owes his ability to pose the question in such a principled and clearcut fashion to his doctrine of the division of labour: in any society based on the division of labour each producer will be fashioning products needed by other members of society.

Thereby, Smith very precisely, and absolutely correctly defines the *object*[1] of his investigation: exchange value. On the other hand, if we ask what is the exact point of view from which Smith studies this object, we find a methodological *duality* in the way that he poses the problem. On the one hand, Smith wishes to uncover the causes that determine first, how much value a commodity possesses and second, any changes in this magnitude; on the other hand, he wants to find a precise, invariable standard which could then be used to measure the value of a commodity. On the one hand he aspires to lay bare the *sources of changes* in value and on the other to find an *invariable measure* of value. It is clear that there exists a fundamental methodological difference between these two ways of posing the question, and that this difference must introduce a dualism into the core of Smith's theory. The theoretical study of real changes in value becomes confused with the practical task of arriving at the best measure of value.[2]

As a result of this confusion, Smith's analysis of exchange value becomes bifurcated and flows along two methodologically different channels: the one the discovery of what causes changes in value, the other the search for an invariable measure of value. Each of these paths leads Smith to a particular conception of labour value or of labour as the basis of value. The first leads him to a concept of *the quantity of labour expended on the production of a given product;* the second to a concept of *the quantity of labour which a given commodity can acquire or purchase through exchange.*

Smith asks, at the outset of his investigation, wherein consists 'the real measure of... exchangeable value'? The quest for such an *invariable measure* occupies the better part of his attention (Book I, Chapter 5). To understand why Smith directs his analysis along such a methodologically incorrect path we ought to recall that Smith had inherited the problem of finding a *measure of value* from his *mercantilist* predecessors. For the mercantilists, inclined as they were to address themselves to practical problems, the theory of value had as its practical task to find a measure of value; we will recall how Petty and Cantillon had sought a measure of value in the 'equation between labour and land.'* It was only slowly and gradually over the course of the 18th century—and largely due to the efforts of Smith himself—that political economy was turned from an agglomeration of practical rules into a system of theoretical propositions, and that the concept of there being theoretical laws behind phenomena ceased to be mixed together with practical prescriptions (as the mercantilists had done) or with 'natural law' (as had the Physiocrats). In Smiths's theory of value this task of theoretically studying the causes of real economic phenomena had still not freed itself from extraneous elements of a practical character.

Smith's general individualist and rationalist approach intruded equally into his search for a measure of value. Earlier we saw that Smith explains the origin of socio-economic phenomena by the utility they possess from the point of view of the isolated economic individual.** He adopts this same approach when dealing with the division of labour and exchange. The division of labour, which is founded upon exchange, makes it possible for each individual to obtain the articles that he needs by exchanging his own product, which thereby acquires special significance for the individual by virtue of his ability to exchange it for other articles. From the *individual's* point of view, the first *practical* question to be posed is how great a *significance does this article hold for him*, i.e., what is the precise measure of exchange value?

What, then, is the measure or index of the value of a given product? It would seem at first glance that we could take as our measure the quantity of other commodities that we get in exchange: the greater their number the higher, obviously, is the value of the commodity

*See above, Chapter Seven.

**See Chapter Twenty.

in question. Smith quite rightly rejects this answer, on the grounds that the value of the commodity that I receive in exchange for my own product is itself subject to constant changes. It is equally impossible to measure the value of a commodity by the quantity of *money* (gold) that it will exchange for, since gold, too, changes in value.

In that case, by what could I measure the value of my product? To answer this question Smith makes recourse to his theory of the division of labour: there he established that a society based on the division of labour is a society of people who labour and who, through mutual exchange of the products of their labour, indirectly exchange their labour. Smith, however, takes what is an extremely valuable objective-sociological conception of exchange value (one which Marx was to use as the basis of his own theory of value) and gives it a *subjective-individualist* interpretation. An exchange society is founded upon the mutual exchange of the labour of its members. Smith then asks, what does this exchange reduce itself to from the standpoint of the isolated individual? His answer: to *the acquisition of the labour of other people* in exchange for his own product. In exchanging the cloth that I have made for sugar or money I am in essence acquiring a definite quantity of other people's labour. My cloth has a greater exchange value the greater the quantity of other people's labour I can dispose over, or 'command', in Smith's expression, in exchange for it. Because of the social division of labour I can obtain what products I need by exchanging products that I have produced, rather than producing these necessities myself, with my own labour. Consequently, I can measure the value of what I have produced by the quantity of other people's labour that I receive when exchanging it. *The quantity of labour which can be acquired or purchased* in exchange for a given commodity is the measure of that commodity's value.*

Although Smith's theory of the measure of value would seem to flow out of his conception of exchange society as a society of labourers, it suffers from the following defect. When we say that in a society of simple commodity producers all of its members exchange the products of their labour, and hence also their labour itself, we are using the term 'exchange' in *two* different ways. The *products* of labour *really are exchanged* and placed on an equal footing with one another in the market; here we have exchange in the literal sense of the word. As regards the 'exchange' of actual labour, we mean essentially a process

*As a secondary measure of a commodity's value Smith takes the quantity of *corn* that it will purchase through exchange (since a given amount of corn will always be able to purchase approximately the same quantity of labour).

through which the labouring activities of individuals are bound to one another and distributed, a process closely associated with the market exchange of the products of labour. Literally speaking there is *no* exchange of *labour*, since it is not actual labour that is bought and sold on the market, but only the products of labour. The labouring activity of people performs a definite *social function*, but it is not an *object of purchase and sale*. When we say that there is an 'exchange' of labour we mean that labours are made socially equal [*uravnenie*] and not that they are equated [*priravnivanie*] on the market.

Thus, when we say that in an exchange society (where people relate to one another as simple commodity producers) I use my cloth to acquire domination over, or to purchase someone else's labour, this says merely that I exert an indirect influence upon the labour of another commodity producer by acquiring what he has made. I exchange my product directly for a product of labour, and not for someone else's labour. In exchange for my cloth I receive sugar, and thereby indirectly the labour of the sugar producer. In other words, I acquire the labour of another person in an already *materialised* form, as a *product* that he has produced. This differs enormously from the *direct* exchange of my cloth for someone's labour, i.e., for the *labour power* of a hired worker. What differentiates these two cases so sharply is not simply the *material form* of the labour being purchased (materialised versus living), but also the type of *social relations* that bind together the participants in the exchange. In the first case they enter into a relation with one another as simple commodity producers; in the second, as capitalist and worker. The first case (i.e., an exchange of one *product for another,* or for materialised labour) constitutes a basic feature of any commodity economy; the second (i.e., the exchange of a *product for living labour*, or of capital for labour power) occurs only within a capitalist economy. Only in the second instance does labour function directly as an *object of purchase and sale* or as a commodity (i.e., labour power).

Smith's mistake was to confuse the *social 'exchange'* (or more properly, equalisation) of labour that takes place in any commodity economy with the *market 'exchange'* of labour as an object of purchase and sale that occurs in a capitalist economy. Smith says that I acquire or purchase with my cloth the labour of other people. But when it is asked whether I am exchanging my cloth for materialized labour (i.e., for the product of someone else's labour) or for the living labour of a hired worker, Smith gives no clear cut answer. He talks about 'the quantity either of other men's *labour*, or, what is the same thing, of the *produce* of other men's labour which it allows him [the owner of the given commodity—*I.R.*] to purchase or command.'[3] Smith carries this confusion of labour with the products of labour right through his analysis. At the beginning of Chapter 5 Smith usually has in mind indirectly disposing over the labour of other independent commodity producers by acquiring the products of their labour. But by the end of this chapter he is already laying greater stress upon the exchange of a commodity for living labour, or *labour power*: the commodity owner appears now as an 'employer' and the commodity surrendered in exchange for labour as 'the price of labour', or the worker's wage.[4] To introduce features inherent in a *capitalist economy* into an analysis of the value of commodities, or of a *simple commodity economy* means to bring into this analysis a terrible confusion. Smith's conception of the labour which is purchased in exchange for a given commodity, and which serves as a measure of that commodity's value, becomes really two concepts: sometimes it appears as the '*materialised labour purchased*', and sometimes as the '*living labour purchased*'.

Smith's conceptual confusion resulted from the fact that having failed from the outset to grasp the social nature of the process of 'exchanging' labour in a commodity economy, he mistook it for the market 'exchange,' or purchase and sale of labour. He took labour as a *social function* to be the same as the labour which functions as a *commodity*. Yet if labour acts as an article of purchase and sale, can it really serve as a measure of value? Does not *the value of labour itself change* thanks to the fact that a given quantity of labour will be able to purchase a greater or lesser amount of commodities (depending upon fluctuations in the wages paid to 'labour')? To get out of this difficulty Smith puts forward his famous proposition that 'equal quantities of labour, at all times and places, may be said to be of equal value to the labourer'.[5] However many commodities the worker may be able to exchange a day of labour for, this day's labour will always mean that he has to sacrifice the same amount of 'his ease, his liberty, and his happiness'.[6] Should he today be able to exchange a day's labour for twice as much cloth as he could last year, this merely shows that the value of cloth has fallen. The value of the labour itself has not changed, and cannot change, since *the subjective assessment of the effort of labouring remains unaltered.* But in that case, the objective quantity of labour purchased in exchange for a given commodity can be taken as an exact measure of that commodity's value. We need only establish that a given commodity previously purchasable with one day's labour can now only be bought with the labour of two days, to be convinced that the value of this commodity has doubled. Two days' labour at all times represents twice the subjective effort and strain compared with the labour of a single day, even if that two days' labour now affords no more commodities (or wages) than one day's labour did before. The distinctive feature of Smith's theoretical confusion between objective and subjective factors (a confusion in which objective factors tend to dominate) is as follows: in order that an objective quantity of labour purchased may preserve its role as the invariable measure of value, Smith has to claim that subjective assessments of the efforts of labouring are also invariable.

Previously Smith had mistakenly turned labour as a social function into labour as a commodity, and had taken 'labour purchased' as an invariable measure of value. Now, in order to be rid of the constant fluctuations in value inherent in labour being itself a commodity, he substitutes for the objective quantity of labour purchased the total subjective strain and effort that this labour elicits. The confusion of labouring activity as a *social function* with labour as a *commodity* (i.e., with 'labour purchased'); the confusion of the '*materialized labour* purchased' with the '*living labour* purchased'; finally, the confusion of the *objective* quantity of labour with the total *subjective* effort and exertion—these conceptual confusions are the price that Smith had to pay for having directed his investigation along the methodologically false path of looking for a measure of value.

Thus far we have been discussing Smith's doctrine of the measure of value. Parallel with this confused and error-ridden train of thought, however, there is another, more valuable and promising theoretical thread which is directed at analyzing the *causes of quantitative changes in the value* of commodities. These two theoretical paths constantly cross one another. Although at the beginning of his analysis, in Chapter 5, Smith's thinking is mostly taken up with the quest for a measure of value, he constantly comes up against the fact that the value of commodities really does change; compelled to inquire further into the causes of such changes, he unhesitatingly deems that cause to be a change in the quantity of labour *expended* on a commodity's production. Especially interesting are Smith's remarks on why money cannot be taken as an invariable measure of value. 'Gold and silver, however, like every other commodity, vary in their value'; it is thus obvious that 'the quantity of labour which any particular quantity of them can purchase or command' also changes. But when the question is put, why has the value of gold and silver (i.e., the quantity of labour which they can *purchase*) changed, the answer forthcoming is unequivocal: because there has been an alteration in the quantity of labour *expended* on their production. '*As it cost less labour* to bring those metals from the mine to the market ... they could *purchase or command less labour.*' It is quite obvious that Smith is combining here the concepts of 'labour purchased' and 'labour expended'. The first is a measure or index of the magnitude of a commodity's value, the second is the cause of quantitative changes in its value.[7]

At the start of Chapter 8, Smith sees changes in the value of commodities as a direct consequence of 'all those improvements in its [labour's—*Trans.*] productive powers, to which the division of labour has given occasion. All things would gradually have become cheaper and cheaper. *They would have been produced* by a smaller quantity of labour; and ... naturally ... would have been *purchased* likewise with the produce of a smaller quantity.'[8] Once a smaller quantity of labour begins to be expended on the production of a certain commodity so, too, must fall the quantity of labour which this commodity will purchase when exchanged. A change in the quantity of '*expended* labour' is consequently a *cause* of changes in the quantity of '*purchasable* labour', hence also of changes in value, of which this latter acts as a *measure* or index. The value of a commodity *is determined* by the labour *expended* on its production, and *is measured* by the labour which it will *purchase* in the course of exchange.

Thus Smith is now determining the value of the commodity in two ways: 1)by the quantity of labour *expended* on its production, and 2) by the quantity of labour which the given commodity can *purchase* through exchange. Do these two definitions not contradict one another? From a *quantitative* point of view there are definite *social conditions* under which the two will *coincide*. Suppose that we have a society of *simple* commodity producers or craftsmen who own their own means of production. Each of them will exchange the product of ten hours of his own labour (e.g., cloth) for the product of ten hours labour (e.g., a table) performed by somebody else. It will be as if he is purchasing a quantity of another person's labour (materialized in the table) exactly equal to the quantity of labour he himself expended on the production of his cloth. In this case we can say that it makes no difference whether the value of the cloth is determined 1) by the quantity of labour expended on its production or 2) by the quantity of labour which it can purchase when exchanged. The quantity of 'expended labour' coincides completely with the quantity of '(materialized) labour that can be purchased'. In a simple commodity economy labour performs a *two-fold* function: 'labour purchased' serves as a *measure* of the value of products while 'labour expended' *regulates the proportions* in which commodities are *exchanged*. 'In that early and rude state of society which precedes both the accumulation of stock and the appropriation of land, the proportion between the quantities of labour necessary for acquiring different objects seems to be the only circumstance which can afford any rule for exchanging them for one another.'[9] In 'early' society, which in essence means simple commodity economy, the exchange of products is subject to the *law of labour value*.

Up to this point these two strands of Smith's analysis—the one leading from the measure of value to purchased labour, the other from the source of changes in value to expended labour—ran parallel and could be reconciled since, under conditions of a simple commodity economy, the (materialized) labour that is purchased is *equal* to the labour that has been expended. Smith, however, did not confine his study to a simple commodity economy, being interested first and foremost in the capitalist economy developing around him. The *'handicraft'* motif in his theory of value is accompanied by a *'capitalist'* motif. If the commodity is a means by which the craftsman can acquire the *product* (or *materialized* labour) of another person, for the capitalist it is a means of acquiring another person's *living labour*. Smith remembers full well that under capitalism the hired labourer receives only a part of the produce of his labour, and that hence a smaller quantity of materialized labour (the commodity) is being exchanged for a greater quantity of living labour (labour power). For the product of ten hours labour the capitalist may receive twelve hours of living labour from the workers. It therefore follows that the quantity of labour *expended* on a commodity's production is no longer *equal* to the quantity of living labour which that commodity will *purchase* in exchange. In a capitalist economy the two determinations of value, which had coincided under conditions of simple commodity production, now sharply diverge. Smith, therefore, now has to make a firm choice: the value of a commodity must be determined *either* by the labour expended on its production, *or* by the (living) labour that it can purchase in exchange. Instead of adopting the first, correct standpoint Smith draws exactly the opposite conclusion. He holds fast to his earlier view that the value of a product is determined (or measured) by the quantity of (living) labour that it will *purchase* when exchanged. But since this quantity of labour *exceeds* the quantity of labour expended on a given product, 'labour *expended*' can *no longer act* as a regulator of the value of products, as it did under a simple commodity economy. *The law of labour value ceases to operate in capitalist society.*

If this is so, what, then, determines a product's value in a *capitalist* economy? Suppose that a capitalist advances a capital of 100 pounds for the hire of labourers (Smith assumes that the entire capital is spent on hiring labour power and ignores outlays on fixed capital*), who in turn produce for him commodities with a value of £120. How is the value of these commodities determined (measured)? As we already know, by the quantity of (living) labour which the capitalist can buy with them when they are exchanged. Out of the total £120 the capitalist can purchase, first of all, *the same amount of the labour* of hired workers as was *expended* on the manufacture of the commodities in question (i.e., £100, or the sum of their wages); second, he can purchase an *additional* quantity of labour with the £20 that are left over and which constitute his profit. As a result, the value of the commodities is no longer determined (measured) by the quantity of labour expended on their production (in fact, Smith now substitutes 'paid labour', i.e., wages or 'the value of labour', for expended labour). The value of the commodities is now large enough to pay in full for the labour *expended* on their production and, on top of this, to yield a certain mass of *profit*. In other words, in a capitalist economy the value of the commodity is defined as *the sum of wages plus profit* (and, in certain circumstances, also plus *rent*), i.e., as the sum of its *'costs of production'* taken in the broad sense of the term. Smith here abandons the terrain of the labour theory of value and replaces it with the *theory of production costs*. Previously Smith defined the value of a commodity by the quantity of labour expended on its production; now he defines it as the sum of wages, profit, and rent. Earlier Smith stated that the value of a commodity *resolves itself* into revenue (wages, profit, and rent); now he says that value *is composed of* revenues, which therefore now function as the 'sources' of a commodity's exchange value. *Revenues* are what is *primary* and given, while the commodity's *value* is seen as *secondary* and derivative, made up by adding together the separate revenues. The *magnitude* of a commodity's *value* depends upon the 'natural rates' of *wages, profit, and rent.*[10]

*See below, Chapter Twenty-Four.

Summing up Smith's trend of thought, one can say that his theory of value suffers from the fundamental defect of a *duality* in his overall methodological approach. His analysis of the *causes* of changes in value leads him to a concept of 'expended labour'; his search for a *measure* of value, deriving as it does from an individualist understanding of the division of labour, leads him to a concept of 'purchased labour'. What is more, these two concepts of labour are each viewed from their objective and subjective aspects, although primarily from the former. In addition, the concept of 'labour purchased' is itself bifurcated, figuring on most occasions as 'materialized labour purchased' (the exchange between simple commodity producers, or an exchange of commodity for commodity), on others as 'living labour purchased' (an exchange between the capitalist and worker, or the exchange of a commodity as capital for labour as labour power). Insofar as it is the first, 'craft' motif which predominates, labour purchased is acknowledged as being equal to the labour expended, and it makes no difference whether the commodity's value be determined by the one or the other. Here Smith is operating with a labour theory of value, so that the parallelism and reconcilability of these two strands of his theory hides his methodological dualism. As soon as the 'capitalist' motif comes to the fore, however, the two analytical paths and the two concepts of labour markedly diverge. In a capitalist economy the labour materialized in the commodity exchanges for a larger quantity of living labour; it is an exchange of non-equivalents, and Smith is unable to explain it from the standpoint of labour value. By preserving for 'labour purchased' its former role as measure of value, Smith must then give up acknowledging 'expended labour' as the regulator of the proportions of exchange. The commodity's value depends now no longer upon the 'labour expended' but on the size of the incomes of the various participants in production (i.e., on wages, profit, and rent). Though the idea of labour value is one of the basic motifs in Smith's thought, he did not take it through to its conclusion, and when applying it to capitalist economy he replaced it with the *theory of production costs*. Smith's labour theory of value was dashed upon the rocks: for it was impossible to make it accord with the *exchange of materialised labour for living labour* (or capital for labour).

So long as Smith kept within the bounds of a simple commodity economy, the contradictory elements which his theory concealed (the regulator of changes in value and measure of value, expended labour and purchased labour, materialised labour purchased and living labour purchased) could still maintain themselves in some sort of unstable equilibrium. As soon as Smith extended his analysis to capitalist economy, however, this unstable equilibrium was destroyed and the dualistic character of Smith's constructs emerged into the full light of day. Each of the different aspects of Smith's doctrine was taken over and developed by later economic schools. Ricardo developed one side of Smith's theory when—with utmost consistency—he defined the value of a commodity by the labour expended on its production. Malthus developed another aspect of the theory and defined the value of commodities by the labour which they can purchase in exchange. The same fate befell Smith's theory (also infused by a dualism) on the relationship between the value of a product and the incomes of those taking part in its production. The idea that the value of a commodity *resolves itself* into wages, profit, and rent formed the basis of Ricardo's theory, who then liberated it from its internal contradictions. Smith's error on this question—his attempt to derive the value of the commodity *from incomes* (wages, profit, and rent)— was taken over by Say, who developed it into the theory of 'productive services'. Here, as elsewhere, the truly valuable kernel in Smith's ideas was subsequently to be developed by Ricardo, Rodbertus, and Marx, while its collateral offshoots were exploited by the so-called 'vulgar' economists.

1 The Russian text reads *'ob'ekt ili predmet,'* both of which in this case mean the object of an investigation or study.

2 At the close of Chapter 4 of Book 1 Smith describes how he will proceed in his ensuing analysis of value:

'In order to investigate the principles which regulate the exchangeable value of commodities, I shall endeavour to shew,

'First, what is the real measure of this exchangeable value; or, wherein consists the real price of all commodities,

'Secondly, what are the different parts of which this real price is composed or made up.

'And, lastly, what are the different circumstances which sometimes raise some or all of these different parts of price above, and sometimes sink them below their natural or ordinary rate; or, what are the causes which sometimes hinder the market price, that is, the actual price of commodities, from coinciding exactly with what may be called their natural price.' *Wealth of Nations*, Book I, Ch.4 p. 46.

3 *Ibid*, Book I, Ch. 5, p. 48. Rubin's italics.

4 *Ibid*, Book I, Ch. 5, p. 51. 'But though equal quantities of labour are always of equal value to the labourer, yet to the person who employs him they appear sometimes to be of greater and sometimes of smaller value. He purchases them sometimes with a greater and sometimes with a smaller quantity of goods, and

to him the price of labour seems to vary like that of all other things. It appears to him dear in the one case, and cheap in the other. In reality, however, it is the goods which are cheap in the one case, and dear in the other.'

5 *Ibid*, Book I, Ch. 5, p. 50.

6 *Ibid*, Book I, Ch. 5, p. 50.

7 The passages quoted in this paragraph are all from *ibid*, Book I, Ch. 5, pp. 49-50. Rubin's italics.

8 *Ibid*, Book I, Ch. 8, p. 82. Rubin's italics.

9 *Ibid*, Book I, Ch. 6, p. 65.

10 The discussion to which Rubin is referring appears in Book I, Ch. 7, p. 72: 'These ordinary or average rates may be called the natural rates of wages, profit, and rent, at the time and place in which they commonly prevail.

'When the price of any commodity is neither more nor less than what is sufficient to pay the rent of the land, the wages of the labour, and the profits of the stock employed in raising, preparing, and bringing it to market, according to their natural rates, the commodity is then sold for what may be called its natural price.'

CHAPTER TWENTY-THREE

The Theory of Distribution

For all the inadequacies and contradictions in Smith's theory of distribution—which it fell to Ricardo and Marx to rectify—it still has one great merit: Smith correctly depicted the *division of classes* and *forms of revenue* characteristic of the capitalist economy. Smith holds that contemporary society is divided into these basic classes: *entrepreneur capitalists, wage labourers,* and *landowners*, a division that is scientifically accepted even in our own day. The basic forms of revenue he takes to be *profit, wages, and land rent*. To fully appreciate the inventiveness of this division of classes and incomes, which today seems common knowledge, we need only compare Smith's doctrine with that of the Physiocrats.

Quesnay had divided society into *three* classes: landowners, cultivators (the productive class), and merchants and industrialists (the sterile class). This scheme confuses class divisions with the difference between branches of production (agriculture and industry). Turgot improved upon this schema substantially by dividing each of these latter two classes again into two. This gave a *five-fold* division of landowners, agricultural entrepreneurs (farmers), agricultural workers, industrial entrepreneurs, and industrial workers.* In Turgot's schema the division of classes coincides with the division between branches of production. Smith took the second and fourth classes and combined them together into a single class of *capitalist entrepreneurs*. In similar fashion he amalgamated the third and fifth classes in a single class of *wage labourers*. Once again we had a *tripartite* division, but one in which the Physiocratic counterposition of agriculture to industry had been removed and the class contradiction between capitalist entrepreneurs and wage labourers became revealed (as it had also been by Turgot) in its full clarity.

Of still greater importance is Smith's systematic *classification of revenue*. The Physiocrats for all intents and purposes knew only two

*See above, Chapter 13.

types of income: land rent (net revenue) and wages.[1] In their constructs entrepreneurial profit does not exist, but is resolved either into a replacement for capital or into the necessary means of subsistence (i.e., wages) of industrialists, farmers, and merchants. Capitalist profit is equated with wages or, to put it more accurately, both these forms of revenue are conceived as being of the same order as the income or 'subsistence' of the independent craftsman.

To *ignore profit* in this way, while it reflected the backward state of capitalist development in 18th century France, would have been impossible in more highly developed England. The English mercantilists had already devoted a great deal of attention to profit, although they knew it primarily as profit on trade. The successes of industrial capitalism found their expression in Smith's scheme, where *industrial profit* taken in the broad sense of the term (including the profit of farmers) figures as the basic form of revenue. The other form of income that had preoccupied mercantilist thinking, *interest on loans*, is subsidiary for Smith: interest is merely that part of profit which the industrialist pays to the lender for the use of the latter's capital.

In singling out *profit* as a special form of income Smith is careful to delimit it from *wages*. He argues against the view that 'profits ... are only a different name for the wages of a particular sort of labour, the labour of inspection and direction.' The volume of profits depends upon the size of the capital invested in a business and not upon the labour that the capitalist might expend on supervision. Hence 'profits ... are altogether different, are regulated by quite different principles' than wages.[2]

On the other hand Smith distinguishes workers' wages not simply from the *profits* of the capitalist, but also from the income of the *craftsman*. Handicrafts were still important in 18th-century England, and it is only natural that the example of the craftsman should often figure in Smith's arguments. Yet Smith was also greatly impressed by the gains made by industrial capitalism (which he tended even to overstate), and he maintained that 'such cases [when an 'independent workman' manufactures a product solely at his own expense—*I.R.*] are not very frequent, and in every part of Europe, twenty workmen *serve under a master* for one that is independent.' Thus 'the wages of labour are everywhere understood to be, what they usually are, when the labourer is one person, and the owner of the stock which employs him another.'[3] In the strict sense, wages are to be understood as the income of the worker who has been deprived of his means of production, and not that of the workman (craftsman) still in possession of them. Obviously Smith is counting as workers not simply the relatively small number at that time working in large-scale manufactories, but also the cottage labourers working on orders from buyers up-putters out: Smith often portrays industrialists as people who supply the workers with 'the materials of their work'.[4]

Smith, then, does not do what Quesnay did and identify profits and wages with the income (subsistence) of the craftsman; his mistake is in the opposite direction. He declares that the revenue of the craftsman (and peasant) includes both wages and profit, when in fact this undifferentiated income of the petty independent producer is unique in character and distinct from these other two forms.

The error that Smith made in transferring the categories of capitalist economy to the forms of economy that preceded it in no way diminishes the merit due to him where the theory of capitalist society is concerned. Smith correctly understood the class structure of that society and its characteristic forms of revenue. By separating profit off as a special form of revenue Smith took a major step towards formulating *the problem of surplus value*. The mercantilists had known surplus value only as *commercial profit*, extracted out of the process of circulation via the non-equivalent exchange of commodities. The Physiocrats, although having sought the origin of surplus value in production, understood it only as *the rent of land*. Because Smith singled out profit and understood that it makes up the capitalist's net income over and above compensation for his costs of production, he linked the problem of *industrial profit* to the problem of *surplus value*.

The Physiocrats were concerned only with the origin of ground rent, since from their point of view this was the one and only form of net income. Smith, by making profit part of revenue, *widened the problem of surplus value*. From a problem of *rent*—which it had been with the Physiocrats—it became a problem of the origin of *all forms of income over and above what goes to labour*: the rent of land, profit, and interest.[5] The question receiving priority was that of the origin of *profit*. Smith correctly regarded *interest* as part of profit. As for *rent*, here Smith was strongly influenced by Physiocratic doctrine, and his explanation was extremely feeble and suffered from glaring contradictions. Smith looked for the source of rent: 1) sometimes in the *monopoly price* of agricultural produce, which price was accounted for by the constantly high demand for such goods; 2) sometimes in the *physical productivity of the land*, which 'produces a greater quantity of food than what is sufficient to maintain ... [and] to replace the stock which employed that labour, together with its profits'; and 3) sometimes in the *labour* of agricultural workers.[6] Rent, therefore figures in Smith sometimes as a 'monopoly' payment or mark-up over and above the value of agricultural produce, sometimes as 'the work of nature which remains after deducting or compensating every thing which can be regarded as the work of man',[7] and sometimes as 'a share of almost all the produce which the labourer can either raise, or collect'[8] and which is given over to the landlord by virtue of his monopoly proprietorship. This last explanation, which accords with the idea of labour value, figures only fleetingly in Smith's theory of rent.

The concept of labour value forcefully asserts itself in Smith's *theory of profit.* The question of the origin of profit as an independent form of revenue had inevitably to lead Smith beyond the bounds of the Physiocratic theory of surplus product. The physical productivity of nature may have still been adequate to explain the origin of rent as a margin of surplus value which agriculture yields over and above total profits, but this explanation was clearly no longer applicable to profit, which is the normal and most often encountered form that surplus value takes. Certainly it is not just within agriculture that profit accrues, but also in industry, where in Smith's view 'nature does nothing; man does all'.* It is obvious that the source of profit must be sought in *human labour*. The problem of *surplus value* (revenue) which had been posed by the Physiocrats, was now tied directly to the *labour theory of value* outlined by the mercantilists. It is one of Smith's greatest merits to have made this synthesis.

Actually, for all the contradictions in his theory of profit and the gaps in his understanding, Smith was quite clearly disposed to the view that profit is that *portion of the value of the product* which the capitalist appropriates for himself. 'In that original state of things, which precedes both the appropriation of land and the accumulation of stock, the whole produce of labour belongs to the labourer.'[9] But once the land has been appropriated as private property and there is an 'accumulation of stock', one part of the product of the worker's labour goes as rent to the landlord and another to the capitalist as profit. Where does this 'accumulation of stock' come from? Smith, in the

*In fact, even industrial labour requires the assistance of the forces of nature. Smith's view to the contrary is characteristic of the manufactory period, when there were no machines and manual labour predominated. However, it seems possible that what is essentially a false notion had a beneficial hand in Smith's development: for it allowed him to transcend Physiocratic doctrine and to locate the source of value and surplus value not in nature, but in human labour. [The quoted phrase is from Book II, Ch. 5, p.364-*Ed.*]

spirit of all the ideologists of the nascent bourgeoisie, offers the following explanation: the more industrious and prudent persons, rather than spending the full produce of their labour, 'saved' part of it and gradually accumulated capital. Capital is what its owner or his forefathers 'saved' out of the product of their labours. 'Capitals are increased by parsimony, and diminished by prodigality and misconduct.' 'Parsimony and not industry, is the immediate cause of the increase of capital.' It was Marx who, with his picture of primitive capital accumulation through commercial monopolies, the plundering of colonies, the displacement of the peasantry from its land, the exploitation of cottage labourers and workers, etc., overthrew the naive myth, so long dominant in bourgeois science, that *the origin of capital lies in 'parsimony'*.

Despite the naivete of Smith's doctrine of the origins of capital, he firmly grasps that in a society where this 'accumulation of stock' has already taken place the mass of the population, deprived of means of production (here taken in the broad sense to include also the means of subsistence to sustain the worker while labouring),[10] becomes immediately dependent on those fortunate individuals whose 'parsimony' has allowed them to accumulate capital. 'The greater part of the workmen stand in need of a master to advance them the materials of their work, and their wages and maintenance till it be compleated. He shares in the produce of their labour, or in the value which it adds to the materials upon which it is bestowed; and in this share consists his profit.'[11] Profit is a *'deduction from the produce of labour'*, which the capitalist appropriates as his own. For their part, the workers are compelled to accede to such a 'deduction', since without a master to invest capital in a business they possess no means either to manage a business of their own or to maintain themselves while they are working.

Smith thereby recognises *labour* to be the source of *value* of the entire product, including that portion of value which accrues to the capitalist as *profit*. As we saw in the preceding chapter, however, Smith proved unable to work the idea of labour value through to the end. It is therefore understandable that his theory of distribution is likewise only incompletely thought out and plagued with major contradictions. We saw that in Smith's view the labour expended on a product's production becomes, in capitalist society, no longer the regulator of that product's value: its value, or 'natural price', is defined as the sum of the natural wage, natural profit, and natural rent. The level of wages, profit, and rent are taken as the primary, or given factors, and the product's value as the result of adding these three quanta of revenue together. The theory of *production costs* is put in the place of the *labour theory of value*.

Smith's *theory of distribution* similarly undergoes a certain change. Previously it had been correctly constructed on the basis of the *theory of value*. Later, however, it is the *theory of value* that is based on the *theory of distribution*. It thus becomes impossible to explain wages and profit as part of the product's value, for the latter can now be explained only after we have determined the level of its 'component parts', i.e., wages and profit. Were Smith fully consistent he would have to conclude (as Ricardo was to do) from his statement that profit is a 'deduction' from the product's value, that the share of profit can rise only when there is fall in the share of wages. Now, however, he maintains that a rise in profit serves only to increase the value of the product, but has no reflection upon wages. With a theory of distribution such as this the investigator must first of all find the natural level of *wages and profit*, so that these can then be used to determine the value of the product. Smith does just that, and attempts to explain wages and profit *independently from the theory of value*—an attempt doomed to failure.

What is it that determines the absolute level of *profit*? Smith does not even venture an answer to this question, and limits himself to trying to explain its *relative* upward and downward fluctuations. Smith distinguishes between the *progressive, stationary, and regressive* states of a nation's economy. The first is characterized by the accumulation and multiplication of the overall mass of a country's capital; in the second total capital maintains itself at its previous level; and in the third the capital is declining and the country is on the road to ruin. In the first situation, capital is abundant, and this causes profits (and interest) to *fall*, while wages *rise* thanks to the competition amongst capitalists for hands. This for Smith explains the *fall in the average rate of profit* observed in Europe from the 16th to 18th centuries. It is only in the young and rapidly advancing colonies with their free virgin land and their shortage of both labourers and capital that wages and profit can *simultaneously* exist at a *high* level. When a society is stationary the market for both capital and labour is completely saturated; thus both profit and wages establish themselves at a very *low* level. Finally, when a society is regressing or in a state of decline, the shortage of capital causes the rate of profit to *rise* and wages to *fall*. The superficiality of Smith's argument limits him to explaining fluctuations in the level of profit from the *abundance or scarcity of capital*.

More successful is Smith's *theory of wages*, which contains a number of apt and accurate remarks and observations. What gives this theory its special appeal is the deeply felt sympathy for the workers that Smith shows on every page. Nevertheless, from a theoretical point of view Smith's theory of wages also suffers from inconsistencies and contradictions.

The so-called *iron law of wages* enjoyed almost universal acceptance among economists of the 17th and 18th centuries. It was enunciated in most clear-cut fashion by the Physiocrats,* who argued that as a general rule the level of wages does not exceed the minimum means of subsistence required to maintain a worker and his family. Smith is reluctant to subscribe fully to this assertion which in his view does not correspond to actual facts. From the 17th to the mid-18th centuries the wages of English workers had been going up, and by Smith's time had reached a level which clearly exceeded what Smith considered the minimum level of means of subsistence. How was this rise in wages to be explained? Smith accounts for it in the same way as he explains the fall in the rate of profit for the period from the 16th to the 18th centuries: economic prosperity and the accumulation of capital create a greater demand for labourers. *The rapid accumulation of capital* (and not its absolute volume) demands a greater number of hands: *high wages* will make it possible for the workers to raise more children, which must in turn cause the level of wages to establish itself at precisely that level at which the rate of population increase more or less corresponds with the rate of growth in the demand for labour. A stagnant economy will be different. When the capital advanced on the hire of workers remains stationary the existing number of workers proves sufficient to satisfy the demand for labour, and 'the masters [would not] be obliged to bid against one another in order to get them'.[12] *Wages will fall to the minimum level of means of subsistence*, the population will reproduce itself at a slower rate, and the size of the working class will hold steady at this particular level. Finally, when a country is in decline and 'the funds destined for the maintenance of labour [are] sensibly decaying', the demand for workers will steadily decline and wages will fall *below* the established minimum 'to the most miserable and scanty subsistence of the labourer'.[13] Poverty, famine, and mortality would reduce the size of the population to what the now reduced volume of capital would require.

*See above, Chapters Three and Thirteen.

Thus the level of real wages will depend on the relationship between the supply and demand for labour, in other words, *upon the rate of growth of capital* or the fund advanced for the hire of workers. Smith, then, is advancing an embryonic version of the theory of the *wage fund*, which was to become so popular among bourgeois scholars.* However, he still confuses the idea of a wage fund with the notion that wages will gravitate towards *the minimum level of means of subsistence*. 'A man must always live by his work, and his wages must at least be sufficient to maintain him. They must even upon most occasions be somewhat more; otherwise it would be impossible for him to bring up a family and the race of such workmen could not last beyond the first generation.'[14] Yet we have seen that Smith believes that wages will really only gravitate towards subsistence level when the volume of capital and the demand for labour are *stationary*. When there is expansion wages will *rise above* this level; when there is a contraction they will fall *below* it. Obviously Smith himself thought that a drop in wages below the subsistence level would be but a temporary and transient occurrence, since poverty and mortality would soon bring the number of workers into correspondence with capital's reduced labour requirements. On the other hand, Smith also believed that there could be a *long-term rise* in wages over and above the minimum of means of subsistence—so long, that is, as high wages did not encourage the workers to reproduce themselves faster than the increased labour requirements of accumulating capital. This faith in the prospect of long-term improvement in the workers' welfare (which was partially evoked by the fact that the wages of English workers had actually risen from the 17th to the mid-18th centuries) distinguished Smith's optimistic world view from the pessimistic views of his followers, for instance, Ricardo.

For all his optimism Smith acknowledged that even when society was advancing, wages would not rise above the minimum required to bring the growth of the working population into line with capital's *demand for labourers*. This is a matter over which the capitalists will show equal concern: because they are *few in number* and hence can easily reach agreement amongst themselves, because they are protected *by the law*, and because the workers *cannot exist* without work for any but the briefest periods, they enjoy in any struggle with the workers a *social superiority of forces* that they can always use to drive down wages to that level beyond which the existing state of capital

*See below, Part V, Chapter Thirty-Four.

and wealth (i.e., whether it is progressing, stagnant, or declining) does not allow them to be driven any further. This recognition of the capitalists' social superiority of forces does not, on the other hand, lead Smith to conclude that the workers must struggle with them to improve their own social position, i.e., utilise strikes, or form trade unions. However much Smith may sympathize with the workers' needs, he does not believe that combinations of workers could improve their lot: in an advancing society they would be superfluous, as purely economic factors would by driving up wages in any case; if society is stagnating or in decline they would not be strong enough to stave off a fall in wages. Smith's *underestimation of the importance of workers' associations* reflected the infant state of the workers' movement during his epoch. At the same time it harmonized with his general views to the effect that economic life had to be left to the free play of individual *personal interests*.

1 We have translated Rubin's term *zemel'naya renta* variously as 'ground rent' (or 'land rent'), which is its more precise meaning, and as 'the rent of land', the terminology actually used by Smith, when dealing with rent as an economic category that specifies the social relation that the landlord class bears to the other classes of society. Smith's specific discussion of ground rent appears in Book V.

2 *Wealth of Nations*, Book I, Ch. 6, p. 66.

3 *Ibid*, Book I, Ch. 8, p. 83, Rubin's italics.

4 *Ibid*, Book I, Ch. 8, p. 83.

5 Rubin's phrase is actually '*chistyi ili netrudovoi dokhod*,' which literally means 'net, or unearned (non-labouring) income.' However, in the context in which it appears this rendering would not convey the full sense of labour being the sole *source* of value.

6 The quotation is from Book I, Ch. 11, pp. 162-63. Of the first source of rent Smith says, 'There are some parts of the produce of land for which the demand must always be such as to afford a greater price than what is sufficient to bring them to market; and there are others for which it either may or may not be such as to afford this greater price. The former must always afford a rent to the landlord. The latter sometimes may, and sometimes may not, according to different circumstances.' (Book I, Ch. 11, p. 162.) What Rubin describes as Smith's third source of rent is discussed by Smith as follows: 'But when by the improvement and cultivation of land the labour of one family can provide food for two, the labour of half the society becomes sufficient to provide food for the whole. The other half, therefore, or at least the greater part of them, can be employed in providing other things, or in satisfying the other wants and fancies of mankind.'

'Food is in this manner, not only the original source of rent, but every other part of the produce of land which afterwards affords rent, derives that part of its value from the improvement and cultivation of labour in producing food by means of the improvement and cultivation of land.' (Book I, Ch. 11, pp. 180 & 182.)

7 *Ibid*, Book II, Ch. 5, p. 364.

8 *Ibid*, Book I, Ch. 8, p. 83.

9 *Ibid*, Book I, Ch. 8, p. 82.

10 Rubin means that workers without their own means of subsistence are deprived of the means of production of the commodity labour power.

11 *Wealth of Nations*, Book I, Ch. 8, p. 83.

12 *Ibid*, Book I, Ch. 8, p. 89.

13 *Ibid*, Book I, Ch. 8, pp. 90-91.

14 *Ibid*, Book I, Ch. 8, p. 85.

CHAPTER TWENTY-FOUR

The Theory of Capital and Productive Labour

Smith, as we have seen, considered profit, rather than rent, to be the primary form of net income (surplus value). But Smith also thought of profit as the 'revenue derived from stock'. Thus it comes as no surprise that Smith had a far broader and more correctly worked out *theory of capital* than did the Physiocrats. His merit is that 1) he broadened the concept of capital beyond the sphere of agriculture to include *industry* as well, and 2) he drew a direct connection between the concepts of capital and *profit*.

Influenced by Rodbertus and Adolf Wagner, bourgeois economists often distinguish between two concepts of capital: a '*national* economy' concept and a '*private* economy' concept.[1] The first refers to the sum total of the produce of society's labour to be used in future production; the second refers to any sum of value that yields its owner a steady unearned income. The first concept of capital derives from a one-sided, material-technical standpoint, namely that capital is the means of production that are in existence, irrespective of their social form; hence the foolish conclusion often encountered in the arguments of the Classical economists and their epigones that the primitive hunter is a 'capitalist' by virtue of his possessing a bow and arrow. In contrast, capital in the second sense separates the concept from the material process of production, thus leaving unanswered the question as to where the capitalist draws his unearned income from.

Here as elsewhere Smith should be considered the progenitor of both concepts of capital. Smith holds that an individual's property (providing it is sufficiently large) will divide up into two parts. 'That part which, he expects, is to *afford him this revenue*, is called his *capital*. The other is that which supplies his immediate consumption.'[2] Capital is property which bears its owner a flow of unearned income, in the form of profit. The main value of this definition is that it links the concept of capital directly to the concept of profit.

Yet Smith understands that he cannot limit himself to defining capital in terms of the '*private economy*.' According to this definition a private house when rented out constitutes capital to its owner; it is equally obvious, however, that when the same house is used directly by its owner 'it cannot yield any [profit] to the publick, nor serve in the function of a capital to it'. [3] In view of this, alongside the aforementioned definition, Smith often talks about capital in terms of the *'national economy'*, i.e., in a material-technical sense, whereby he understands it as an *'accumulated stock of produce'* for use in *future production*, namely 1) the raw materials needed for the work, 2) the implements of production and 3) means of subsistence for the workers.

Smith is *unable to reconcile* these two definitions of capital because, owing to the confusions within his own theory of surplus value, he cannot trace out how the capital invested in agriculture, industry, and trade (Smith mistakenly places the capital invested in commerce and exchange on an equal footing with productive capital invested in agriculture and industry) possesses the ability to bear a steady income in the form of profit. The duality of Smith's views on capital reveals itself clearly in the fact that he sometimes understands capital correctly as the *total value* that the entrepreneur spends on purchasing machinery, raw materials, etc., but at other times mistakenly takes it to be the actual machines, raw materials, and the like *in natura*. This confusion of the material and technical elements of production (means of production as such) with their given social form (i.e., with their function as capital) is both a distinctive feature of Smith's theory of capital and a characteristic of the Classical school in general.

This lack of clarity in Smith's theory of capital was reflected in his view that capital is divided into two types, *fixed* and *circulating*. We have already met up with the embryonic form of this theory in Quesnay, who made the distinction between *avances primitives* and *avances annuelles*.* Smith generalized these categories beyond agricultural capital to industrial capital (which was correct) and to commercial capital (which was wrong, inasmuch as the division between fixed and circulating capital applies only to productive and not to commercial capital). [4]

Now circulating capital differs from fixed capital according to the length of time it takes for it to circulate: the value of circulating capital (e.g., raw materials) is wholly restored to the factory owner out of the price of his product upon the completion of *a single production period*; the value of fixed capital (e.g., machinery), on the other hand, is restored only in part, being fully cancelled out only after *several*

*See above, Chapter Thirteen.

production periods have expired. Smith remained vague about this distinction. His attention was devoted to the *material* aspect of phenomena as things, to the actual machinery *in natura*, and not to their value. While the entire *value* of a machine enters into circulation, albeit slowly and bit by bit, the actual *machine* remains at all times in the possession of the factory owner until it has completely depreciated. Smith, noticing this, comes to the strange conclusion that no part of fixed capital *passes into circulation*: unlike circulating capital (raw materials, for example), which 'is continually going from him [its owner—*Trans.*] in one shape, and returning to him in another', fixed capital yields a profit 'without changing masters, or circulating any further'. [5] The incongruities to which such a definition leads Smith are visible from the way he is compelled to classify the value of the seed which the farmer keeps on hand for later sowing as fixed capital simply because it stays in the farmer's possession. Using the same definition Smith deems the commodities held by traders as circulating capital, though generally speaking they constitute commodity, or commercial capital, and not productive capital at all.

In his theory of capital Smith came very close to the problem of *reproduction*, including that of the relationship between *capital and revenue*. He formulated it in much broader terms than had the Physiocrats, understanding that the formation of *net income*—in the form of *profit*—also occurs within industry. However, the rest of his analysis of reproduction is full of the most flagrant errors.

As we have seen, according to Smith's theory, a portion of capital is expended on the purchase of implements of production (fixed capital) and raw materials (circulating capital). From this it would seem to follow that the value of the annual product of society as a whole must first and foremost go to replace the total capital expended; it is only what remains over and above this sum that constitutes society's revenue, which is then divided up between the three social classes as wages, profit, and rent (whereas wages figure simultaneously as a portion of the circulating capital, profit and rent make up surplus value, or net income). In certain passages Smith actually arrives at just such a correct understanding of the problem: 'The gross revenue of all the inhabitants of a great country, comprehends the whole annual produce of their land and labour; the neat revenue, what remains free to them after deducting the expence of maintaining; first, their fixed; and, secondly, their circulating capital; of what, without encroaching upon their capital, they can place in their stock reserved for immediate consumption, or spend upon their subsistence, conveniencies, and amusements.'[6] Thus, the value of society's annual product contains ***not simply the revenue*** going to each of society's classes (i.e., wages, profit, and rent), but also the fixed and circulating *capital* that is being reproduced.

After coming so close to formulating the problem of reproduction correctly, Smith then begins to have his doubts. What confuses him is the fact that a value which represents *capital* for one person, represents *revenue* for another. For the owner of a cloth-making factory the textile machinery he purchases represents fixed capital. Yet what he pays to the machine maker for it, and what the latter then disburses to his workers as wages constitutes income for the workers and a replacement of circulating capital for the machine maker. Marx analyzed the complex intersection of these relatons between capital and revenue in Volume II of *Capital*. There he examines the process of reproducing the social product from two aspects: that of its material elements (means of production and means of consumption), and that of the component parts of its value (the reproduced constant capital, wages, and surplus value). Smith, as we know, confused these two aspects—the material and the social—of the process of production; in his theory of surplus value he vacillates between various points of view, having no knowledge of the division between constant and variable capital that Marx was to introduce into science. As a result, Smith proved unable to provide a correct solution to the problem of reproduction and, to get around the doubts that confounded him, resorted to a very simplistic approach. He merely assumes that the value of the constant capital, textile machinery, for instance, can be *resolved in its entirety into revenue*, i.e., into wages plus profit (and rent). Granted, the value of the constant capital necessary to the manufacture of this machinery (e.g., iron) must in turn enter into that machinery's value; but the value of the iron once again consists of the wages of the workers who extracted and processed it, plus the profit of the entrepreneur, etc. What this argument actually shows is that at every stage of its production the value of the product contains not simply the incomes going to the participants in production (i.e. wages, profit, and rent), but equally a replacement of constant capital (machinery, raw materials, and the like). Smith, however, comes to precisely the opposite conclusion. He thinks that the value of constant capital resolves itself in the last instance purely into revenue: wages, profit, and rent. Consequently, the price 'of all the commodities which compose the whole annual produce of the labour of every country, taken complexly, must resolve itself into the same three parts, and be parcelled out among different inhabitants of the country, either as the wages of their labour, the profits of their stock, or the rent of their land.' [7] While Smith has previously understood that a portion of society's annual product is designated to replace constant capital, he now arrives at the absurd conclusion, that *the entire value of the social product resolves itself exclusively into revenue*, entering, in other words, into the personal consumption of the individual members of society.

This mistaken theory became ruling doctrine among the economists of the Classical school: Ricardo accepted it, Say turned it into a dogma, and John Stuart Mill was repeating it even in the middle of the 19th century.*

For Smith, then, the value of a product consists of wages, profit, and rent. Now wages constitute what, in Marx's terminology, is variable capital; we can thus reformulate this statement as follows: *the value of the product consists of variable capital plus net revenue* (profit and rent). *The entire capital* is assumed to consist solely of *variable capital*. That part of a product's value making up the reproduced constant capital is totally forgotten. Yet how can the reproduction of the social product be understood if one ignores the reproduction of constant capital, which has such a great, and constantly growing importance in a capitalist economy? Clearly, Smith's erroneous notion that the value of a product breaks down into revenue mars his entire theory of reproduction. On this question he even lags behind Quesnay, who never for a moment forgot that part of the annual product goes to restore the depreciated portion of fixed capital.

The errors that Smith made in analyzing the process of reproduction-in-general could not fail to find reflection in his understanding of expanded reproduction, that is, of capital *accumulation*. If the entire capital is spent as variable capital, on hiring labourers, the process of accumulation will obviously take place as follows: there is a part of the capitalist's revenue (i.e., his profit) that he does not spend on personal consumption, but adds to his capital, that is, he advances it for the hire of labour. *All capital that is accumulated is expended on the hire of labour*. This position is simply wrong, and once again ignores the fact that the capitalist must lay out part of his additional capital on the purchase of machinery, raw materials, etc.

Two important conclusions could have been drawn from this mistaken theory of accumulation. The first is that, because the entire capital is expended on the hire of labour, 'every increase or diminution of capital, therefore, naturally tends to increase or diminish the real quantity of industry, the number of productive hands.'[8] Consequently, *any addition to capital*, by calling forth a proportional increase in the

*See the chapter on Sismondi in Part V, below.

demand for labour, *works fully to the advantage of the working class.* The proponents of this argument have forgotten that in reality the demand for labour grows only in proportion to the rise in capital's variable portion, and not to the growth of capital as a whole. The second conclusion is that *the accumulation of capital does not imply a cut in personal consumption for the members of society.* If a capitalist accumulates half of a profit of £1,000, he is using £500 to hire workers. *The capitalist is foregoing this much of his own personal consumption in favour of the personal consumption of the workers.* 'What is annually saved is as regularly consumed as what is annually spent, and nearly in the same time too; but it is consumed by a different set of people,' i.e., workers: 'The consumption is the same, but the consumers are different.'[9] Insofar as Smith was directing these words against the primitive petty-bourgeois or peasant notion that capital accumulation means hiding gold coins away in a sock or a money box, he was correct. Accumulated capital is certainly spent. But it is spent not simply on hiring workers, but equally on the purchase of machinery, raw materials, etc. Overall *personal* consumption falls in favour of *productive consumption*; the production of means of production rises *at the expense of* means of consumption. Disregard for this fact laid the basis for the Classical *theory of markets* of Say and Ricardo; even opponents of this theory, like Sismondi, shared Smith's mistaken doctrine that the entire annual product of society goes to the personal consumption of its members.*

Closely tied to Smith's theory of capital and revenue is his extremely interesting and valuable theory of *productive and unproductive labour.* It was Smith's view, as we already know, that the entire capital is spent on hiring workers, i.e., is made up of wages. Does this mean that every single worker has his wages paid out of capital? No, says Smith, workers can receive their wages either from *capital* or from *net income* (profit and rent). A capitalist uses his capital to hire workers, who by means of their labour not only restore their wages, but provide on top of this a profit (surplus value). The capitalist can use his net income (i.e., profit) either to buy various commodities or to purchase the labour of different workers to be used directly for his own consumption (a maid, a cook, a domestic tutor, etc.). The labour of these people provides the capitalist with a definite use value yet yields no exchange value or surplus value. This constitutes the basis for distinguishing between productive and unproductive workers. *Productive* workers are those who *exchange their labour directly against capital; unproductive* workers are those who

*See the chapter on Sismondi in Part V, below.

exchange their labour directly against revenue. To be sure, the capitalist can spend part of his revenue on hiring additional productive workers. But in that case he is converting a portion of his revenue into capital; he is accumulating or capitalizing it. As capital must yield a surplus value, we can formulate this statement another way: *productive* workers are those whose labour *yields surplus value*; *unproductive* workers are those whose labour is devoid of this property. 'Thus the labour of a manufacturer adds, generally, to the value of the materials which he works upon, that of his own maintenance, and of his master's profit. The labour of a menial servant, on the contrary, adds to the value of nothing.'[10]

We can see how the concept of *productive labour* has changed with the evolution of the concept of *surplus value* (or net income). The only form in which the mercantilists had known surplus value was as *commercial profit* earned from foreign trade, flowing into the country as gold or silver. Hence for them the most productive labour was that of the *merchants and seamen* involved in foreign trade. The Physiocrats understood that surplus value was created in the process of production, but, by ignoring profit and identifying surplus value with *rent*, they came to the erroneous conclusion that only the labour of the *agricultural* population was productive. Smith, expanding the concept of surplus value to include also profit, thereby transcended the restricted concept of productive labour held by the Physiocrats. According to Smith's theory, *all wage labour*, be it agricultural or industrial, is productive when it is exchanged directly for *capital* and earns the capitalist a *profit*.

Smith is here deriving the distinction between productive and unproductive labour from their different *social forms*, rather than from their material properties. On the basis of the above definition, the labour of a servant ought to be deemed unproductive if a capitalist hired him for his personal services, and productive when employed by a capitalist running a large restaurant. In the first instance the employer relates to the servant as a consumer buyer, in the second as a capitalist buyer. Although materially speaking the servant's labour is identical in both cases, they each entail different social and production relations between people, productive in the one case and unproductive in the other. Here, however, Smith fails to reach such a correct conclusion and proves unable to differentiate labour's social form from its material content. Looking at what is actually going on around him Smith sees that the entrepreneur sometimes uses his capital to hire workers whose labour is embodied in material objects, or commodities, but at other times he uses his revenue to purchase personal services where this property of materiality is absent. From there he comes to the conclusion that productive labour is that which 'fixes and realizes itself in some particular subject or vendible commodity, which lasts for some time at least after that labour is past ... The labour of the menial servant, on the contrary, does not fix or realize itself in any particular subject or vendible commodity. His services generally perish in the very instant of their performance, and seldom leave any trace or value behind them, for which an equal quantity of service could afterwards be procured.' [11]

As we see, Smith is here giving us a *second* definition of productive labour, the defining characteristic of which is its ability to create *material objects*. Smith is obviously unaware that he is putting forward two definitions that do not fully concur with one another. From the standpoint of the first, correct definition, the labour of the servant in a restaurant run on capitalist lines is *productive*; from the point of view of the second, incorrect definition, this labour will always be considered *unproductive*, since it is not embodied in any material objects. By way of contrast, the labour of a gardener whom a capitalist keeps at his summer home to tend his plants is by the first definition *unproductive*, since that labour is purchased out of the capitalist's revenue and not out of his capital—in short, it is put towards his personal consumption and not to the production of surplus value. According to the second definition, the gardener's labour, because it leaves behind 'material' results in the form of flowers and plants, would always have to be considered *productive*.

On this, as on other questions, we see Smith (and this is typical of the Classical school) *confusing the material-technical aspect of the production process with its social form*. Wherever Smith is studying the social form of the economy he is discovering new perspectives and is one of the founders of contemporary political economy. When he confuses the social form of the economy with its material-technical content he falls into innumerable errors and contradictions, of which his two definitions of productive labour offers but one example.

The epigones of the Classical school, who directed their attention towards the material-technical side of production, paid no regard whatsoever to the first definition that Smith gave of productive labour, and embraced only his second, mistaken one. Some of them shared Smith's view of unproductive labour as that which is not embodied in material objects. Others objected to it on the grounds that the labour of officials, soldiers, priests, etc., had also to be considered productive. Yet neither the partisans nor the opponents of Smith's view in the least understood his truly valuable social definition of productive labour, which it fell to Marx to develop further.

1 The Russian text here reads '*chistokhozyaistvennoe*', which means 'purely economic'. On the following page it reads '*chastnokhozyaistvennoe*,' or 'private economic.' As the first of these seems to make little sense in the context in which Rubin is using it, we have—perhaps boldly—assumed it to be a misprint, and have translated it as 'private economic', to conform with the second term that appears in the text.

2 *Wealth of Nations*, Book II, Ch. 1, p. 279. Rubin's italics.

3 *Ibid*, Book II, Ch. 1, p. 281.

4 In Volume II of *Capital* Marx distinguishes three different forms assumed by industrial capital, each characterized by its own formula of circulation: *Money capital*, whose basic formula is M—C ... P ... C'—M', i.e., money (M) is transformed into commodities (C—means of production and labour power), which function as *productive capital* (P), and out of which appear commodities of greater value which are finally transformed again into money (M', i.e., now a greater sum than before, because it contains an increment of surplus value). Second, there is *productive capital*, which refers specifically to the form assumed by capital within the process of production. Its circuit is P ... C'—M'—C ... P. That is, the process of production yields commodities augmented by surplus value and which are then sold for money. If all of the surplus value is to go for the capitalist's personal consumption (i.e., is consumed as revenue) the commodities purchased to renew production (means of production and labour power) will be of the same value as before, and so we have C ... P (this is simple reproduction). If part of the surplus value is capitalized, and used to purchase a greater value of means of production and labour power than represented by the original P at the beginning of the circuit, we will, as a result of this accumulation have at the end of the formula C ... P'. Finally, there is *commodity capital*, whose formula is C'—M'—C ... P—C'. Here we start with the total commodity-product as it emerges out of the process of production, that is, containing both the original value of P plus surplus value. This is then transformed into money capital, which is used to purchase anew means of production and labour power. These, after functioning in the process of production yield a new commodity product, C', which also contains both the value of the original productive capital plus surplus value. Marx's entire discussion of fixed and circulating capital revolves upon these distinctions, for, as Marx emphasizes, the distinction between fixed and circulating capital only has relevance *within the process of production*. Smith's error, as Rubin discusses here, was to confuse the circulation of *value* with the circulation of the material objects embodying that value. Circulating capital is capital whose *value* completes the entire circuit of *productive capital* within a single production period. Fixed capital is capital whose *value* traverses this same circuit only over a protracted period of time, i.e., over several production periods. Smith was thus led into the confusion of *circulating capital* (which is necessarily part of P) with *capital in circulation*, that is, with *commodity capital* (or what Rubin refers to here as commercial capital).

5 *Wealth of Nations*, Book II, Ch. 1, p. 279.

6 *Ibid*, Book II, Ch. 2, pp. 286-87.

7 *Ibid*, Book I, Ch. 6, p. 69.

8 *Ibid*, Book II, Ch. 3, p. 337. Other passages on the same page make a similar point. 'Whatever a person saves from his revenue he adds to his capital, and either employs it himself in maintaining an additional number of productive hands, or enables some other person to do so, by lending it to him for an interest ... ' 'Parsimony, by increasing the fund which is destined for the maintenance of productive hands, tends to increase the number of those hands

whose labour adds to the value of the subject upon which it is bestowed.'

9 *Ibid*, Book II, Ch. 3, pp. 337-38.

10 *Ibid*, Book II, Ch. 3, p. 330.

11 *Ibid*, Book II, Ch. 3, p. 330.

Part Four

David Ricardo

CHAPTER TWENTY-FIVE

The Industrial Revolution In England

In the England [and Scotland!—*Ed.*]* of Adam Smith industrial capitalism was still in its early stages. Agriculture held first place, while handicraft and cottage industry continued to prevail within industry. Industrial capitalism could begin its victorious progress only after the *factory*, with its extensive application of machinery and steam engines, had supplanted the manual labour of the *manufactory*. This transition from manufactory to factory took place during England's *industrial revolution*; embracing the latter quarter of the 18th century and the first quarter of the nineteenth. This is precisely the lapse of time that separates Ricardo's activity from that of Smith. If we can call Smith the economist of the *manufactory* period, Ricardo's writings arose against the background of rapidly developing *factory*, machine production.

The beginning of the industrial revolution is usually set at 1769, the jumping off point for a rapid succession of inventions which completely transformed production technology. It would be a great mistake, however, to see the industrial revolution as the result of the accidental appearance of fortuitous inventions. Machines to replace human labour had been invented before. But during the guild period, when the crafts were working for a restricted local market, such machinery was unnecessary, and could only spell ruin to the handicrafts. It is therefore understandable that the guilds used every means they could to oppose their introduction, secure their prohibition, destroy the prototypes made by audacious inventors, and have the

*Throughout, apart from this addition, we have retained Rubin's constant references to 'England' and 'English' rather than changing these to 'Britain', 'The United Kingdom', 'British', etc. 'Britain' and 'British' would obviously be more accurate in most cases, but for several reasons (the industrial revolution's locational priority in England, the barely consolidated nature of the entity 'The United Kingdom' which was formed only in 1801, the lack of centralization of the State in many spheres, as well as Rubin's own preference) we have retained his 'England' and 'English' [Ed.].

latter banished from a town or put to death. Thus the use of the ribbon loom was banned in the 16th century, that of a machine for manufacturing needles at the beginning of the 17th century, and so on.

During the 17th and 18th centuries—the epoch of the decline of the guilds, the strengthening of *merchant capital*, the growth of mass (cottage industry) production for *export*, and the birth of the *manufactories*—the situation altered. An immediate objective for entrepreneurs was now to lower production costs. The urge to make technological improvements and economies in *costs of production* gave rise during the 17th century to a feverish pursuit of inventions. The innovations of the 17th century—the extensive use of any and every type of water mill, technical innovations in mining and metallurgy (the use of machines to pump water out of mines, the construction of blast furnaces), improved methods of transmitting power (cog-wheels and fly-wheels, transmissions)—all prepared the way for the enthusiastic acceptance of the machine within industry. Yet prior to the middle of the 18th century these different inventions were incapable of revolutionizing an industry which remained dependent upon power sources (man, animals, and water) that were either weak or could be driven by machine power only in specific localities.

The stimulus for the industrial revolution at the close of the 18th century came, as we know, from inventions 1) in the *cotton textile* industry, 2) in *metallurgy*, and 3) the invention of the *steam engine*. Each of these was merely the end result of a long line of preceding inventions, the outcome of quests that had extended over decades.

It was no accident that this rapid succession of inventions took place in the youngest branch of England's textile industry, *cotton textiles*. Making its appearance in England only late (in the 17th century) it had not been subjected to guild regulations. Cotton textiles could only win out in its intense struggle with the older woollen industry by relying on new technical improvements. In the middle of the 18th century looms were both improved and made bigger in size. But as the spindles used in spinning continued to retain their medieval construction, spinners were unable to provide the weavers with enough thread. This thread 'famine' compelled inventors to start looking for new methods of spinning. In 1769 Arkwright took out a patent on his 'water' machine, an improved version of the *spinning machine* that he had invented in the 1730's. Within a year Hargreaves had taken out a patent on his spinning 'Jenny'. Finally, in 1779, Crompton combined the achievements of these two inventions into his 'mule', which began rapidly to drive out hand spinning. A spinner using this machine could prepare 200 times as much thread as he could without it. Now it was the weavers who could no longer keep up with all the thread supplied by the spinners: there was an urgent need for an improvement in weaving methods. In 1785 Cartwright invented the *mechanical loom*, but it was not used extensively until further improvements had been made to it. From 1813 onwards it began to drive out hand weaving.

Gradually the spinning and weaving machines spread into the wool industry as well.

A second field of technical inventions was *metallurgy*. Up until the middle of the 18th century both iron and cast iron had been produced using wood fuel. Blast furnaces were set up near forests, moving to new sites when the supply of wood became exhausted. By the 17th century England was already beginning to record a shortage of forests. At the start of the 18th century the scarcity and rising price of wood fuel caused metallurgy to pass through a severe crisis and recession. It was essential to find new forms of fuel. Such fuel existed in the form of hard coal, but prior to the mid-18th century the numerous attempts that had been made to coke coal and use it in the processing of iron had all met with no result. Only after the mid-18th century was pig iron extensively produced using mineral fuel (Derby's method, invented in 1735); beginning in the 1780's, rolled iron started to be produced using hard coal, thanks to the new method of 'puddling' invented by Cort in 1780. The *combination of iron and coal* that was to be so important for capitalism had now taken place.[1]

Finally there was the most important and universal invention of this period: in 1769 James Watt built his famous *steam machine*, a pump for removing water from mines. The artificial removal of water from mines had begun as early as the 16th century. In 1698 Severi had invented for this purpose the first steam engine which, in the improved version given it by Newcomen in the early 18th century, had become widely used in mining. However, Newcomen's machine could not cope with very deep shafts or a strong head of water. Watt's new invention eliminated this defect. His initial machine was intended only for the extraction of mine water. In 1781, however, after additional improvements, Watt converted his machine from a pump into a *universal* steam engine applicable to all branches of industry. Following its initial introduction into textile and metallurgical production, the steam engine seized one branch of industry after another. At the start of the 19th century the steam engine was applied to *transportation* (the steam ship, railways). England entered *the age of steam.*

The inventions just described could not have exerted the swift and revolutionary impact they did had there not existed the *socio-economic conditions* necessary for the extensive development of factory industry. By the end of the 18th century these conditions were already present in England. On the one hand, the epoch of commercial capital had already seen a significant accumulation of capital in the hands of traders, financiers, industrialists, etc.; the new factory industry presented these free capitals with a wide-open field for investment. On the other hand, landless peasants, ruined craftsmen and cottage labourers, and paupers of various sorts provided in abundance the *human material* that capital could employ for its own needs. The ancient guild restrictions that had stood in the way of capitalist development had already fallen into decay by the end of the 18th century. In the 1780's Tucker could say 'the privileges of the guilds and the trading corporations in the towns have at the present moment only insignificant power and are incapable of causing a great deal of harm, as was formerly the case.'[2]

Under these conditions *factory industry* grew at an extraordinarily rapid rate. In the words of one contemporary, 'a new race of factory owners rushed to set up factories wherever the opportunity presented itself: they began to fix up old barns and sheds, punched windows in bare walls, and transformed these premises into weaving workshops.' 'Any who had capital, however small it might be, threw it into a business: shop keepers, inn keepers, goods ferrymen, all became factory owners. Many of them met with failure, but others attained their objectives and acquired fortunes.'[3] The period from 1788 to 1803 was called the 'golden age' of cotton textiles, with production increasing three-fold during that time. This type of rapid growth in production was made possible only by the introduction of machinery which cut *production costs* and caused the *price* of cotton cloth *to fall* considerably. The introduction of the spinning machine brought down the production costs of thread from twelve shillings to three shillings in 1800, and even to 1 shilling in 1830. With the fall in the costs of production came a cheapening of commodities: the price of a pound of thread fell from thirty-five shillings to nine shillings in 1800, and to three shillings in 1830. Production costs and prices on many industrial commodities fell between ten and twelve times. Cheap cotton cloth began to displace more expensive woollens; thanks to their cheapness they managed to force their way into the remote countryside and onto foreign markets. In the 17th and 18th centuries the fate of England's economy had depended primarily on its wool industry; from the beginning of the 19th century onwards, it was the cotton industry that played this role.

The feverishly quick advance of factory production brought profound changes to the English economy. It was only now that the centre of gravity shifted from agriculture to industry. On the eve of the industrial revolution (1770) England's population was divided about equally between town and country; a half century later (1821) agriculture employed only 33% of the population. A flight from the countryside had begun: the population of the factory towns grew with incredible speed. Between 1760 and 1816 the population of Manchester increased from 40,000 to 140,000; that of Birmingham from 30,000 to 90,000; that of Liverpool from 35,000 to 120,000. England was on the way to becoming '*the workshop of the world*,' providing factory-made goods for the rest of the world. Its foreign trade grew rapidly. Between 1760 and 1815 imports into England went from ten million to thirty million pounds sterling, its exports from fifteen million to fifty-nine million. Having previously had the export industry subordinate to it, the export trade now itself became subsidiary to a powerfully developed industry. The leading role gradually passed from *commercial* to *industrial* capital.

The industrial revolution opened up vast prospects for a great forward surge of England's productivity of labour and national wealth. Yet even in these first stages of its development, industrial capitalism revealed with utmost clarity its negative, as well as its positive aspects. The colossal rise in the nation's production did not reduce the poverty of its masses in the least. Machinery which was intended to save on human labour frequently gave a further push to the *deterioration* in the labourers' working conditions. Introduced at a feverish pace, it displaced hand spinners, weavers and other workers, who were threatened with either death by starvation or an existence as paupers. Understandably, the workers looked upon the machine as the most evil of their enemies. 'The machine' wrote one worker, 'has left us in rags and without a living, the machine has driven us into a dungeon, locked us up in a prison worse than the Bastille. I look upon any improvement which tries to reduce the demand for human labour as the most dreadful curse that can fall on the head of the working class, and I consider it my obligation to oppose the introduction of machinery, this scourge, into any branch of industry whatsoever.'[4] This passionate protest expressed a feeling widely held by the working masses. The introduction of machines often provoked workers' riots: they burned down factory buildings, smashed the machinery, and tried to have it proscribed. These spontaneous movements, however, were powerless to halt the process of bringing in machinery.

The machine meant the utter ruin of hand spinners and weavers, put an end to the *cottage* industries that had provided the peasant family with a second means of income, and made adult workers compete for work by drawing women and children into the factory. Although it is true that female labour had also been used in cottage industries, the woman had previously been working at home on her own, whereas now her departure for the factory meant leaving the children unattended unless they, too, came along. Engels, in his famous book, *The Condition of the Working Class in England*, painted a shocking picture of the conditions under which workers laboured in the final period of the industrial revolution (the 1830's and 1840's): five year old children working in factories, women and children performing heavy labour down the pits, children of seven spending twenty hours a day underground. Parish orphanages used to hand over whole flocks of children to factory owners, ostensibly for 'training', but in reality for forced labour. The factory owners would pass them from one to another like slaves.

Conditions were no less difficult for adult workers. Factory legislation was as yet non-existant; the law placed no restrictions on the exploitation of labour, while workers' trade unions were banned and subject to government prosecution. The working day averaged 13 to 14 hours, but was often even longer. The lack of hygiene in the factories was horrific. As for wages, in *monetary* terms these on the whole rose throughout the second half of the 18th century,* but in *real* terms they fell due to the sharp rise in the price of corn and other means of subsistence (meat, butter, etc.). According to Barton, in 1790 the weekly wages of a skilled worker would buy 169 pints of corn, in 1800 only 83.

The sharp fall in real wages is accounted for by the swift rise in the prices of grain and other agricultural produce which began in the last decade of the 18th century and ended in 1815, with the conclusion of the Napoleonic war. In the 1770's, when the industrial revolution began, the average price of corn stood at about forty-five shillings per quarter. In the 1790's it was fifty-six shillings, rose to eighty-two shillings during the first decade of the 19th century, and to 106

*In those branches of industry (such as spinning and weaving) where the displacement of manual labour by machinery was very rapid, money wages also fell.

shillings in the period 1810-1813. That corn prices rose so rapidly is explained first by the growth of England's urban industrial population, which heightened the demand for corn, and second, by the shortfall in the supply of corn coming from agrarian countries (e.g., Prussia and Poland) during the war with Napoleon. It was not simply the war and Napoleon's declaration of the continental blockade that slowed down the flow of cheap corn into England: the English government, acting in the interests of the landlords, did all it could to hinder the import of foreign grain through the imposition of *high customs duties*. By a law of 1791, the import of foreign grain into England became possible only if the latter's price on the domestic market was raised to fifty-five shillings per quarter. In 1804 this base price was raised—in the interests of the landowners—to sixty-four shillings, and in 1815 to eighty-two shillings. The combined effect of a number of factors (the country's rapid industrialization, the war with France, harvest failures, and agricultural protectionism) acted to produce a colossal rise in grain prices over the period 1790-1815.

At the sight of such a vertiginous increase in corn prices, farmers and landowners rushed to utilize every spare plot of land. The 'enclosure' of common lands took on vast proportions. Large capitalist farms increasingly displaced peasant holdings. Poor lands, waste lands, bogs—all of which were deemed unprofitable when corn prices were lower—now began to be cultivated. The drawing of *inferior lands* into production, the associated increase in the *cost of producing* corn, and the rise in *grain prices* were all features of English agriculture at the start of the 19th century and all found their precise reflection in Ricardo's theory of rent.

A second consequence of the advance in corn prices was a rapid rise in the ground rent that farmers paid to the landlords. From the 1770's up until the end of the war with France rental payments rose on average by 100% to 200%, and not infrequently by four or five times. In Scotland the total amount of ground rent in 1795 was £2,000,000; in 1825 it was £5,250,000. A farm in Essex which had been leased in 1793 at ten shillings an acre rented in 1812 [5] for fifty. The war, high prices, and bad harvests had made the landlords stupendously rich.

> Safe in their barns, these Sabine tillers sent
> Their brethren out to battle—why? for rent! [6]

When Byron, the famous poet, hurled these indignant lines at the aristocracy he was expressing the sentiments of the most diverse sections of the population.

Indeed, dissatisfaction with high corn prices and with protective legislation on behalf of the gentry had spread throughout the country. The industrial bourgeoisie assumed the leadership of the movement against the *corn laws*. Industrialists remarked with dismay that the lion's share of the profits brought by England's industrialization were slipping right through their own hands into those of the land magnates. The industrialists' dream was to shower the entire world with cheap goods from their own factories; but for this *cheap hands* were necessary. The high price of corn made it impossible to lower money wages. Further, high corn prices undermined the *purchasing power* of the workers and urban petty bourgeoisie, thus reducing the domestic market for industrial products. Periods of poor harvests and high grain prices often coincided with severe trade and industrial crises.

The broad mass of *workers* suffered not simply from expensive corn, but also from the introduction of machinery, unemployment, and low wages. The early ideologists of the proletariat had already grasped that the root of these evils lay not in the corn laws, but in the capitalist system. Yet the propaganda of the first utopian socialists (Owen for example) affected but a narrow circle. The broad mass of workers still lent a sympathetic ear to the agitation against the corn laws. The first decades of 19th-century England were passed in an atmosphere of bitter struggle between the landowning class and the commercial and industrial bourgeoisie supported by the broad mass of workers and petty bourgeois. In 1815 the agrarians still held the upper hand, and the protective tariffs on corn were increased. In 1820 the London merchants presented their famous petition to Parliament, in which they demanded the introduction of free trade as the only means by which the products of England's factories could gain broad access to the world's markets. In 1822 the merchants of Manchester put the same demand in their own memorandum. Manchester, the centre of cotton textile production, had become the fortress of the partisans of free trade, who hence became known as the '*Manchester*' school. With the industrial crisis at the end of the 1830's the struggle for free trade took on greater dimensions. The Manchester chamber of commerce presented a petition to Parliament in which it explained that 'without the immediate repeal of the corn duties the ruin of factory industry [would be] inevitable, and that only the broad application of the principle of free trade [could] assure the future prosperity of industry and the peace of the country.'[7] The anti-Corn Law League, founded by Cobden and Bright, enlisted hundreds of thousands of supporters and conducted a powerful agitation over the entire country. In 1846 the long decades of struggle finally ended in victory for the bourgeoisie: the *corn laws* were repealed, and England went definitively over to a system of *free trade*.

The bourgeoisie secured its victory only in the period following Ricardo's death, although the historic debate between the commercial-industrial bourgeoisie and the landlord class was already well alight during his lifetime. All Ricardo's literary activity took place in this atmosphere of struggle between social classes. The fundamental socio-economic phenomena of his day—the rapid growth of industry and the successes of machine production, the menacing rise in corn prices and ground rent, and the bourgeoisie's dissatisfaction with the corn laws—left a deep imprint on the whole of his theoretical system. In economic policy Ricardo stood as a leader of the industrial bourgeoisie: he demanded that the corn duties be repealed and free trade introduced. His theoretical system, for all its abstractness and apparent separation from the real economic conditions of his day, was in fact closely tied to them. Its two central components—the theory of value and the theory of distribution—both reflect the economic conditions of early 19th-century England. In his *labour theory of value* Ricardo summed up the many and varied factors which caused technical improvements and increases in labour productivity to lower the price of factory products. The extensive application of machinery had compelled Ricardo to ponder on the extent to which the use of machines (fixed capital) might modify the law of labour value. The raging struggle between the bourgeoisie and landowners and the more distantly perceptible battle between bourgeoisie and proletariat concentrated Ricardo's thoughts on to the *theory of disribution*. Ricardo made the impetuous rise in corn prices and ground rent the basis of his *theory of rent*. The grievous distress of the workers, notwithstanding rising nominal wages, found theoretical reflection in the Ricardian *theory of wages*. The struggle between the landowners and the bourgeoisie caused Ricardo to think in terms of an irreconcilable *conflict of interests* between these two classes: the idleness of the aristocracy and the rise in corn prices that were typical features of a capitalist economy were for him the main reason for the fall in profits and the primary threat to capital accumulation and the ability of the capitalist economy to grow.[8] Ricardo owes to his epoch both the strong and weak points of his theoretical system. Insofar as the English economy at the start of the 19th century had already managed to develop those features that are typical of a capitalist economy, Ricardo succeeded in making theoretically ingenious generalizations that have entered permanently into economic science. Wherever he took transient or temporary contemporary phenomena to be inevitable characteristics of capitalist economy in general, he fell into errors and biases that later economic schools, and above all Marx's, were to correct.

1 A detailed and interesting study of technological change during the industrial revolution, including the events Rubin is talking about here, is David Landes, *The Unbound Prometheus* (Cambridge University Press, 1969), Chapter 2, 'The Industrial Revolution in Britain'.
2 Translated from the Russian.
3 Both quotations have been translated from the Russian.
4 Translated from the Russian.
5 The text reads 1912, which is obviously a misprint.
6 The quotation is from Byron's poem, *The Age of Bronze*.
7 Translated from the Russian.
8 A phrase is missing here from the Russian text. The passage from 'the idleness ...' to the end of this sentence is interpolated from the apparent meaning as indicated by what is printed in the Russian original and by Rubin's argument in later chapters.

CHAPTER TWENTY-SIX

Ricardo's Life

David Ricardo (1772-1823) was born in London into the family of a wealthy Jewish banker. By the age of fourteen Ricardo was assisting his father in his stock exchange operations, but a few years later he broke with his family when he was converted to Christianity.[1] He became an independent jobber on the stock exchange, where, thanks to his remarkable ability to foresee the price movements on securities, he amassed a huge fortune in just a few years. At the age of twenty-five Ricardo was already enjoying a reputation in London as a millionaire and famous banker.

Apparently, however, playing the market soon ceased to afford Ricardo any satisfaction: his spirit harboured a passionate thirst for knowledge. At twenty-five he abruptly altered his style of life, gave up speculating on the exchanges, purchased an estate, and devoted his time to self-education. At first he studied mathematics and natural science, set up his own laboratory and collected minerals. Two years later he was so impressed by Smith's book as to give himself wholly over to the study of economic questions, which could get quite a grip on the mind of a man familiar with the secrets of stock-jobbing.

At the beginning of the 19th century economic questions had once again become the subject of animated discussion in England. The long war with France had thrown English economic life into profound disarray. This disorder showed up particularly in the depreciation of England's currency (the bank notes issued by the Bank of England, whose convertibility into gold had been suspended during the war) and in the exorbitant rise in the *price of corn*. These were practical questions, which touched the vital interests of different social groups, and gave rise to tremendous discord. Nor was this an academic debate among students in the quiet of some study; it was accompanied by bitter polemics in Parliament and the press. Such a fierce conflict of opinions and interests prompted the modest Ricardo, who had little confidence in his own abilities, to embark upon a literary career. In 1809, some ten years after he had set about his study of economic matters, he published some articles and a pamphlet, *On the High Price of Bullion*, in which he gave an outline of his *quantity theory of money*.[2] He explained the depreciation of the bank notes by their excessive emission and demanded that a certain portion of them be withdrawn from circulation if the currency was to be brought back to health.

In the years that followed Ricardo issued a number of short polemical works also dedicated to questions of monetary circulation. In 1815 he published *An Essay on the Influence of a Low Price of Corn on the Profits of Stock*. In this work Ricardo was already acting as a defender of *industrial capitalism* and had come to the conclusion that the interests of the landowning class conflicted with those of the other classes of society. At this time, as a letter of 1815 makes clear, Ricardo had no ambition to publish a work embracing the fundamental theoretical questions of economics. 'Thus you see', he wrote, 'that I have no other encouragement to pursue the study of Political Economy than the pleasure which the study itself affords me, for never shall I be so fortunate however correct my opinions may become as to produce a work which shall procure me fame and distinction.'[3] However, just two years later, in 1817, influenced by the persistent advice of his friend, James Mill, Ricardo published the book that was to earn him immortal fame, his *Principles of Political Economy and Taxation*. Although most of the chapters in the book are devoted to discussions of practical questions, mainly taxation, the few theoretical chapters guaranteed Ricardo permanent fame as one of the great economists. His book marks the highest point that the Classical school was able to attain—after that it went through only agony and a period of decay.

Although Ricardo himself at one time said that no more than twenty-five men in the whole of England had understood his book, it nevertheless earned him tremendous fame among his contemporaries and made of its author the head of an entire school. Ricardo stood at the centre of the vital economic discussions of his day. He was in constant personal contact or in correspondence with all the outstanding economists of his day. Some of them became his closest disciples and followers (James Mill, McCulloch), the first apostles of the orthodox *'Ricardian'* school. Yet even those of his opponents who created their own economic systems (Malthus, Say, Sismondi) could not fail to defer to his great intellect and scientific candour. Malthus, who was his constant opponent and a fierce defender of the landowning class, called the day Ricardo died the unhappiest day of his life.

Ricardo loved to hold domestic gatherings of friends and famous economists for uninhibited chats and discussions about topical economic subjects. These meetings of friends formed the basis of the

London *Political Economy Club*, which was founded in 1821 and stayed in existence for 25 years. The club's members were in the main practical people, merchants and industrialists, political figures; only a few were academic scholars. At its monthly meetings they discussed the most important questions of the day, the debates usually revolving around questions of monetary circulation and the duties on corn—questions that were uppermost in Ricardo's mind. Up to the day of his death, which came unexpectedly in 1823, Ricardo was the central figure in the club's meetings, the majority of whose members ardently defended—and did a great deal to implement—the ideas of *free trade*.

Ricardo successfully championed the ideas of *economic liberalism*—not only in his pamphlets and books, at gatherings of friends, and at meetings of the Political Economy Club, but also from the tribune of Parliament. Chosen as a member of Parliament in 1819, he delivered speeches, despite his shyness and dislike for oratory, during the debates on monetary circlation, parliamentary reform, etc., in which he declared himself in favour of bourgeois-democratic reforms (extension of the suffrage, the secret ballot). His teaching on *monetary circulation* had enormous influence both on the parliamentary commissions debating this issue and on subsequent English legislation.

Ricardo's literary and parliamentary declarations in defence of economic and political liberalism inevitably made him an object of attack, primarily from the representatives of the *landowning* class. They accused him of defending the narrow interests of the monied and industrial bourgeoisie, and even, on occasion, of having a personal interest in the passage of this or that measure. With unshakeable tranquility and dignity Ricardo repudiated these personal suspicions, and even refused to acknowledge himself as defending the interests of a single social class. Indeed, Ricardo was subjectively correct to see himself as a defender of 'true' economic principles and of the interests of all the 'people'(which he counterposed in one of his works to the interests of the aristocarcy and the monarchy), since what he invariably championed was the need for the rapid development of the productive forces, which in his epoch could occur only in the form of capitalist economic development. The high duties on corn, the poor laws, the rule of the landowning oligarchy all retarded the growth of the productive forces, and thus Ricardo consistently came out against them. On the other hand, it is true that he never imagined that the growth of the productive forces might be possible in a form other than a capitalist economy, and so he rejected Owen's communist schemes (on this see the following chapter).

Ricardo's horizons never extended beyond *capitalist* economy. Yet if he ardently defended capitalism's interests it was because his researches, being infused with the utmost scientific honesty and candour, led him to see it as the only form of economy that would provide sufficient scope for a powerful growth of the *productive forces* and the wealth of society as a whole. In Marx's words, 'Ricardo's conception is, on the whole, in the interests of the *industrial bourgeoisie*, only *because* and *in so far as*, their interests coincide with that of production or the productive development of human labour. Where the bourgeoisie comes into conflict with this, he is just as *ruthless* towards it as he is at other times towards the proletariat and the aristocracy.[4]

1 To the extent that Ricardo had any religious attachments at all, these were with the Unitarians.

2 *On the High Price of Bullion, A Proof of the Depreciation of Bank Notes* (1810), in *The Works and Correspondence of David Ricardo*, edited by Piero Sraffa with the collaboration of M.H. Dobb, Volume III (Cambridge University Press, 1951).

3 Ricardo, letter to Trower of 29 October 1815, in *Works* (Sraffa edition), Vol. VI (CUP, 1952), p. 315.

4 Marx, *Theories of Surplus Value*, Part II (Progress Publishers English edition), p. 118 (Marx's italics).

CHAPTER TWENTY-SEVEN

The Philosophical and Methodological Bases of Ricardo's Theory

In the great historical contest between the landed aristocracy and the industrial bourgeoisie Ricardo stood decisively on the side of the latter. It would be a great mistake, however, to accept Held's statement that 'Ricardo's doctrine was dictated simply out of the money capitalist's hatred for the landlord class.'[1] In Ricardo's time the industrial bourgeoisie still played a progressive historical role, and its ideologues still felt themselves leaders of the entire 'people' in a struggle against the aristocracy and monarchy.[2]

Ricardo was an ardent champion of the bourgeois capitalist order because he saw it as the best means for guaranteeing, 1) the greatest *individual happiness*, and 2) the maximum growth of the *productive forces*.

Bourgeois economic science had already raised the demand for free competition and individual economic initiative in the 18th century. Both the Physiocrats and Smith consecrated this demand by making reference to the eternal, *natural right* of the individual. By the beginning of the 19th century the role of natural right as the bourgeoisie's main spiritual weapon in its struggle for a new order had played itself out. The foundations of the capitalist order had already been laid, and the greater its successes the more were the ideologists of the bourgeoisie themselves prepared to abandon their naive faith in the impending realization of a 'natural order' of universal equality and brotherhood. The bitter disappointments of the French revolution, the desperate state of the labouring masses during the time of the industrial revolution, and the first portents of the budding struggle between the bourgeoisie and working class left little room for the illusions of yesteryear. From the beginning of the 19th century demands for equality and brotherhood alluding to the natural right of the individual were mostly coming from the mouths of the first defenders of the *proletariat*, the early utopian socialists. Henceforth, the antethesis previously made between bourgeois natural right and feudal tradition became impossible and inadequate. The ideologists of the bourgeoisie were faced with a new and difficult problem: to justify the bourgeois order at one and the same time against both feudal tradition and the demands for natural equality being raised by the socialists. Called upon to solve this problem was the new philosophical system of '*utilitarianism*' developed by Bentham, which gained great currency from the 1820's onwards. If the theory of natural right had served as philosophical basis for the doctrines of the Physiocrats and Smith, Ricardo and his closest disciples were fervent adherents of utilitarianism.

Although utilitarianism denied the doctrine of natural right, on one point it continued in the same direction: it gave definitive formulation to the *Weltanschauung* of *individualism*. For the Physiocrats the demand for individual freedom followed from the character of their ideal social system (the 'natural order' of society); in this sense society still had domination over the individual, in effect itself determining the degree of freedom that the latter was allowed. In the writings of Adam Smith the individual and society are equal entities, existing in complete harmony with one another: the 'invisible hand' of the creator ensures that they are in complete accord.[3] Finally, in the utilitarian system, society is completely subordinate to, and dissolved into the individual. Society is nothing but a *fictitious body*, a mechanical sum of the *individuals* who comprise it. In Bentham's words, 'the interest of the community ... is ... the sum of the interests of the several members who compose it. It is vain to talk of the interest of the community, without understanding what is the interest of the individual.'[4] 'The interest of individuals, it is said, ought to yield to the public interest. But what does that mean? Is not one individual as much a part of the public as another? ... *Individual* interests are the only *real interests*.'[5] What does this interest of the individual consist of? The enjoyment of pleasures and security from pains, i.e., to attain for himself the greatest benefit. The '*principle of utility*' forms the cornerstone of the entire utilitarian system (the name derives from the Latin *utilis*, or useful). To evaluate the utility of a given action we must sum up all its beneficial effects, on the one side, and all its harmful effects, on the other; we then deduct the sum of the pains from the sum of the pleasures (or vice versa) to obtain a balance that is either positive or negative.[6] By using this '*moral arithmetic*'[7] we know what actions will be capable of assuring the 'greatest happiness' for the individual.

By what means can we construct a bridge from the happiness of the *individual* to the well-being of *society*? Since society is itself a mechanical sum of constituent individuals it follows that social well-being is nothing more than the result of *mechanically adding up* these individuals' happiness. The well-being of society means '*the greatest happiness for the greatest number.*' And since a sum increases only with increases in its components, social progress is possible only as a rise in the welfare or *happiness of the individual*. 'Everything that conforms to the utility or interest of the community increases the total welfare of the individuals who compose it.'[8] But how do we increase this general sum of individual welfares? Very simply: care for this should be left to the *individuals themselves*, since 'each is his own judge of what is useful for him.'[9] 'Here we have a general rule: grant people the greatest possible freedom of action in all those circumstances where they can do harm to noone but themselves, since they themselves are the best judge of their own interests.'[10] Thus the social ideal that Bentham, as founder of the utilitarian school, constructs out of the principle of utility is *maximum freedom of the individual* and *limitation of the state's functions* to the purely negative task of keeping its citizens from doing damage to one another. This system of *bourgeois individualism* is preferable to *feudalism* and the 'inconveniences of its useless burden' because it guarantees the individual the greatest possible freedom of action and hence also the opportunity to attain maximum happiness. It is preferable to *socialism* because the latter deprives the individual of the opportunity to attain the greatest utility or happiness through the agency of his own labour. 'When security and equality are in conflict, it will not do to hesitate a moment. Equality must yield ... The establishment of perfect equality is a chimera; all we can do is to diminish inequality.'* While the thinkers of the 18th century had been filled with a magnanimous enthusiasm for universal equality and brotherhood, the voice of the sober bourgeois now declared equality a chimera. While in the 18th century the duty of the bourgeois order had been to realize the sacrosanct rights of the individual, it now faced a more modest task: to guarantee to each individual the freedom to select what was most profitable ('useful' or affording the 'greatest happiness') from amongst those undertakings left open to him by the social system as it was.

Ricardo became a philosophical adherent of utilitarianism via James Mill, a man who on economic questions had been Ricardo's pupil. Bentham had said, 'I was the spiritual father of Mill, and Mill

*This quotation, along with those preceding are taken from Bentham's works. [The passage quoted here is from *The Theory of Legislation*, p. 120—*Ed.*]

was the spiritual father of Ricardo: so that Ricardo was my spiritual grandson.'[11] Like Bentham, Ricardo was firmly convinced that 'where there is free competition, the interests of the *individual* and that of the *community* are never at variance.'* The interest of society can reside nowhere but in the optimal realization of the interests of its constituent members. That which 'is less profitable to individuals [is] therefore also less profitable to the State.' Ricardo believes it impossible for there to be employments 'which, while they are the most profitable to the individual, are not the most profitable to the State'.[12] 'The pursuit of *individual advantage* is admirably connected with *the universal good of the whole*. By stimulating industry, by rewarding ingenuity, and by using most efficaciously the peculiar powers bestowed by nature, it [the pursuit of personal advantage—*I.R.*] distributes labour most effectively and most economically: while, by increasing the general mass of *productions*, it diffuses *general benefit*, and binds together by one common tie of interest and intercourse, the universal society of nations throughout the civilized world.'[13] In Ricardo's eyes, to give free reign to the principle of *'individual advantage'* (or, what is the same thing, to Bentham's 'principle of utility') is the best guarantee of increasing the 'general benefit', which consists of augmenting 'the general mass of products', i.e., developing the productive forces. Conversely one need only remove or impede the activity of the personal-interest principle for there to be an inevitable deterioration of the productive forces, a reduction in general welfare, and a decline in the total happiness of society's members. It was on this basis that Ricardo rejected Owen's projects to set up communist communities. 'Owen is himself a benevolent enthusiast, willing to make great sacrifices for a favorite object', wrote Ricardo in one of his letters. '... Can any reasonable person believe, with Owen, that a society, such as he projects, will flourish and produce more than has ever yet been produced by an equal number of men, if they are to be stimulated to exertion by a regard to community, instead of by a regard to their private interest? Is not the experience of ages against him?'[14]

The ideal society for Ricardo, therefore, is capitalism, where *competition between individuals*, each of whom is out to attain the greatest possible personal *advantage*, assures that there will be maximum growth of the productive forces. In this sense Ricardo was heir to the Physiocrats and Smith. Unlike his predecessors, however, he had

*This quotation, as with all ensuing ones, are taken from Ricardo's works. [The passage here is from *The High Price of Bullion, A Proof of the Depreciation of Bank Notes*, in Ricardo's *Works* (Sraffa edition), Vol. III, p. 56 (Rubin's italics)—*Ed.*]

before him a capitalist economy at a higher stage of development and was therefore able to formulate more correctly and more fully its characteristic economic laws. The Physiocrats had lived in a France that was still semi-feudal; Adam Smith had been part of the age of manufactories. Ricardo, because he was witness to the rapid growth of large-scale capitalist machine production, was better able to make note of is fundamental technical and socio-economic features.

Smith's theoretical horizons had been completely bounded by the technology of the manufactory. When he spoke about machinery he in essence understod it as the specialized instruments employed by the manufactory workers. It was Smith's assertion that 'in agriculture ... nature labours along with man', while in industry 'nature does nothing; man does all'.[15] Only the era of the manufactory, where production was based on manual labour, could have spawned such a naive conception of industry. With the progress of machine production and the advance of technology such a conception became clearly outmoded. 'Does nature nothing for man in manufactures? Are the powers of wind and water, which move our machinery, and assist navigation, nothing? The pressure of the atmosphere and the elasticity of steam, which enable us to work the most stupendous engines—are they not the gifts of nature? to say nothing of the effects on matter of heat in softening and melting metals, of the decomposition of the atmosphere in the process of dyeing and fermentation. There is not a manufacture which can be mentioned, in which nature does not give her assistance to man, and give it too, generously and gratuitously.'[16] While Smith explains industrial progress almost exclusively by the development of the *division of labour*, Ricardo adduces such factors as 'the improvements in *machinery* ... the better division and distribution of labour ... and the increasing skill, both in science and art, of the producers.'[17]

Ricardo expected the introduction of machinery to make products cheaper and to bring a rise in output. True enough, he did not close his eyes to the disastrous situation of the workers whom the machines had ousted. The defenders of capitalism argued that the introduction of machinery was incapable of causing even the slightest deterioration in the workers' condition since those displaced would immediately find employment in other branches of production. At first Ricardo, too, ascribed to this 'theory of compensation', but later on he acknowledged—with his great, and characteristic honesty and scientific candour—'that the substitution of machinery for human labour, is often very injurious to the interests of the class of labourers.'[18] This view notwithstanding, Ricardo remained a fervent advocate of the introduction of machines as a necessary condition for the development of the productive forces. He rejected the petty-bourgeois utopianism of Sismondi, who wanted to reverse the wheel of history and go back to the patriarchal economy of independent petty producers (craftsmen and peasants) that had existed prior to large-scale machine production.

This rejection of the Smithian counterposition of agriculture to industry made it possible for Ricardo to overcome the *residua of Physiocratic ideas* in Smith. In starting out from the view that nature assists man in agriculture but not in industry Smith was assuming that agriculture (rather than industry) was where society could most profitably invest its capital. This view was understandable in the middle of the 18th century, when England was still feeding its population with its own grain and agriculture played the dominant role in the country's economy. Although at the start of the 19th century it still held this honoured position, and Ricardo was still unable to conceive of England's transformation into a onesidedly industrial state, he nevertheless maintained a firm course in favour of England's *industrialization*, even if this was to be at the expense of a curtailment in agriculture. Heated debates on this issue flared up between Malthus and Ricardo once the war with France had ended. The defenders of the landowning class, including Malthus, were demanding high import duties on corn so as to keep corn prices from falling and agriculture (which had been intensively developed during the war years under the impact of high grain prices) from being cut back. Malthus labelled as 'extravagant' schemes to turn England into an industrial state feeding on imported corn. Ricardo foresaw that it would be necessary to import cheap foreign corn and that *English capital would have to flow out of agriculture and into industry*. The prospect that 'the corn of Poland, and the raw cotton of Carolina, will be exchanged for the wares of Birmingham, and the muslins of Glasgow'[19] not only failed to frighten him—he hailed it. He saw the 'unusual quantity of capital ... drawn to agriculture'[20] as an abnormal phenomenon that had been created by the war and which was leading, as a result of its high costs of production, to excessively expensive corn. Ricardo welcomed the import of cheap foreign corn and a reduction in the capital invested in English agriculture: cheaper corn would lead, he thought, to a rise in profits and a tremendous flowering of the country's industrial life.

Thus, in Ricardo's constructs we have a country at a much higher stage of technical development than that described by Smith, one that is rapidly proceeding towards *industrialization* by going through a feverish period of *introducing machinery*. Ricardo advances our understanding of capitalism's *social* characteristics noticeably less than Smith; yet, for all that, these acquire much sharper outlines with Ricardo than with the earlier economist, for whom a *'capitalist'* point of view is still able to coexist with a *'handicraft'* one: in his descriptions we often encounter, besides the capitalist economy, an economy of petty producers; the figures of the capitalist and farmer at times alternate with those of the craftsman and peasant. In Ricardo the social background to capitalist economy is far more *homogeneous*: to judge from his constructs of society we could well think that England's handicraftsmen, cottage labourers, and peasants had already completely disappeared by the beginning of the 19th century (when in fact they still existed, and in healthy numbers). The entire stage is occupied by *capitalists* (including farmers), *wage labourers*, and *landlords* (capitalist landlords, that is, renting their land to farmers). This is a *'pure'* or *'abstract'* capitalism, freed from the admixtures and debris of pre-capitalist forms of economy. Ricardo presupposes that the tendencies inherent in a capitalist economy act with full force, encountering no delays along their way. If Smith is prepared to describe in great detail the innumerable obstacles that interfere with the equalization of the rate of profit and wages in different branches of production, Ricardo cites them merely in passing.

Ricardo conceives of capitalist economy as an enormous mechanism whose error-free functioning is ensured by the capitalists' desire for maximum profit; this desire results in the equalization of the rate of profit in all branches of production (differences in the rate of profit being maintained only so far as it is necessary to balance out the advantages held by some branches of production over others). *The striving to obtain the greatest profit* is the basic, motive force of capitalist economy, and *the law of equalization of the rate of profit* is its basic law. By grasping the central role of this law Ricardo once again proves himself superior to Smith. It is true that Smith had already presented a magnificent picture depicting how labour and capital pour from some branches of production into others consequent upon deviations in the market prices of commodities from their 'natural prices' (values). Yet it was still not clear to Smith that the capitalist entrepreneur plays the central role in this process of redistributing the productive forces. Smith still thought that the entrepreneur was joined in his function of prime mover in this process by the wage-labourers and landowners. Ricardo correctly identified the *capitalist entrepreneur* as the prime mover in this redistribution of the productive forces between branches. 'This restless desire on the part of all the employers of stock, to quit a less profitable for a more advantageous business, has a strong tendency to equalize the rate of profits of all.'[21] The flow of capital out of less profitable branches and into those that are more lucrative (in consequence of the greater credit granted to the latter by the banks and the expansion of their production) rectifies imbalances in the supply and demand of commodities. The movement of the entire capitalist economy is subordinated to the law of an equal rate of profit, this 'principle which apportions capital to each trade in the precise amount that it is required.'[22]

Ricardo has thus 'purified' the capitalist economy from its pre-capitalist admixtures and alloted the central role in this 'pure' capitalism to the capitalist. Ricardo studies each tendency within capitalist economy in its *'pure'* or *'isolated'* form, on the presupposition that the force of its action will be undiluted by counteracting tendencies. This is Ricardo's *'abstract'* method which provoked such censure from his opponents (especialy from economists of the historical school). Often Ricardo's 'abstract' or 'deductive' method is counterposed to the 'experimental' or 'inductive' method of Smith, which is deemed more correct. The contrast is itself false. Wherever Smith is seeking to discover the laws or tendencies of economic phenomena he, too, utilized the method of isolation and abstract analysis, without which any theoretical study of complex social phenomena would be impossible. With Smith, however, the train of his theoretical analysis is broken (and at times distorted) by a superfluity of descriptive and historical material. In Ricardo the sturdy skeleton of *theoretical* analysis is freed of the living flesh of concrete material culled from real life. An iron chain of syllogisms rapidly and inexorably carries the reader forward, supported only by hypothetical examples (usually beginning with the words, 'let us suppose that ... ') [23] and arithmetical calculations. Instead of Smith's vivid and captivating descriptions, the reader can look forward to an abstact, dry exposition, the difficulty of which is made all the greater by the fact that he cannot for a minute let slip from view the multitude of premises that the author either explicitly or tacitly assumes. Ricardo's method of abstract analysis is precisely what gives his theoretical thinking its consistency and intrepidity and endows him with the power to trace the workings of each tendency of economic phenomena through to its very end. This method allowed Ricardo to overcome Smith's innumerable contradictions and to construct a logically more integral and cohesive theory of value and distribution.

If Ricardo is to be reproached it is not for having applied an abstract method, but for having forgot that the theoretical positions arrived at by using it are *contingent*. Above all Ricardo, as with the other representatives of the Classical school, lost sight of the one basic historical condition for the correctness of all theoretical economic propositions: the existence of a determinate *social form of economy* (i.e., capitalism). That this social form of economy should appear to Ricardo as given and intelligible in its own right is a feature that he shared in common with all the ideologists of the young bourgeoisie, who in place of the old feudal system had posited a new social order that they saw as natural, rational, and eternal. 'The real laws of political economy do not change', wrote Ricardo. It is therefore understandable that even this thinker who, by differentiating value from riches and who, with his doctrines of labour value and rent did so much to transform political economy into a social science, readily sought the ultimate explanation for socio-economic phenomena in the action of 'immutable' *natural* laws (the biological law of population and the physico-chemical law of the declining fertility of the soil).

Besides ignoring the basic socio-historical precondition to his investigation, Ricardo often forgot, or lost sight of those *partial* premises that formed the basis of his theoretical propositions. He forgot that every economic tendency only fully manifests itself in the *absence* of counteracting tendencies, or as we say, 'all other conditions being equal'. By underestimating the multitude of tendencies that intermingle with one another in real life, Ricardo was inclined to explain real phenomena, created by *many* different factors, in terms of the activity of *a single* abstract law. One such abstract Ricardian law, for example, states that when farmers begin to cultivate inferior lands this will raise the value of a unit of corn (providing technique and other conditions remain the same). The author then hastens to apply this law to actual situations, declaring that the real rise in the price of corn is explained by the fact that farmers are now cultivating inferior land. Ricardo takes another such abstract law—that a general rise in wages necessarily lowers the rate of profit (all other things being equal) and rashly (and erroneously) uses it to explain the historical fact of the fall in the rate of profit. This tendency to attribute *unconditional* validity to *conditional* conclusions and to detect the immediate activity of '*pure*' laws in *concrete*, historical phenomena led Ricardo into a number of errors. These mistakes did not, however, prevent him from grasping (precisely through using the method of abstraction) the basic tendencies whose *continuous*, though at times concealed *operation* lie at the very *basis of capitalist economy*. It is for this reason that Ricardo's theoretical constructs, once altered and corrected, retain their validity even today, and we are justified in acknowledging his work as one of the great monuments of human thought.

1 Adolf Held, a German bourgeois economist who lived from 1844-1880.

2 In general, Rubin's discussion of Ricardo's views on the conflict between the landlords and the other classes of society requires some qualification, especially in light of the way Rubin presents Ricardo's theory of rent (Chapter Twenty Nine). Ricardo made a number of statements similar to this passage from *An Essay on the Influence of a Low Price of Corn on the Profits of Stock*: 'It follows then, that the interest of the landlord is always opposed to the interest of every other class in the community. His situation is never so prosperous, as when food is scarce and dear: whereas, all other persons are greatly benefited by procuring food cheap.' [Ricardo, *Works*, Sraffa edition, Vol. IV (CUP, 1951), p. 21.] In the very same paragraph and the discussion following, however, Ricardo immediately qualifies the context in which he makes this statement: 'High rent and low profits, for they invariably accompany each other, *ought never to be the subject of complaint, if they are the effect of the natural course of things.*

'They are the most unequivocal proofs of wealth and prosperity, and of an abundant population, compared with the fertility of the soil. The general profits of stock depend wholly on the profits of the last portion of capital employed on the land; if, therefore, landlords were to relinquish the whole of their rents, they would neither raise the general profits of stock, nor lower the price of corn to the consumer. It would have no other effect, as Mr. Malthus has observed, than to enable those farmers, whose lands now pay a rent, to live like gentlemen...' (*ibid*, pp. 21-22, *our emphasis*).

The *Essay on the Low Price of Corn* was a comparatively early pamphlet (1815). In his correspondence following publication of the *Principles* Ricardo clarified his position still further. 'He [Malthus] has not acted quite fairly by me in his remarks on that passage in my book which says that the interest of the landlord is opposed to that of the rest of the community. I meant no invidious reflection on landlords—their rent is the effect of *circumstances over which they have no control,* excepting indeed as they are the lawmakers, and lay restrictions on the importation of corn.' [Letter of 2 May 1820 to McCulloch, in Sraffa's edition of the *Works*, Vol. VIII (CUP, 1952), p. 182; our emphasis.] In a letter of 21 July that same year to Trower, Ricardo elaborated still further: 'He [Malthus] represents me as holding the landlords up to reproach, because I have said that their interests are opposed to those of the rest of the community, and that the rise of their rents are at the expence of the gains of the other classes. The whole tenor of my book shews how I mean to apply those observations. I have said that the community would not benefit if the landlords gave up all their rent—such a sacrifice would not make corn cheaper but would only benefit the farmers.—Does not this shew that I do not consider landlords as enemies to the public good? They are in possession of machines of various productive powers, and it is their interest that the least productive machine should be called into action—such is not the interest of the public—*they* [i.e., the public—*Ed.*] must desire to employ the foreign greater productive machine rather than the English productive one. Mr. M. charges me too with denying the benefits of improvements in Agriculture to Landlords. I do not acknowledge the justice of this charge, I have more than once said, what is obvious, that they must ultimately benefit by the land becoming more productive ...

'... I contend for free trade in corn on the ground that while trade is free, and corn cheap, profits will not fall however great be the accumulation of capital. If you confine yourself to the resources of your own soil, I say, rent will in time absorb the greatest part of that produce which remains after paying wages, and consequently profits will be low ...' (*Ibid*, Vol. VIII, pp. 207-208; Ricardo's italics.)

3 See Rubin's discussion in Chapter Twenty, above, especially note 9, p. 176.

4 Jeremy Bentham, *The Principles of Morals and Legislation* (New York, Hafner, 1965), p. 3.

5 Bentham, *The Theory of Legislation*, edited by C.K. Ogden (London, Kegan Paul, Trench, Trubner & Co., 1931), p. 144. Rubin's italics.

6 'Sum up all the values of all the *pleasures* on the one side, and those of all the pains on the other. The balance, if it be on the side of pleasure, will give the *good* tendency of the act upon the whole, with respect to the interests of that *individual* person; if on the side of pain, the *bad* tendency of it upon the whole.' (*Principles of Morals and Legislation*, p. 31; Bentham's italics.)

It is worth at this point recalling Marx's assessment of Bentham. 'Bentham is a purely English phenomenon ... [I]n no time and in no country has the most homespun manufacturer of commonplaces ever strutted about in so self-satisfied a way. The principle of utility was no discovery made by Bentham. He simply reproduced in his dull way what Helvétius and other Frenchmen had said with wit and ingenuity in the eighteenth century ... [H]e that would judge all human acts, movements, relations, etc. according to the principle of utility would first have to deal with human nature in general, and then with human nature as historically modified in each epoch. Bentham does not trouble himself with this. With the dryest naiveté he assumes that the modern petty bourgeois, especially the English petty bourgeois, is the normal man. Whatever is useful to this peculiar kind of normal man, and to his world, is useful in and for itself ... This is the kind of rubbish with which the brave fellow, with his motto "*nulla dies sine linea*" [no day without its line] has piled up mountains of books. If I had the courage of my friend Heinrich Heine, I should call Mr. Jeremy a genius in the way of bourgeois stupidity.' *Capital*, Volume I (Penguin edition), pp. 758-59, fn.

7 The expression is from *The Theory of Legislation*. A similar concept which he frequently used is that of a 'hedonistic calculus.'

8 Translated from the Russian.

9 Translated from the Russian. 'Now as there is no man who is so sure of being *inclined*, on all occasions, to promote your happiness as you yourself are, so neither is there any man who upon the whole can have had so good opportunities as you must have had of *knowing* what is most conducive to that purpose. For who should know so well as you do what it is that gives you pain or pleasure?' (*Principles of Morals and Legislation*, p. 267; Bentham's italics.)

10 Translated from the Russian.

11 Cited by Sraffa in his introduction to Volume VI of Ricardo's *Works*, p. xxviii, fn.

12 The two quotations are both from Ricardo's *On the Principles of Political Economy and Taxation*, Volume I of the Sraffa edition of the *Works* (CUP, 1951), pp. 349-50, fn.

13 *Principles*, pp. 133-34. Rubin's italics.
14 Letter to Trower, 8 July 1819, in *Works* (Sraffa edition), Vol.VIII, p. 46.
15 Smith, *Wealth of Nations*, Book II, Ch. 5, pp. 363-64. See above p. 201.
16 *Principles*, p. 76, fn.
17 *Ibid*, p. 94, Rubin's italics.
18 *Ibid*, p. 388. 'It is incumbent on me to declare my opinion on this question [the effect of machinery on each of the different classes in society], because they have, on further reflection, undergone a considerable change; and although I am not aware that I have ever published any thing respecting machinery which it is necessary for me to retract, yet I have in other ways given my support to doctrines which I now think erroneous...

'Ever since I first turned my attention to questions of political economy, I have been of opinion, that such an application of machinery to any branch of production, as should have the effect of saving labour, was a general good accompanied only with that portion of inconvenience which in most cases attends the removal of capital and labour from one employment to another... The class of labourers also, I thought, was equally benefited by the use of machinery, as they would have the means of buying more commodities with the same money wages, and I thought that no reduction of wages would take place, because the capitalist would have the power of demanding and employing the same quantity of labour as before, although he might be under the necessity of employing it in the production of a new, or at any rate of a different commodity ... As ... it appeared to me that there would be the same demand for labour as before, and that wages would be no lower, I thought that the labouring class would, equally with the other classes, participate in the advantage, from the general cheapness of commodities arising from the use of machinery.

'These were my opinions, and they continue unaltered, as far as regards the landlord and the capitalist; but I am convinced, that the substitution of machinery for human labour, is often very injurious to the interests of the class of labourers.

'My mistake arose from the supposition, that whenever the net income of a society increased, its gross income would also increase; I now, however, see reason to be satisfied that the one fund, from which landlords and capitalists derive their revenue, may increase, while the other, that upon which the labouring class mainly depend, may diminish, and therefore it follows, if I am right, that the same cause which may increase the net revenue of the country, may at the same time render the population redundant, and deteriorate the condition of the labourer.' *Principles*, pp. 386-88.

19 *Ibid*, p. 267, fn. The passage is not, in fact, Ricardo's, but is quoted by him from an article by McCulloch in the *Encyclopaedia Britannica*.
20 *Ibid*, p. 266.
21 *Ibid*, p. 88.
22 *Ibid*, p. 90.
23 It is interesting that Gramsci made an identical observation about Ricardo's contribution to Marx's analytical method: 'In order to establish the historical origin of the philosophy of praxis ... it will be necessary to study the conception of economic laws put forward by David Ricardo. It is a matter of realising that Ricardo was important in the foundation of the philosophy of praxis not only for the concept of "value" in economics, but was also "philosophically" important and has suggested a way of thinking and intuiting

history and life. The method of "supposing that ...", of the premiss that gives a certain conclusion, should, it seems to me, be identified as one of the starting points (one of the intellectual stimuli) of the philosophical experience of the philosophy of praxis. It is worth finding out if Ricardo has ever been studied from this point of view.' Gramsci, *Selections From the Prison Notebooks* (London, Lawrence and Wishart, 1971), p. 412.

CHAPTER TWENTY-EIGHT

The Theory of Value

1. Labour Value

Smith, as we know, had left behind a number of unresolved problems and contradictions (see chapter Twenty-two above). Let us briefly recall the most important:

1) Smith's theory suffered from a methodological dualism in the very way that he posed the problem: he confused the *measure* of value with the causes of *quantitative changes* in value.

2) Because of this he confused the labour *expended* on the production of a given product with the labour that that product *will purchase* in the course of exchange.

3) Smith's attention focused sometimes upon the *objective* quantity of labour expended and at others upon the *subjective* assessment of the efforts and exertions that go into it.

4) Smith confused the labour *embodied* in a particular commodity with *living labour* as a commodity, i.e., with labour power.

5) Smith came to *deny* that the law of labour value operates in a *capitalist* economy (in which labour nevertheless retains its function as a measure of value).

6) Together with a correct point of view, which sees the *value* of a product as the primary magnitude which then resolves itself into separate *revenues* (wages, profit, and rent), Smith sometimes mistakenly derives value from *revenue*.

It is fair to say that on each of these questions Ricardo adopted the correct standpoint and did away with Smith's contradictions. It must be added, however, that he worked through only the first three of these problems to a successful completion. As for the rest, although his stance was formally correct and he appeared on the surface to have eliminated Smith's inconsistencies, he was unable to genuinely resolve Smith's underlying difficulties and contradictions.

Above all, Ricardo decisively rejected any and all attempts to find an *invariable measure* of value, returning time and again to show that such a measure could not be found. The *method* that Ricardo consistently applied to the theory of value is that of the *scientific study of causality*, which the Classical school did so much to establish as part of political economy. Ricardo was looking for the *causes of quantitative changes* in the value of products, and wished to formulate the laws of those changes. His ultimate aim was 'to determine the laws which regulate the distribution' of products between the different social classes.[1] To do this, however, he first had to study the laws governing changes in the value of these products.

By posing the problem unambiguously in terms of scientific causality, Ricardo frees himself from the contradictions that befell Smith when he was defining the *concept of labour*. Ricardo starts out his work with a critique of the way Smith confused 'labour *expended*' with 'labour *purchased*', a question that he returns to in other chapters. Ricardo consistently bases his entire investigation upon the concept of the labour *expended* on a commodity's production, and sees changes in the quantity of this labour as the constant and most important reason for quantitative fluctuations in value.[2]

In this sense Ricardo makes the monistic principle of *labour value* the foundation of his theory (he makes certain exceptions to this, which we will discuss below in Section 3 of this chapter). Like Smith, Ricardo at the very outset excludes utility, or *use* value from the field of his enquiry, allocating to it a role as a *condition* of a product's exchange value. It is true that he talks here of 'two sources' of exchange value: the *scarcity* of articles and the quantity of *labour* expended on their production; this has led some scholars to speak of a dualism in his theory as well. This view is mistaken, since scarcity determines the value (or more accurately, the price) only of individual articles *not subject to reproduction*. Ricardo, however, is studying the process of production and the laws governing the value of products that are *reproduced*—and their value is determined by the quantity of *expended labour*. What is more, Ricardo shows the genuine maturity of his thought when he limits his investigation to 'such commodities only as can be increased in quantity by the exertion of human industry, and on the production of which competition operates without restraint.'[3] 'This in fact means that the full development of the law of value presupposes a society in which large-scale industrial production and free competition obtain, in other words, modern bourgeois society.'* In Chapter IV of his book Ricardo reveals this same clear understanding that the essential premise of the law of labour value is the existence of *free competition* between producers.

*Marx, *A Contribution to the Critique of Politial Economy* [Progress Publishers edition, London: Lawrence & Wishart, 1970, p. 60].

There he shows that any deviation between market prices and 'natural price' (value) is eliminated by capital flowing out of certain branches of industry into others.* If Ricardo is to be faulted, it is not for having made free competition (and hence the possibility of the reproduction of products) his starting point, but, to the contrary, for having grasped with insufficient clarity the social and historical conditions of the emergence of free competition and for having assumed these to be present even in the primitive world of hunters and fishermen.

Thus the value of products subject to reproduction is determined by the quantity of labour expended on their production. On analysis, this formula raises a number of questions: 1) when examining expended labour, do we do so from its *objective* or its *subjective* aspect; 2) do we take only the labour *directly* expended on a product's manufacture, or do we include the labour previously expended on manufacturing the *means of production* used in its production; 3) do we consider only the *relative*, or the *absolute* quantity of expended labour; 4) is the value of a commodity determined by the quantity of labour *actually* expended on its manufacture, or by the quantity of labour that is socially *necessary?*

As to the *first* of these questions, it should be noted that Ricardo rigourously adopts the *objective* point of view, doing away once and for all with the question of the individual's subjective assessment of the efforts that go into his labours (here again showing his superiority over Smith).** In receiving the products of labour the capitalist market shows scant regard for the personal vicissitudes of the producers who stand behind them. These impersonal, inexorable laws of market competition find reflection in Ricardo's system, which is so pervasively objective as to verge on detachment.

To the *second* of these questions Ricardo dedicated a special section —Section III of chapter I. Its heading maintains that 'not only the labour applied immediately to commodities affect their value, but the labour also which is bestowed on the implements, tools, and buildings, with which such labour is assisted.'[4] Implements, tools, and machinery *transfer* their value (either wholly or, where they depreciate only slowly, in part) to the product in whose manufacture they assist, but in no way do they *create* any new value. At the beginning of the 19th century, economists such as Say and Lauderdale, who were enraptured with the high productivity of machines, attributed the

*Here he even identifies the mechanism (expansion or contraction of the credit accorded a given branch) by which this expansion or contraction of production takes place.
**See the third of Smith's contradictions enumerated at the start of this chapter.

ability to create new value, the source of capitalist profits, to the machines themselves. Ricardo understood perfectly well that machines and the forces of nature which they set in motion, though they may raise the technical efficiency of labour and thereby augment the quantity of *use* values that this labour can manufacture per unit of time, nevertheless create no *exchange* value. Machines will only transfer their own value to the product 'but these natural agents, though they add greatly to *value in use*, never add exchangeable value, of which M. Say is speaking, to a commodity: as soon as by the aid of machinery, or by the knowledge of natural philosophy, you oblige natural agents to do the work which was before done by man, the exchangeable value of such work falls accordingly.'[5] By making a sharp distinction between 'riches' (use value) and 'value' Ricardo revealed the absurdity of the theory that *nature* creates value—a theory developed with greatest consistency by the Physiocrats and carried over by Smith in his theory of the exceptional productivity of agricultural labour.

On the *third* question, the view is often expressed that Ricardo, because he was concerned only with the *relative* value of different commodities and with the *relative* quantities of labour expended on their production, ignored the problem of *'absolute'* value. Indeed, Ricardo does study the problem of value primarily from its quantitative aspect and is looking to find the causes of *quantitative changes* in the value of products. If the relative value of two products A and B is expressed by the proportion 5:1, Ricardo accepts this fact as given and spares it no further consideration. A phenomenon holds his attention when he can see in it indications of change; for example, when the above-mentioned proportion of exchange gives way to a new one of 6:1. This does not, however, mean that Ricardo confines himself simply to observable alterations in the relative values of two commodities or in the relative amounts of labour required for their production. If the relative value of two commodities changes, he asks himself whether this is because the *'real'* ('actual', 'positive') value of commodity A has risen, or because the *'real'* value of commodity B has fallen? A change in a commodity's 'real' value is for Ricardo the result of changes in the quantity of labour needed to produce it. 'Labour is a common measure, by which their real as well as their relative value may be estimated.'[6] Ricardo is here affirming that his theory is not to be restricted simply to the study of the relative value of commodities.

The last question relates to the attributes of *value-forming labour*.

Marx accorded this question a great deal of attention, characterizing this labour as *social, abstract, simple*, and *socially necessary*. Ricardo, given his over-riding concern for the quantitative side of value, devoted his attention to those aspects of labour which influence the magnitude of value. Thus we find Ricardo commenting upon both *skilled* and *socially necessary* labour.

Ricardo, following Smith, acknowledges that one hour of *skilled* labour, e.g., that of a watch-maker, can create twice the value of one hour's labour by a spinner. This inequality is to be explained by 'the ingenuity, skill, or time necessary for the acquirement of one species of manual dexterity more than another.' The fact that this is so does not, in Ricardo's view, invalidate the law of labour value. Ricardo assumes that once the scale between these two types of labour (here taken at 2:1) becomes fixed it will show almost *no variation* over time. Once this is so the only change that can occur in the relative value of the two given products is that produced by changes in the relative *quantities of labour* necessary to their production.

Similarly we find in Ricardo a concept—albeit not fully developed—of *socially necessary* labour. Value is determined by the labour *necessary* for production. In his theory of rent Ricardo derives his famous law that the value of products is regulated not by the labour expended by the given individual producer, but 'by the *greater* quantity of labour necessarily bestowed on their production' by producers working under the most unfavourable circumstances.[7] Where Ricardo went wrong was to have derived this law from differences in the natural conditions of agricultural production and then advanced it as a general law applicable to all situations and to all products, be they from agriculture or industry. Marx rectifed Ricardo's error here with his own theory of *average* socially necessary labour.

Ricardo contrasted his own labour theory of value to others which attempted to explain the magnitude of a product's value by the extent of its utility or by the relationship between supply and demand. He was scathingly critical of Say's theory of *utility*: 'When I give 2,000 times more cloth for a pound of gold than I give for a pound of iron, does it prove that I attach 2,000 times more utility to gold than I do to iron? certainly not; it proves only as admitted by M. Say, that the cost of production of gold is 2,000 times greater than the cost of production of iron. If the cost of production of the two metals were the same, I should give the same price for them; but if utility were the measure of value, it is probable I should give more for the iron.'[8]

Ricardo rejected the vapid theory of *supply and demand* no less decisively: 'It is the cost of production which must ultimately regulate the price of commodities, and not, as has been often said, the proportion between the supply and demand: the proportion between supply and demand may, indeed, for a time, affect the market value of a commodity, until it is supplied in greater or less abundance, according as the demand may have increased or diminished; but this effect will be only of temporary duration. Diminish the cost of production of hats, and their price will ultimately fall to their new natural price, although the demand should be doubled, trebled, or quadrupled.'[9]

To judge from these quotations one might hink that Ricardo subscribed to a theory of *production costs*. This is not so. The vulgar theory of production costs holds that a rise in wages will automatically call forth a rise in the product's value. Ricardo expressed his dissent from this view in the very first words of his book: 'The value of a commodity ... depends on the relative quantity of labour which is necessary for its production, and not on the greater or less compensation which is paid for that labour.'[10] Although there were occasions when Ricardo failed to properly differentiate between costs of produc-tion and outlays of labour, his entire system is geared towards establishing the law of labour value and surmounting the theory of production costs which Smith, owing to his own inconsistencies, had fallen prey to (see sections 2 and 3 of this chapter).

Thus we can see that Ricardo contributed greatly towards improving the theory of value. He freed the idea of labour value from the wealth of contadictions that we find in Smith. Ricardo fundamentally reformed the *quantitative* side of the theory of value. He discarded the search for a constant measure of value—that deceptive mirage that economic thinkers had been pursuing from Petty to Smith—and presented a doctrine on how *quantitative changes in the value* of products are *causally dependent* on changes in the *quantity of labour expended on their production*. Ricardo sees the development of the *productivity of labour* as the ultimate cause behind changes in the value of commodities: but more than that, he is also looking in this direction to find the key to the riddle of how the different branches of production (agriculture and industry) and the different social classes (landlords, capitalists and workers) inter-relate with one another. Ricardo explained the progressive *cheapening* of industrial manufactures and the progressive *rise in price* of agricultural produce—both characteristic phenomena of early 19th century England— in terms of the workings of one and the same *law of labour value*. The value of industrial wares falls as a result of *technical progress*—the introduction of machinery and rising productivity of labour. The rise in value of agricultural produce is accounted for by the greater outlays of labour needed for its production, occasioned in turn by the increasing cultivation of *inferior land*. This downward trend in the value of industrial products and upward movement in the value of agricultural produce will provide the key to understanding the tendencies behind the distribution of the nation's revenue between *classes*. The rise in corn prices, which results from poor land being brought under cultivation, brings in its train a sizable increase in *ground rent*, and hence also a simultaneous need to raise *money wages* (real wages remaining unchanged, however). This rise in wages inevitably provokes *a fall in the rate of profit*. In this fashion Ricardo dervies his entire theory of distribution *from the law of labour value*.

While Ricardo's analysis of value's *quantitative side* represented an enormous advance over that of Smith, the *qualitative* or *social* dimensions of value remained outside his field of vision. Here we find the achilles heel of a theory whose horizons fail to extend beyond those of capitalist economy. Ricardo takes phenomena that belong to a specific form of economy and ascribes them to any economy. The *social forms* that things acquire inside the context of determinate production relations between people are taken by Ricardo as *properties of things in themselves*. He does not doubt that each and every product of labour possesses 'value'. It never occurs to him that value is a specific social form, which the product of labour acquires only when social labour is organized in a definite social form. Changes in the magnitude of value of products are conditional upon changes in the quantity of labour necessary for their production. This is Ricardo's basic law. His attention is riveted to the quantitative side of phenomena, upon the *'magnitude of value'* and the *'quantity of labour'*. He evinces no concern for the qualitative or social 'form of value', which is nothing but the material expression of social and production *relations between people* as independent commodity producers. Nor does Ricardo show any interest in the qualitative or social form in which *labour is organized*: he provides us with no explanation as to whether he is talking about labour as a technical factor of production (*concrete labour*), or about social labour organized as an aggregation of independent, private economic units connected to each other through the generalized exchange of the products of their labour (abstract labour). Certainly, we find in Ricardo the embryonic shoots of a theory of skilled and socially necessary labour, but it was left to

Marx to develop the theory of both socially abstract labour and the social 'form of value'.* Ricardo's great reform of the theory of value affected only its *quantitative* aspect. To him the existing social (i.e., capitalist) form of economic phenomena was given in advance, was already known and therefore required no analysis. As to the *qualitative* side of value, only a thinker who had taken as his object of enquiry the social form of *economy* (i.e., production relations between people), the social *form of labour*, and the social '*form of value*' could reform that aspect of the theory. Such a thinker was Marx.

The failure of Ricardo to recognize that the social form of an economy is historically conditioned did him little harm so long as he restricted his investigation to those phenomena that corresponded to the *existing* production relations between people (for example, to the law of labour value of commodities, which is premised upon production relations between people as commodity producers). But as soon as Ricardo passed onto the exchange of capital for labour power (an exchange predicated upon production relations between people as capitalists and wage labourers) or to the exchange of products produced by capitals of different organic compositions (an exchange which presupposes production relations between capitalists in different branches of production), his lack of a sociological method led him into the most basic analytical errors, as we shall see below.

2. Capital and Surplus Value

Ricardo's inability to grasp the social nature of value as an expression of the production relations between people created enormous difficulties for him even in his theory of labour value; when it came to his theory of *capital and surplus value* the difficulties only increased. Nevertheless, Ricardo did improve upon the existing theory of surplus value, ridding the quantitative analyisis of these phenomena of a number of the mistakes that had been present in Smith's account.

Smith's theory of value came to ruin, as we know, when it moved from petty commodity production to capitalist production. The very fact that a commodity (as capital) could exchange for a greater

*This disregard for the form of value led Ricardo, as it did the other representatives of the Classical school, to misapprehend the social function of *money*. Ricardo subscribed to a '*quantity*' theory of money and, apart from his doctrine on the movement of precious metals between countries, added nothing new in principle to what Hume had already formulated (see Chapter Eight on Hume above).

quantity of labour (labour power) than was embodied in it appeared to Smith as a violation of the law of labour value (see Chapter Twenty-Two above). Smith's only recourse was to declare that the law of labour value ceased to operate with the appearance of capital (profit) and the private ownership of land (rent).

Ricardo directed his entire efforts to showing that the law of labour value could operate even where there is *profit and rent*. But surely the working of this law is nullified by the fact that the value of a product (corn) is sufficient to cover not simply the remuneration of labour (wages) and the capitalist's profit, but also to yield an additional margin (*rent*) originating, as it would appear, not in labour but in the forces of nature? Not at all replies Ricardo in his theory of rent. The value of corn is determined by the quantity of labour needed to produce it on land of the most inferior quality. The value of corn produced on such land divides up only into wages and profit. The better lands receive a differential rent, comprised not of a mark up on top of the value of the commodity, but only of the difference between the labour value of the corn produced on better land and its social labour value as determined by the conditions of production on lands of the poorest quality. Rent is not a component part of price. By taking this position Ricardo simplified the entire problem of the relationship between value and revenues (we will have more to say about this in Chapter Twenty-Nine) such that it merely remained to explain the relationship between wages and profit.

Let us continue: the value of the product is sufficient not only to remunerate the labour expended on its production but also to yield a *profit* over and above this—surely this must invalidate the law of value as well? Surely the fact that the value of the product breaks down into wages and profit must conflict with a law which states that the product's value is determined only by the quantity of labour expended on its production? To resolve this problem in full one would have to discover the laws behind the exchange of capital for living labour (labour power), an exchange premised on production relations between capitalists and wage labourers. But Ricardo's thinking was, as we know, a long way from investigating the production relations between people. The social attributes of capital, on the one hand, and of labour power (wage labour), on the other, are simply missing. For Ricardo *capital and labour* confront one another as different *material* elements of production. Ricardo defines capital in *material-technical* terms, as 'that part of the wealth of a country which is employed in production, and consists of food, clothing, tools, raw materials, machinery, &c. necessary to give effect to labour.'* *Capital*, then, is means of production, or '*accumulated labour*,' so that even the primitive hunter possesses some capital. Ricardo turns the confrontation between *capital and labour power* from a conflict between *social classes* into a *material-technical* counter-position of *'accumulated'* labour to *'immediate'* labour. Hence capital has a *dual* function in Ricardo's arguments. On the one hand, the emergence of capital (in the sense of means of production) does not in the least invalidate the law of labour value: the value of the means of production (machinery, and the like) is simply *transferred* to the product that they help to manufacture. On the other hand, the value of products contains not simply the previously existing 'accumulated' value of the machinery and other means of production, which is reproduced on the same scale as before, but an *additional* margin of determinate size in the form of profit. Where does this *profit*, or *surplus value* come from? Ricardo provides no clear answer to this question.

To reveal the laws which govern the exchange of embodied labour (as capital) for living labour (as labour power) we must understand that, in addition to the production relations that exist between people as commodity producers, there appears in society a new, more complex type of *production relation*: that between capitalists and wage labourers. However, the method of distinguishing and gradually studying the different forms of production relations between people was alien to the Classical economists. Smith had come to conclude that the exchange of capital for labour** overturns the laws by which commodities exchange for one another. Ricardo was able to avoid this conclusion only because he studiously delimited these two types of exchange. Feeling powerless to explain the exchange of *capital for labour* in a way which would be consistent with the law by which commodity is exchanged for *commodity*, he confined himself to a more modest task: to demonstrate that the laws governing the mutual exchange of commodities (i.e., the law of labour value) is not abolished by the fact that capital exchanges for labour.

Let us suppose, says Ricardo, that a hunter expends the same quantity of labour on hunting a deer as does a fisherman in catching two

*Following Smith's example, Ricardo divides capital into *fixed* and *circulating* portions, differentiating them according to their 'durability.' By circulating capital Ricardo usually has in mind the capital which is laid out on *hiring* workers ('variable capital', in Marx's terminology). [The passage quoted here is from the *Principles* (Sraffa edition), p. 95—*Ed.*]

**In fact, as Marx made clear, capital is not exchanged for labour, but for labour power. The economists of the Classical school, however, remained unaware of this distinction and spoke about an exchange of capital for labour.

salmon, and that the means of production that each of them uses (the bow and arrow of the hunter, the boat and implements of the fisherman) are products of identical amounts of labour. In this case one deer will exchange for two salmon, completely independently of whether or not the hunter and the fisherman are independent producers or capitalist entrepreneurs conducting their business with the help of hired labour. In the latter case the product will be divided up between capitalist and workers, 'but it [the proportion of the product going to wages—*Trans.*] could not in the least affect the relative value of fish and game, as wages would be high or low at the same time in both occupations. If the hunter urged the plea of his paying a large proportion, or the value of a large proportion of his game for wages, as an inducement to the fisherman to give him more fish in exchange for his game, the latter would state that he was equally affected by the same cause; and therefore under all variations of wages and profits ... the natural rate of exchange would be one deer for two salmon.'[11] In other words, no matter by what principle capital is exchanged for labour, the exchange of *one commodity for another commodity* still takes place on the basis of the *law of labour value*: the proportions in which commodities mutually exchange for one another are determined exclusively by the relative quantities of labour required for their production.

We can now see the error in Smith's view, where in a capitalist economy *revenues* (wages and profit) appear as the basic sources of value, the primary magnitudes which, when altered, entail changes in the *value* of the commodity. 'No alteration in the wages of labour could produce any alteration in the relative value of these commodities; for suppose them to rise, no greater quantity of labour would be required in any of these occupations but it would be paid for at a higher price, and the same reasons which should make the hunter and fisherman endeavour to raise the value of their game and fish, would cause the owner of the mine to raise the value of his gold. This inducement acting with the same force on all these three occupations, and the relative situation of those engaged in them being the same before and after the rise of wages, the relative value of game, fish, and gold, would continue unaltered.'[12] From here we get Ricardo's famous rule: *a rise in wages,* contrary to the view of Smith, does not cause *the value of the product to go up, but rather causes profits to fall.* A fall in wages makes profits rise. The value of the product can rise or fall only in consequence of changes in the amount of labour demanded for its production, and not because wages have gone up or down.

This proposition, which runs like a red thread through the whole of Ricardo's work, is of cardinal importance. In the first place, by adopting it Ricardo took a correct position on the question of the relationship between *value and revenue*, an issue over which Smith had observed his own helplessness and inconsistency. Smith had incorrectly maintained that the value of a product is composed of the sum of wages, profit, and rent (and hence that the size of these revenues determines the amount of a commodity's value). This was completely alien to Ricardo's view. His standpoint is that the size of a product's value—as determined by the quantity of labour expended on its production—is the *primary*, basic magnitude that then *breaks down* into wages and profit (rent for Ricardo is not a component part of price). It is obvious that once the entire magnitude (the value of the product) is given *in advance* as a fixed entity (being dependent on the quantity of labour needed to produce it), any increase in one of its parts (i.e., wages) will invariably lead to a fall in the other (i.e., profit).

Secondly, the proposition under discussion is testimony that Ricardo saw profit as that part of the *value* of the product—created by the *labour of the worker*—which remains after deducting wages, and therefore moves *inversely* to the latter. Ricardo's position here definitively disproves any and all attempts to interpret his doctrine as a theory of production costs. If Ricardo's view had been that value is determined in conformity with production costs, i.e., by what is actually paid to labour in the form of wages, changes in the latter would elicit a corresponding change in the product's value. However, this is the very view that Ricardo is so forthrightly rebelling against. His assertion that wages and profits change inversely to each other is comprehensible only under one condition: if profit has its source in the surplus value created by the worker's labour. We are compelled, therefore, to acknowledge that *the idea of surplus value* (as viewed in its quantitative aspect) lies at the very basis of Ricardo's system, and that he applied it with greater consistency than did Smith. The fact that Ricardo concentrated his attention mainly on the exchange of commodities for other commodities and refrained from directly analyzing the exchange of capital for labour in no way refutes this statement; nor does the fact that Ricardo's specific mentionings of surplus value are less frequent than we find in Smith, who often makes reference to the 'deductions' made from the worker's product on behalf of the capitalist and the landlord. For Ricardo the existence of profit—and even an equal rate of profit—is presupposed in the very first pages of his study, providing, so to speak, a permanent background to the picture he is going to paint. Although

Ricardo does not inquire directly into the origins of profit, the general direction of his thinking leads him to the concept of surplus value. The value of the product is a *precisely fixed* magnitude, determined by the *quantity of labour* necessary for its production. This magnitude divides up into two parts: wages and profit. Of these, *wages* are firmly fixed, being determined by the value of the worker's customary means of subsistence (see below, Chapter Thirty)—that is, by the quantity of labour needed to produce corn on land of the poorest quality. What is left after wages (i.e., the value of the worker's means of subsistence) have been deducted from the product's value constitutes *profit*.

Like Smith, Ricardo analyzed profit and rent as separate entities, rather than bringing them together under the general category of surplus value. He confused surplus value with profit, mistakenly extending to it the laws applicable to surplus value.

Ricardo ignores the social nature of profit, riveting his entire attention on its quantitative aspect. The state of the productivity of labour *in agriculture*, the value of the worker's *means of subsistence*, the size of *wages*, and, depending upon fluctuations in the latter, *the size of profits*, are the causal connections and *quantitative relationships* that Ricardo studies. Ricardo makes *the size of profits* depend exclusively on the magnitude of wages and hence, in the last instance, on changes in the *productivity of labour* within agriculture. This is far too unilinear and narrow. Insofar as we are dealing with the *mass* of profits, this depends not simply on the size of wages, but on many social factors as well (the length of the working day, the intensity of labour, the number of workers). Insofar as we are dealing with the *rate* of profit, this depends to a very large degree upon the size of the total capital on which the profit is being calculated. Ricardo's disregard for these various factors is a weak point in his theory of profit; yet at the same time it graphically reveals one of its valuable strengths: Ricardo's overriding interest in the growth of the *productivity of labour* as the factor which ultimately determines *changes in the value of products and the revenues* of the different social classes.

3. Prices of Production

Up to this point Ricardo has been more or less successful in avoiding the reefs on which Smith's theory of value ran aground. True, he did not really resolve the problem of the exchange of capital for labour which had been so theoretically troublesome for Smith. But by pushing it to one side he neturalized, as it were, its inherent dangers and was able to show that the distribution of the product's value between capitalist and worker in no way affects the *relative* values of the products being exchanged. Of course, this argument conceals its own pitfalls. It assumes, for example, that a rise in wages (and a corresponding fall in profits) affects each of the two commodities being exchanged *to the same degree*. This assumption, however, is justified only under one condition: that the producers of the two commodities either advance their *entire* capital on the purchase of labour power (i.e., on the hire of workers) or divide it up between constant and variable capital in *exactly the same* proportions (Ricardo talks about fixed and circulating capital, but this has no effect on the problem). If each of them expends £1,000 on constant capital (machinery, raw materials, etc.) and £1,000 on hiring workers, then a rise in wages (say, by 20%) will have the same effect on both our entrepreneurs and have no influence on the relative values of their commodities. It is a different matter if, while one entrepreneur divides up his capital in the proportions we have stated here, the other lays out his entire capital of £2,000 purely and simply on hiring workers. Obviously a 20% rise in wages is going to be felt more sensibly by the second entrepreneur; and his rate of profit will fall below that earned by entrepreneur number one. In order to equalize the rate of profit in the two branches of production the relative value of the products in the second branch must rise in comparison to the value of the products of the first so as to compensate it for the greater loss suffered from the increase in wages. [13] We arrive, then at an *exception* to the rule that a change in wages does not affect the relative value of the products that are being exchanged: should exchange take place between branches of production with *different* organic compositions of capital, any increase in wages will be accompanied by a *rise* in the relative value* of the products of the branch of production with the *lower* organic structure of capital (i.e., the branch with the greater proportion of living labour) and a *fall* in the relative value of the products in the branch whose capital structure is *higher*. Consequently, the *relative values* of products (produced either by capitals with different organic compositions, by fixed capitals of unequal lifespans, or by capitals having unequal turnover periods) can alter not only because of changes in the relative quantities of labour necessary for their production, but also from a *change in the level of wages* (which means a corresponding change in the rate of *profit*). This is the famous '*exception*' to the law of

*In fact it is the 'price of production' that changes, and not the product's value. However, Ricardo did not differentiate prices of production from value.

labour value that Ricardo examines in Sections IV and V of the first chapter of his *Principles*. The heading to Section IV reads, 'The principle that the quantity of labour bestowed on the production of commodities regulates their relative value, considerably modified by the employment of machinery and other fixed and durable capitals.'* The law of labour value retains full validity only when the products being exchanged are produced by capitals that have equal *organic compositions*, are of the same *longevity*, and are advanced for equal *periods of time*.[14]

Ricardo illustrates his idea with the following example. Farmer A hires 100 workers, each of whom he pays a wage of £50 a year. His total circulating (variable) capital is £5,000 pounds. We assume that he makes no outlays on fixed capital. Given an average rate of profit of 10% the farmer's corn will at year's end have a value of £5,500. At the same time cloth manufacturer B also hires 100 workers, investing in his business a circulating capital of £5,000. However, to manufacture the cloth these workers use machinery with a value of £5,500 pounds. This means that B is investing in his business a total capital of £10,500. If, for the sake of simplification, we assume that the machinery does not depreciate, the cloth that has been manufactured in the course of the year will have a value of £6,050: £5,000 as replacement for circulating capital, plus £500 (= 10% of this circulation capital), plus £550 (= 10% of the fixed capital). Although both the corn and the cloth have been produced with equal quantities of labour (100 men),** the cloth is worth more than the corn: into the price of the cloth there enters an additional sum of 550 pounds, which is *profit on the fixed capital*. Where does this additional profit come from if no more labour has been expended on producing the cloth than on the corn? Ricardo does not ask this question. He states and then accepts as given the fact that the ratio of the corn's value to the cloth's is 5500:6050.

*Ricardo always speaks of fixed and circulating capitals, but by the latter he essentially means capital advanced for the hire of workers (i.e., variable capital, in Marx's terminology). [This quotation is from the *Principles* (Sraffa edition), p. 30.—*Ed.*]

**Since we have assumed that the machinery used in cloth manufacturing does not depreciate, it does not transfer any of its value to the cloth. [Rubin might more properly have said here that it does not transfer any of its value *to the value* of the cloth. Although Marx and virtually every Marxist economist since have talked of value being transferred or imparted directly to the commodity, one does not want to lose sight of the fact that value is a *social*, and not a *material* property of the product. For a truly excellent discussion of the problems caused by the 'mental materialization of human relations' (the latter being the proper subject of political economy) amongst students of Marxism, see E. A. Preobrazhensky, *The New Economics* (Oxford University Press, 1965), pp. 147-50. From the point of view of their method, especially their philosophical treatment of the categories of political economy, Preobrazhensky and Rubin shared a great deal in common—*Ed.*]

From here Ricardo goes on to examine what effect *a change in wages* will exert on the value of these two commodities. Assume that wages rise, thus causing the average level of profits to fall from 10% to 9%. The value of the corn *will not change*, but will remain at its old figure of £5,500: whatever the fall in the farmer's profits, his total wage bill will increase by the same amount, so that the sum of wages plus pofit will still be equal to £5,500. Similarly, the sum of cloth maker B's circulating capital (i.e., his workers' wages) plus the profit derived from it is unchanged to £5,500. What does alter is the additional profit on his £5,500 of fixed capital. Previously he had added on 10% (£550), thus making his cloth worth £5,500 + £550, i.e., £6,050. Now he charges only 9% (£495), so that the price of the cloth becomes £5,500 + £495, i.e., £5,995. The ratio of the value of the corn to the value of the cloth, which before had stood at 5,500:6,050, stands now at 5,500:5,995. Consequently, *a rise in wages* (or, what is the same thing, *a fall in profits*) *lowers* the relative value of those commodities being produced using *fixed capital* (or using a larger amount of fixed capital). The reason for this is that the price of these commodities contains *an additional amount of profit* charged on the fixed capital which declines with the fall in the rate of profit.

The example we have anlayzed poses the investigator not only with the problem of how changes in wages affect the value of different commodities, but also with the much more profound and basic problem of how to reconcile *the law of labour value* with the law of *the equalization of the rate of profit* on capital. We saw that prior to there being any change in wages—and completely independent of this change— the value of corn stood to the value of cloth in the ratio of 5,550:6,050, even though equal quantities of labour had been expended on their production. Here before us we have two commodities, produced with equal quantities of labour (100 workers), but where the capitals advanced are unequal (£5,500 compared with £10,500). From the point of view of the theory of labour value the labour value possessed by the two commodities is *equal*. From the point of view of the law of an equal rate of profit, the price of the latter commodity must be *higher*, since it contains a profit on a *larger capital*. How do we resolve this contradiction? It was to answer this question that Marx constructed his theory of *'prices of production'*. According to Marx's theory, in a capitalist economy, with its tendency towards an equalization of the rate of profit, commodities are sold not at their labour values, but at their 'prices of production', i.e., production costs plus average profit. The total mass of surplus value produced in society is divided up between all of its capitals in proportion to the size of each. If some commodities are sold at prices above their labour value, others are sold at prices below it. A branch of production with a high capital structure receives the average profit, which exceeds the total surplus value that this branch has produced. These 'additional' sums of profit are taken, however, out of the general reserve of surplus value created by all of the branches of production together.

Ricardo was not only unable to *resolve* the problem of 'prices of production' he could not even *pose* it in all its scope. True, he understood that with two branches of production having different organic structures of capital the prices of their products must deviate from their labour values to allow their rates of profit to be equalized. Ricardo started out grasping a firm hold of the idea that the governing tendency within capitalist economy was for *for profits to be equalized.* He had no doubt that cloth must cost more than corn, despite their equal labour values, so that its owner could earn a profit on his larger capital investment. The cloth manufacturer's right to receive a profit corresponding to the size of his capital appeared to Ricardo so natural that the question of where this additional £550 profit (on fixed capital) originated from did not concern him. By assuming an *average rate* of profit from the very outset, i.e., that commodities sell not at their labour values but at their prices of production, he avoids the *basic* problem of *how the average rate of profit is formed* and how labour value is transformed into *prices of production.* Rather, his attention is focused specifically on the effect that *changes in wages* have on the relative prices of commodities produced by capitals with unequal organic compositions, independently of alterations in labour value. Ricardo, in establishing that changes in wages and profit do influence the relative values of commodities, acknowledges that here we have a '*modification*' or '*exception*' to the law of labour value. He consoles himself that this 'exception' is of no great significance: the effect that changes in wages (and profit) exert on the relative values of commodities is *insignificant* compared to the impact of changes in the quantity of labour necessary for their production. By analyzing the quantitative changes that take place in the value of commodities the growth in the productivity of labour preserves its former role as the predominant factor. On this basis Ricardo considers himself justified in pushing aside his exception and considering 'all the great variations which take place in the relative value of commodities to be produced by the greater or less quantity of labour which may be required from time to time to produce them.'[15] Exceptions notwithstanding, the law of labour value *retains its validity* in his eyes, and he subsequently constructs his entire *theory of distribution* upon it.

Although Ricardo continues to hold fast to the law of labour value, the exceptions to it in fact punch a gaping hole in his formulation of the theory of value. To the question, where does the *profit on fixed capital* come from?, Ricardo gives no answer. Instead of demonstrating that the product of one branch of production will sell as much above its labour value as the product of another branch sells below its own, Ricardo makes another, totally unintelligible assumption: corn sells at its *full* value (5,500), but cloth sells *above* its value (£5,500 + £550). Instead of demonstrating the process by which the average rate of profit *is formed*, Ricardo takes the rate of profit to be 10% in advance, without any explanation. The source of the profit on *circulating* (variable) capital is the *labour value* of £5,500 created by the labour of 100 men; it therefore falls with every increase in wages (and vice versa): the sum of wages (circulating capital) plus the profit on circulating capital is assumed to remain steady at £5,500. The profit on fixed capital is mechanically *added* to the labour value created by the workers' labour at the defined rate of 10% (that is, a profit of unknown origin equal to £550, or 10% of the fixed capital, is added to the £5,500 value that the 100 workers have created). This mechanical adding together of the profit on fixed capital and the profit on circulating (variable) capital illustrates clearly the way in which Ricardo had *mechanically combined the law of labour value and the law of an equal rate of profit on capital*. Ricardo did not abandon the first, but he was unable to make it accord with the second. Smith's theory of value came to ruin over the problem of exchanging *capital for labour*; Ricardo's theory, on the other hand, was unable to resolve the problem of how *prices of production and an equal rate of profit* are formed. Ricardo himself acknowledged that his exceptions had introduced a contradiction into the theory of value. He says in his correspondence that the relative value of commodities is regulated not by one, but by two factors: 1) the relative *quantity of labour* necessary for their production, and 2) the size of the *profit* on capital up to the time when a product of labour can be put on the market (or, what is the same thing, the relative periods of *time* required in bringing a product to market).[16] Here *profit on capital* (or the time over which capital is advanced) functions as an independent factor which regulates—*along with labour*—the value of commodities.

This *contradiction* in Ricardo's doctrine served as a starting point for subsequent scientific developments. Ricardo's followers (James Mill and McCulloch) did their best to maintain that unstable equilibrium between the theory of labour value and the theory of production costs (or between the law of labour value and the law of an equal rate of profit) which was to be found in Ricardo. Freedom from these contradictions could be had either at the price of abandoning the labour theory of value or by fundamentally reworking it. Malthus, a severe critic of Ricardo, called for the first of these when he argued that the many 'exceptions' allowed for by Ricardo sapped the law of labour value of any definitive validity. The second line was pursued by Marx, whose theory of 'prices of production' resolved those contradictions which, though latent and confused, had made themselves felt in Sections IV and V of the first chapter of Ricardo's book, and which were to become the subject of lively debates in post-Ricardian literature (see Chapter Thirty-Three below).

1 'The produce of the earth—all that is derived from its surface by the united application of labour, machinery, and capital, is divided among three classes of the community; namely, the proprietor of the land, the owner of the stock or capital necessary for its cultivation, and the labourers by whose industry it is cultivated.

'But in different stages of society, the proportions of the whole produce of the earth which will be allotted to each of these classes, under the names of rent, profit, and wages, will be essentially different; depending mainly on the actual fertility of the soil, on the accumulation of capital and population, and on the skill, ingenunity, and instruments employed in agriculture.

'To determine the laws which regulate this distribution, is the principal problem in Political Economy ... ' Ricardo, Preface to the *Principles*, p. 5.

2 It is interesting to note just how closely Rubin's critique of Smith's theory of value (see Chapter Twenty-Two above) parallels the critique offered by Ricardo. 'Adam Smith, who so accurately defined the original source of exchangeable value, and who was bound in consistency to maintain, that all things became more or less valuable in proportion as more or less labour was bestowed on their production, has himself erected another standard measure of value, and speaks of things being more or less valuable, in proportion as they will exchange for more or less of this standard measure. Sometimes he speaks of corn, at other times of labour, as a standard measure; not the quantity of labour bestowed on the production of any object, but the quantity which it can command in the market: as if these were two equivalent expressions, and as if because a man's labour had become doubly efficient, and he could therefore produce twice the quantity of a commodity, he would necessarily receive twice the former quantity in exchange for it.

'If this indeed were true, if the reward of the labourer were always in proportion to what he produced, the quantity of labour bestowed on a commodity, and the quantity of labour which that commodity would purchase, would be equal, and either might accurately measure the variations of other things: but they are not equal; the first is under many circumstances an invariable standard, indicating correctly the variations of other things; the latter is subject to as many fluctuations as the commodities compared with it. Adam Smith, after most ably showing the insufficiency of a variable medium, such as gold and silver, for the purpose of determining the varying value of other things, has himself, by fixing on corn or labour, chosen a medium no less variable ...

'It cannot then be correct, to say with Adam Smith, "that as labour may sometimes *purchase* a greater and sometimes a smaller quantity of goods, it is their value which varies, not that of the labour which purchases them;" and therefore, "that labour *alone never varying in its own value*, is alone the ultimate and real standard by which the value of all commodities can at all times and places be estimated and compared;"—but it is correct to say, as Adam Smith had previously said, "that the proportion between the quantities of labour necessary for acquiring different objects seems to be the only circumstance which can afford any rule for exchanging them for one another;" or in other words, that it is the comparative quantity of commodities which labour will produce, that determines their present or past relative value, and not the comparative quantities of commodities, which are given to the labourer in exchange for his labour.' *Principles*, pp. 13-17 (Ricardo's italics).

3 *Ibid*, p. 12.

4 *Ibid*, p. 22.

5 *Ibid*, pp. 285-86; Ricardo's italics.

6 *Ibid*, p. 284.

7 *Ibid*, p. 73. 'The exchangeable value of all commodities, whether they be manufactured or the produce of the mines, or the produce of land, is always regulated, not by the less quantity of labour that will suffice for their production under circumstances highly favorable, and exclusively enjoyed by those who have peculiar facilities of production; but by the greater quantity of labour necessarily bestowed on their production by those who have no such facilities; by those who continue to produce them under the most unfavorable circumstances; meaning—by the most unfavorable circumstances the most unfavorable under which the quantity of produce required, renders it necessary to carry on the production.'

8 *Ibid*, p. 283.

9 *Ibid*, p. 382.

10 *Ibid*, p. 11.

11 *Ibid*, p. 27.

12 *Ibid*, p. 28.

13 As Rubin notes later on in this discussion, it is not really the *relative values* of the two commodities that are changing (and we must at all times keep in mind that Ricardo is talking about their *relative* standing to each other, and not their *absolute* values—although as Meek points out in his *Studies in the Labour Theory of Value*, p. 104, there are special conditions under which a rise in wages can cause absolute price to fall as well), but their *prices of production*.

In Volume III of *Capital* Marx noted the seeming conflict between the theory of value, which, as we will illustrate, can have capitals of equal size earning unequal rates of profit, and the clearly observable realities of every day economic life, where such inequalities in the rate of profit do not exist but for exceptional cases. Let us take two capitals, A and B, each with total capitals of 100 (we have taken the example from Chapter IX of *Capital*, Volume III):

A. 80c + 20v + 20s = 120
B. 70c + 30v + 30s = 130

The two capitals are of identical size, but create products of unequal value, owing to different proportions between constant capital, which simply transfers its value to that of the final product, and variable capital, which is the only value-creating element. What is more, though of equal size, these capitals have unequal rates of profit. The rate of profit, which is defined as the ratio of surplus value to the *total capital*, equals for capital

A: $\frac{20s}{80c + 20v} = 20\%$; for capital B, $\frac{30s}{70c + 30v} = 30\%$.

Marx resolved the problem by noting that commodities do not actually sell at their simple labour values, but at *prices of production* which *deviate* from these labour values but which nevertheless *are based upon them*. We know that the two capitals must have equal rates of profit. This rate is determined by the relationship between society's *aggregate* surplus value and its *aggregate* capital. The total capital (assuming that capitals A and B are the only two capitals in society) here equals 200; the total surplus value equals 50. The rate of profit, p', therefore equals 25%. Each of these capitals will sell at a price of production determined by its '*costs of production*,' i.e., total capital, plus the profit on that capital, which is the *average rate of profit for society as a whole*, or 25%. Thus, capital A will have a price of production for its product of

80c + 20v + 25p = 125

and capital B a price of production on its product of:

70c + 30v + 25p = 125.

Now the two capitals have equal selling prices and equal rates of profit; their selling prices are the same only because these are capitals of equal gross size earning the average rate of profit. What has happened is that the total surplus value of society as a whole has been *apportioned according to the size of the total capital of each of its constituent capitals*. This means that capital A sells *above* its value and capital B *below* its value. However, total surplus value remains the same; it is merely redistributed so as to equalize rates of profit. Also total price equals total value (250 in both cases).

In the example that Rubin has given here we have two capitals of equal size, but with different apportionments between constant and variable capital. We do not know the rate of profit, but it is assumed to be equal in the two cases, let us say 30%.

A. 1000c + 1000v + 600p = 2600
B. 0c + 2000v + 600p = 2600

On the assumption that a rise in wages comes out of profit, a 20% rise in wages for capital A will raise them to 1200; if this comes out of profits (since the actual labour expended does not alter), capital A stands at:

A. 1000c + 1200v + 400p = 2600

Similarly, a 20% rise for capital B will raise them to 2400; reducing profit by the same amount, capital B will be:

B. 0c + 2400v + 200p = 2600

They still have equal prices, but now they have unequal rates of profit; capital A's rate of profit equals 400/2200 = 18%; capital B's rate of profit equals 200/2400 = 8.3%. To equalize its rate of profit with that of capital A, capital B would have to raise its price (by raising its total profit) from 200 to 432. Then, with a rate of profit of 18%, its price would be:

B. 0c + 2400v + 432p = 2832.
Its price of production (since that is really what we are dealing with here) has risen *relative* to the price of production for capital A.

It is important to recognize why this has happened. A 20% rise in wages has affected the two capitals unequally by ***changing the size of their total capital***. Given the existence of an average rate of profit, once their capitals were unequal in size their selling prices had to diverge. It is equally important to note that this example already presumes the existence of an average rate of profit; i.e., values in terms of labour values in no way figure into it. In the example given, if we assume that the two capitals function with equal rates of exploitation (s/v), they would in value terms look as follows (assuming that s/v equals 40%).

A. 1000c + 1000v + 400s = 2400
B. 0c + 2000v + 800s = 2800

In other words, the very assumption of an equal rate of profit in this example hides the fact that they have *unequal* labour values. On Marx's premises these two capitals could not have had equal rates of profit and sold at their values in the first place, except by assuming either that the rate of exploitation in capital A is double that in capital B, so that they each produced 800 in surplus value, or that A's capital circulated *twice as fast as* B's (in that case its 1000v would circulate twice in a year, earning a *total annual* surplus value of 800). Were either of these exceptions permitted (the last one being quite plausible) the two capitals would be equal in size, produce equal surplus values, have equal rates of profit, and hence the values of their products and their prices of production would be identical. On the effects of times of turnover on the annual rate of surplus value and the rate of profit, see *Capital*, Volume II, Chapter XVI, and Volume III, Chapter VIII. An excellent and lucid explanation of the problem of prices of production and its relation to Marx's theory of value (discussed by Marx in Part II of *Capital*, Vol. III) is Rubin's chapter 'Value and Production Price', in his *Essays on Marx's Theory of Value*.

14 The question of the longevity of fixed capital can be illustrated very simply. Suppose that we have two capitals of equal size each earning equivalent surplus values, and hence having equal rates of profit, but experiencing unequal rates of depreciation on their fixed capital. Suppose that capitals A and B each have a stock of fixed capital of 1000, and that they use no circulating constant capital. Their fixed capital, however, depreciates at different rates: the fixed capital of capital A wears out in ten years; that of capital B wears out in five. In value terms the value of A's annual product will contain a constant capital component (which, after all, represents only the value transferred by the means of production in that particular year) of 100, the value of B's product a constant capital component of 200.

A. Total capital = 1000 fixed capital stock + 100v
Value of product = 100c + 100v + 100s = 300
B. Total capital = 1000 fixed capital stock + 100v
Value of product = 200c + 100v + 100s = 400

Here the *total* capital equals 1100 for both A and B; their rates of profit are also equal, being 1/11 in both cases. However, the *value* of their annual product is different, because of the faster depreciation of fixed capital in B.

Similarly, if they have unequal periods of turnover (what Rubin means when he says they may be advanced for unequal periods of time), their values can also differ, as we have shown in the previous note. In the example given here, if their fixed capitals depreciated at the same rate, so that both their total capital and the annual constant capital value were identical in both A and B, but A's capital turned over at twice the speed of B's, A's annual surplus value would equal 200, as compared to B's surplus value of 100. Their values would now be unequal (400 for A's annual product value versus 300 for B's), as would their rates of profit (A would earn a higher rate of profit than B).

An interesting variation of this example would be if A's capital turned over twice as fast as B's, but B's fixed capital depreciated at twice the rate of A's (i.e., we combined the two sets of assumptions in this illustration). Their values would be

A. 100c + 100v + 200s = 400

B. 200c + 100v + 100s = 400

The value of their annual product would now be equal, but A's *rate of profit* would be higher.

15 *Principles*, pp. 36-37.

16 In other words, a producer at greater distance from the market will require greater time to realize his product, and hence his capital will have a longer turnover period.

CHAPTER TWENTY-NINE

Ground Rent

Ricardo's theory of *differential rent* has suffered far fewer alterations during the ensuing development of economic thought than have all his other theories. At present it is generally accepted by nearly all economists of the most diverse tendencies. Marx was to incorporate its basic features into his own theory of rent.

The second chapter of Ricardo's book, devoted to rent, is, by virtue of its simplicity and the clarity of its basic ideas, one of the most brilliant examples of the application of the method of abstraction in the history of economic literature. From a few initial propositions and the application or implication of a number of simplifying conditions, Ricardo derives his entire theory of rent* which abuts directly upon his theory of value, developed by him in Chapter I of his book. He asks, at the very outset, whether the fact that the price of agricultural produce (in the broad sense) includes rent does not contradict the theory of value?

Prior to Ricardo, queries as to the origin of rent had received the following answers. The *Physiocrats* (see Chapter Fourteen) had said that rent originates in the superior productivity of agricultural labour which, in collaboration with the forces of nature, yields a 'net product' over and above the produce consumed by the workers themselves: rent is created *by nature*. In Smith (See Chapter Twenty-Three), as usual, we find several embryonic solutions to the problem. In the first he partially takes over the physiocratic idea that rent results from the special productiveness of agricultural, as compared to industrial labour; secondly, in his idea that profit and rent are both 'deductions' from the value created by the worker's labour, he reduces rent to labour; finally, there is his idea that the value of the product is defined as the sum of wages, profit, and rent, by virtue of which he opened the way for those theories that attribute the higher value of agricultural

*The forerunner to Ricardo's theory of differential rent was a writer from the end of the 18th century, Anderson. The law of 'diminishing fertility of the soil' was formulated in 1815, practically simultaneously by West, Malthus, and Ricardo. [On Anderson—and Malthus's alleged plagiarism of Anderson's theory—see *Theories of Surplus Value* (Progress Publishers English edition), Part II, pp. 114-20—*Ed.*]

produce to the need to pay rent to the proprietor of the land. If taken to its logical conclusion this last idea turns into a theory that explains rent by the landowner's 'monopoly' status which results in the sale of agricultural poducts at prices which exceed their value by the amount of the rent.

Thus, from the point of view of the *Physiocrats*, rent is an *in natura* surplus of products over and above those consumed by the workers. According to the *'monopoly'* theory, rent is an increment added onto the price of the agricultural product, which is then sold above its value. The first solution tears the theory of rent from the theory of value, the second sees rent as an exception to the principle of labour value.

Ricardo's theory was directed against both these viewpoints. As an objection to the Physiocrats he points out that the exceptional productivity of agricultural labour—assuming that it actually exists—is accompanied by a rise in the number of use values or *in natura* produce and hence ought to result in a decline, and not a rise in their exchange value. The source of rent must be sought not in the surplus of products *in natura*, but in their greater exchange value, which to the contrary, arises from the difficulty of producing them. Ricardo shifts the entire problem out of the sphere of *use* value and into that of *exchange* value. 'When land is most abundant, when most productive, and most fertile, it yields no rent; and it is only when its powers decay, and less is yielded in return for labour, that a share of the original produce of the more fertile portions is set apart for rent.'[1]

Hence we have Ricardo's first thesis: rent comes not from the special productivity of agriculture, but on the contrary, results from the *deterioration of the conditions under which labour is applied, or the transfer of production from superior land to land of poorer quality.* The value of corn is determined by the quantity of labour expended to produce it on the worst land.* *Rent* is the *difference* between the value of this corn (its 'socially necessary' or 'market value', to use Marx's terminology) and the *'individual* value' of a given bushel of corn produced on land of prime quality. This rent is called, therefore, *'differential rent'*; and arises where expenditures of labour**

*Ricardo mistakenly generalized this law to apply to the exchange value of all products.

**Ricardo talks about expenditures of labour and capital, but makes no distinction between a simple commodity economy, where labour is expended and the product sold at its labour value, and a capitalist economy, where what is expended is capital and the product is sold at its price of production (or, in agriculture, at its price of production plus absolute rent).

have different productivities, either by virtue of being made on pieces of land of unequal fertility (rent of *fertility*) or at different distances from a common market (rent of *distance*)* or by having been successively applied to one and the same plot of land (rent of *intensity*).

The theory that rent is the margin between the *individual* value and the *socially necessary* value of products** links the theory of rent immediately and inseparably to the theory of value, making the phenomena of rent akin to other economic phenomena, especially to 'differential profit', or '*superprofit*'. The latter accrues to those capitalist entrepreneurs who carry out production using new improvements, particularly new methods of production, etc. The difference between superprofit and rent is as follows: 1) superprofit is a *temporary* phenomenon, which disappears as soon as the improvement in question becomes universally applied and thus lowers the product's socially necessary value, whereas differential rent, because it depends on permanent differences in the fertility or location of plots of land or in the productivity of successive expenditures of labour, is *constant*;† 2) superprofit is earned by the *capitalist*, whereas rent goes to the *landlord*. Let us consider this point further.

Why is it that the *superprofit* which the farmer receives from employing more advanced machinery stays in his pocket, while the superprofit accruing from the greater fertility of the land he is cultivating has to be paid over to the landlord and turned into *rent*? Should a portion of this rent remain with the farmer he would be receiving a superprofit (i.e., a profit greater than the average rate of profit) solely by virtue of the fact that he is producing on a plot of land that is more fertile. In this case all other farmers would want to lease this plot, upping what they would pay as rent until the entire superprofit (the rent) was passing into the hands of the landowner and the farmer was left with only an average rate of profit. Thus, to explain why the whole of the differential rent is transferred to the landlord, Ricardo puts forward a second premise which states that *there are*

*Here it is a question of differences in expenditures not on production, but in transporting the produce to where it will be sold. Ricardo mentions this form of rent only in passing. The doctrine of rent of distance was developed by Thünen in his famous book, *Die isolierte Staat* (1827).

**Because there is no explanation in Ricardo of the social process by which individual labour is transformed into socially necessary labour, he was unable to give his theory precise formulation, even though he had developed it in its essentials.

†Even though this difference is constantly present its magnitude nonetheless fluctuates, thus giving rise to changes in the volume of differential rent.

sufficient capitals in the country looking to invest in agriculture wherever they can be assured of receiving the average rate of profit.

Rent, therefore, is received not because the price of corn exceeds its value, but because the value of the particular corn in question is below the socially necessary value. With this explanation Ricardo resolutely rejects the second of the theories that we earlier referred to, namely the '*monopoly*' theory, which sees rent as an increment added onto the value of the product. 'The reason then, why raw produce rises in comparative value, is because more labour is employed in the production of the last portion obtained, and not because a rent is paid to the landlord. The value of corn is regulated by the quantity of labour bestowed on its production on that quality of land, or with that portion of capital, *which pays no rent.* Corn is not high because a rent is paid, but a rent is paid because corn is high.'[2] Rent *does not enter into the product's value*, which is determined by the amount of labour (or capital) expended on poor quality land. Land of this quality earns the farmer only an average profit on capital, but will provide nothing extra that could be payed over as rent to the landlord. Yet how can the farmer get hold of such a tract for cultivation without paying rent to a landowner? Ricardo is obviously presuming the existence of poor quality land *freely accessible* to anyone who wishes to work it. In other words, Ricardo is ignoring just those limitations that private property in land—including very poor land—places in the way of capital investment in agriculture. Only in this way could Ricardo arrive at the conclusion that *inferior tracts of land yield no rent.*

Ricardo's theory of rent gives us, then, the following three propositions: 1) there is no such thing as *absolute* rent (i.e., rent paid for cultivating land of the poorest quality); 2) the only rent that exists is *differential* rent, which equals the difference between *individual* and *socially necessary* expenditures of labour (or capital) and arises because farmers are gradually bringing land of increasingly *inferior* quality under cultivation; 3) the whole of the differential rent goes to the *landowner*. Ricardo's first thesis, as we will see, is wrong and needs correction. His doctrine of differential rent is on the whole correct. It is still true that the theory of differential rent as Ricardo developed it contains a number of non-essential elements that need to be expunged. Ricardo had tied his theory of rent to the mistaken idea that, because farmers would be tilling land of poorer and poorer quality, the quantity of labour needed to produce a bushel of corn would go up and there would be an inevitable and progressive rise in the price of corn. Indeed, Ricardo does acknowledge that progress in agricultural technology reduces the quantity of labour required to produce corn, but it is his opinion that these technical advances can only momentarily retard or attenuate the operation of this so-called *law of 'diminishing fertility of the soil'* and not abolish it.

Ricardo's erroneous idea that technical progress in agriculture tended in a direction opposite to that of industrial development was simply a theoretical reflection of *fortuitous* economic phenomena that *temporarily* appeared in England at the beginning of the 19th century. English *industry* in Ricardo's time was marked by the rapid introduction of machine production and the cheapening of commodities. In his theory of *value* Ricardo generalized this phenomenon: he was convinced that 'alterations in the quantity of labour necessary to produce commodities are of daily occurrence. Every improvement in machinery, in tools, in buildings, in raising the raw material, saves labour, and enables us to produce the commodity to which the improvement is applied with more facility, and consequently its value alters.'[3] Industry develops in an atmosphere of non-stop *technical progress*, growth in the *productivity of labour*, and *cheapening of products*. *Agriculture* develops in a different direction—and here again Ricardo is generalizing from the previously described features of early-19th-century English agriculture (the tillage of new land of poorer quality, the rising costs of producing corn, and an awesome rise in corn prices). These were the historically transient conditions of English agriculture during the period 1770-1815, but Ricardo in effect incorporated them *in toto* into his theoretical conceptions. According to Ricardo, agriculture develops under the inexorable necessity to move from better land to inferior, with a *rise* in the quantity of labour needed to produce a bushel of corn on land of decreasing quality. Ricardo's famous law of 'diminishing fertility of the soil' was formulated (and this was also done by his contemporaries, West and Malthus) as a hurried and mistaken generalization of the temporary phenomena that he was witnessing. Because of the operation of this law, corn 'has a tendency to become *dearer* from the greater difficulty of producing it'.[4] The development of labour productivity in industry and agriculture is subordinated to *different* laws, the result of which is that the values of industrial and agricultural products move in *opposite* directions: 'manufactured commodities [are] always *falling*, and raw produce always *rising*, with the progress of society.'[5]

Ricardo moves on from here to draw a number of conclusions as to how society's revenue will be distributed between its different classes. With the price of corn constantly rising money wages will grow (although real wages will remain unchanged). The growth of money wages and the rise (both real and monetary) of rent create a tendency for the rate of profit to fall. The lion's share of the benefits of economic progress go to the landlords, to the detriment of the capitalists, and to a lesser extent of the workers as well. Thus, in terms of the *distribution* of society's revenue, the tendencies that Ricardo depicts are these: first, a colossal rise in the price of corn and ground rent, second, an increase in money wages while real wages remain stationary or even fall, and third, a declining rate of profit (this will be discussed further in the next chapter). This entire theory of distribution proceeds from the assumption that corn prices will inevitably rise owing to the operation of the law of 'diminishing fertility of the soil'.

Every one of these conclusions is premised on a precipitous generalization of a few facts taken from the history of English agriculture at the start of the 19th century. In the first place, it is *historically* incorrect that the best land was always cultivated before inferior areas. Carey shows, using historical examples, that farmers frequently began by cultivating land that was poorer in quality but more easily accessible, and started the cultivation of higher quality land only later (see the chapter on Carey and Bastiat in Part 5 below). Secondly—and this is Ricardo's decisive mistake—it is untrue that a gradual transition to cultivating inferior land inevitably leads to a progressive rise in the price of corn. Once new technical improvements are introduced corn can be produced on inferior land at a lower production cost than it could previously on land of better quality. The brilliant successes of agricultural technology in the mid-19th century progressively lowered the outlays of labour and capital required to produce a unit of corn and overthrew the pessimistic forebodings of Ricardo and Malthus. Thirdly, it is incorrect that rent only rises when there is a rise in the price of corn. If the *difference* in productivity of expenditures made on different land widens and *the number* of bushels of corn harvested per acre increases, rent can go up even if the price of corn *falls*. No less mistaken was Ricardo's attempt to explain the *falling rate of profit* on the basis of a rise in the price of corn: its explanation in fact lies in the rising organic composition of capital (see next chapter). Each and every one of these assertions falls as soon as we remove the basic premise of an inevitable and progressive rise in the price of corn.

However false Ricardo's predictions about the tendencies of revenue movements may have been, this in no way detracts from the *theoretical validity* of his doctrine on diferential rent. Let us accept that Ricardo was historically inaccurate when he maintained that farmers always begin by cultivating the best lands and only later shift to poorer ones; let us allow that his certitude that the price of corn must progressively rise was misplaced. Independent of these facts, that is, no matter what the *order in which we transfer* from some tracts of land to others and no matter what the price of corn is, even if it be a *low* one, it remains beyond dispute that labour (and in a capitalist economy, capital as well) will be *simultaneously* expended on lands of *different* fertility and geographical location (or on one plot of land at different points in time). It follows, then, that there will be long-term *differences in the individual amounts* of labour (or capital) *expended* per unit of product, e.g., per bushel of corn (and not temporary differences, as in industry). Given that in a commodity economy products are exchanged according to their socially necessary expenditures, producers operating under more favourable conditions will inevitably receive from the sale of agricultural produce a *surplus* quantum of value over and above costs of production and the average profit on capital (i.e., over and above their prices of production). Given that the capitalists (farmers) and landowners are separate classes, this surplus quantum, or superprofit, goes to the latter and is transformed into *rent*, that is, into the specific form of income of a definite social class. Thus, for all the corrections that have to be made in Ricardo's theory of differential rent, it remains on the whole fully valid.

His theory of rent needs to be supplemented, however, by the doctrine of *absolute* rent. So long as all land is privately owned Ricardo is wrong to assume that the worst lands under cultivation yield no rent: the landowner would prefer to let this poorest plot of land lie fallow rather than gratuitously give it over to the farmer for cultivation merely so that the latter might earn an average profit on his capital. Where all land is held as private property and farmers and landlords exist as separate classes, even the worst lands under cultivation will yield some rent, even if it is very small. This is what is referred to as *absolute* rent. The best lands will yield both absolute rent and a differential rent (the size of the latter depending on the quality of the land in question, that is, on its fertility or its proximity to a market). Development of the theory of absolute rent belongs to Rodbertus and to Marx.

1 *Principles*, p. 74.
2 *Ibid*, p. 75 (Rubin's italics).
3 *Ibid*, p. 36.
4 *Ibid*, p. 93 (Rubin's italics).
5 *Ibid*, p. 97 (Rubin's italics).

CHAPTER THIRTY

Wages and Profit

Although Ricardo's doctrine on wages was to gain wide currency under the title '*the iron law of wages*' (given it by Lasalle), from a theoretical standpoint it is one of the weakest and least satisfactory parts of his system.

Worst of all, Ricardo—and this is in accord with his general method—paid no regard to the *qualitative* or *social* side of wages. Under what socio-economic conditions do wages arise, what relationships between social classes do they presume, on the basis of what laws does the exchange of wages for labour power take place? Ricardo asks none of these questions. Because he fails to distinguish labour power from labour, he is unable to explain how it is that 'labour' (i.e., labour power) possesses less value than the value that it creates. To explain this Ricardo would have had to differentiate the social characteristics of labour as a commodity (i.e., the labour of the wage worker, or labour power) from the social characteristics of the labour that creates the commodity (i.e., the labour of the commodity producer). Yet we have already noted Ricardo's disregard for the social characteristics of labour and capital (see Chapter Twenty-Eight, Section 2).

Ignoring the qualitative or social side of wages, Ricardo focuses his entire attention on their *quantitative* dimension. Ricardo's writings on the magnitude of wages possess both significant merits and enormous deficiencies.Their greatest merit is that Ricardo persistently strives to define wages as a magnitude that is *precisely fixed*. Ricardo rejects the superficial explanation of the level of wages in terms of the relationship between the supply of, and demand for labour—an explanation that we have already encountered in Smith and which was developed in the 1830's by the proponents of the 'wages fund' theory (see Chapter Twenty-Three and the chapter below on the wages fund). In Ricardo's view demand and supply influence only the '*market price* of labour' i.e., 'the price which is really paid for it, from the natural operation of the proportion of the supply to the demand'. 'However much the market price of labour may deviate from its natural price, it has, like commodities, a tendency to conform to it.'[1] As with commodities, the market price of labour fluctuates around a stably determined centre, which forms its 'natural price' (or value).

By what is labour's '*natural price*' determined? 'The natural price of labour', says Ricardo, 'is that price which is necessary to enable the labourers, one with another, to subsist and to perpetuate their race, without either increase or diminution.' 'The natural price of labour, therefore, depends on the price of the food, necessaries, and conveniences required for the support of the labourer and his family. With a rise in the price of food and necessaries, the natural price of labour will rise; with the fall in their price, the natural price of labour will fall.'[2] The natural price of labour (or the value of labour power, in Marx's terminology) is determined by the value of *the necessary means of subsistence* of the worker and his family. Lasalle was later to give this theory of '*minimum means of subsistence*' the name '*the iron law of wages*', which he used as an agitational device to demonstrate to the workers the impossibility of achieving any fundamental improvement in their situation within the capitalist system.

Even though we can find embryonic versions of the 'iron law' among economists of the 17th and 18th centuries, it was Ricardo who gave it its classical formulation. Among the mercantilists (see Chapter Three) the iron law bore the character of a practical prescription: wages *had to* be limited to the necessary minimum of means of subsistence in order to cut the costs of production and expand the export of domestic commodities. The Physiocrats (see Chapter Thirteen), among whom Turgot is often deemed to be the author of the iron law, made no clear distinction between the wages of the worker, on the one hand, and the subsistence of the craftsman, or even the profit of the entrepreneur, on the other: according to Physiocratic doctrine all these forms of revenue were restricted to the necessary means of subsistence. Ricardo's merit is to have: 1) formulated the iron law as applying specifically to the wages of *wage labourers*, 2) endeavoured to uncover—albeit unsuccessfully, as we shall see—the *mechanism* which explains how this law works, and 3) tied the theory of wages to the theory of *profit*. For all its failings, Ricardo's theory of wages has enormous advantages over the theory of supply and demand, as formulated by Smith (where it intermingles with the theory of means of subsistence), Malthus, and proponents of the 'wages fund'.

As we know, we can find among economists two variants of the *means of subsistence* theory: one is the theory of a '*physiological minimum*', the other a theory of a '*cultural minimum*'. Proponents of the former say that workers' wages are confined to the sum total of means of subsistence physiologically needed to sustain the worker and his family. Partisans of the second theory justifiably extend the concept of a minimum of means of subsistence to include all those means needed to maintain the worker at his customary standard of living in conformity with the social and cultural conditions of a given population during a particular historical period. At first glance Ricardo seems to be closer to the broader and more flexible formulation of a cultural minimum. He grasps that the 'natural price' of labour 'varies at different times in the same country, and very materially differs in different countries. It essentially depends on the habits and customs of the people'.[3] Further on, however, he usually forgets these qualifications and comes close, when substantiating the iron law, to a physiological minimum theory.

How does Ricardo *substantiate* his iron law? In other words, how does he account for the fact that wages will gravitate towards a level which corresponds to the value of the worker's necessary means of subsistence? In Ricardo's view the mechanism which keeps the market price of labour from straying very far or for very long from its natural price is *changes in the population level.* When wages *exceed* the natural price of labour 'the condition of the labourer is flourishing and happy' and he is able 'to rear a healthy and numerous family. When, however, by the encouragement which high wages give to the increase of population, the number of labourers is increased, wages again fall to their natural price.'[4] They cannot fall below that level for very long, for if they did the workers would be deprived of their essential means of subsistence, 'privations [would] reduce their number', and wages would again go up. The workers' rapid multiplication prevents wages from rising for any length of time above the natural price of labour; when they multiply slowly or die off this keeps wages from falling for too long below it. If, because of ensuing deprivations, a drop in wages below the natural price of labour causes the number of workers to be reduced, it is obvious that the 'natural price' of labour includes only that aggregate of means of subsistence as is unconditionally needed to keep the worker and his family alive. Here Ricardo's teaching comes close to the physiological minimum theory.

Ricardo thus substantiates his iron law of wages by having recourse to the unvarying, biological *law of human reproduction* formulated by Malthus. Once the movement of wages is regulated by 'natural' changes in the population, any and all attempts to raise wages by artificial means, e.g., through strikes or factory legislation, become doomed to failure. Ricardo did not understand that the workers, by intensifying their economic struggle—itself a reflection of their rising social needs—can bring about a rise in wages. Nor did he grasp the significance of factory legislation (which in his day was still non-existant). In accord with other ideologists of the bourgeoisie, he proclaimed that 'wages should be left to the fair and free competition of the market, and should never be controlled by the interference of the legislature.'[5] The only possibility of a more or less long-term improvement in the workers' condition that Ricardo admitted would be if the law of population was unable to assert its influence. This could happen either because the workers, in seeking to preserve the high level of subsistence that they had obtained, consciously *abstained* from reproducing or because of new colonies with an abundance of fertile land, where the rate of growth of capital *outstrips* the rate of increase in population. On the first point Ricardo was conceding to Malthus, on the second to Smith. Nevertheless, Ricardo nurtured no great faith in the workers' conscious abstention, and considered a rapid growth of capital to be but a temporary phenomenon. Thus, these exceptions notwithstanding, Ricardo continued to hold to his iron law and to take a pessimistic view towards the prospect of a protracted rise in real wages.

Because his theory of wages suffers, as we have already noted, from its approximation to the theory of a *physiological* minimum, it acquires traits of unreality and ahistoricism. These features of the iron law are intensified still further by the *false grounds* on which Ricardo justified it. Especially false is the idea that one can look to the speed or slowness with which the workers *reproduce themselves* as a cause of upward or downward movements in wages. The appearance or disappearance of a surplus working population depends, in capitalist economy, not on the absolute increase or decline in the number of workers, but on the periodic expansion and contraction of capitalist production. The reserve army of unemployed is a necessary appurtenance of capitalist economy, which in no way stems from the fact that the workers are reproducing themselves with exceptional rapidity. In periods of expansion capitalist industry recruits new hands from this reserve army: to do so it does not have to wait the twenty years it would take, on Ricardo's assumption, for a rise in wages to encourage the workers to multiply and bring forth genuinely 'new' labourers into the world. If we are to look for that mechanism which forces wages to gravitate towards the level of customary means of subsistence it should not be in the workings of a Malthusian '*absolute* law of population,' but in a '*relative* law.'

Ricardo's doctrine on the 'static' level of wages, then, despite the healthy kernel that it conceals, was marred by the biological or 'natural' basis that he gave to it. His interesting doctrine on the '*dynamics*' of wage movements suffers from exactly the same defect. Here Ricardo seeks the ultimate cause of phenomena in the workings of natural laws: the 'physico-chemical' *law of diminishing fertility of the soil*, and the 'biological' *law of population*. We saw above, in our chapter on rent, that Ricardo, basing himself on a mistaken belief in the permanence of the former law, considered it inevitable that the prices of corn and other agricultural produce would progressively rise. Since the worker requires a determinate quantity of food stuffs to sustain life, any rise in their price will invariably boost the 'natural price' of labour, or money wages (even though real wages will remain unaltered or even fall, as we shall see below). 'The same cause which raises rent, namely, the increasing difficulty of providing an additional quantity of food with the same proportional quantity of labour, will also raise wages.'[6] 'But there is this essential difference between the rise of rent and the rise of wages.'[7] The landlord's rent will increase both in terms of corn (because of the extension of cultivation to inferior lands and the growing disparity between the fertility of superior versus poor plots) and even more so in terms of money (as a consequence of the rise in both value and price of each bushel of corn). 'The fate of the labourer will be less happy; he will receive more *money* wages, it is true, but his corn wages will be reduced.'[8] To understand why it is that, according to Ricardo, corn or *real* wages will decline it is necessary to look at the tendencies behind *movements in profits*.

We have already encounterd Ricardo's theory that profits always move inversely to changes in wages. 'Profits would be high or low in proportion as wages were low or high',[9] says Ricardo, confusing here—as everywhere—the rate of profit with the rate of surplus value (for the rate of profit can in fact fall even with a fall in wages, providing that the total amount of advanced capital rises at the same time). From here it follows that if money 'wages should rise with the rise of corn ... profits would necessarily fall',[10] since with the labour value of commodities remaining unchanged manufacturers will sell them at their former price, despite wages having gone up. '*The natural tendency* of profits then is *to fall*; for in the progress of society and wealth, the additional quantity of food required is obtained by the sacrifice of more and more labour.'[11] Although this tendency will from time to time be arrested owing to advances in agricultural technique and the free import of cheap foreign corn, it casts its gloomy shadow over the entire future of the capitalist economy: it threatens to bring economic progress to a total halt and to reduce society to a state where 'the very low rate of profits will have arrested all accumulation, and almost the whole produce of the country, after paying the labourers, will be the property of the owners of land.'[12]

Even though capitalist society had not yet reached this position, the pace of its economic progress was progressively decelerating with the fall in profit. 'The farmer and manufacturer can no more live without profit than the labourer without wages. Their motive for accumulation will diminish with every diminution of profit.'[13] Thus the natural law of diminishing fertility of the soil results in *a slow down in the rate of capital accumulation*. By virtue of our natural law, however, i.e., the biological law of population, the workers will continue to increase their numbers at the same rate as before. If the number of workers rises at 2% per year while the rate of capital accumulation drops from 2% to 1%, the *demand* for labour power will obviously lag behind its *supply*, in other words, real wages will *fall*. Admittedly, 'instead ... of the money wages of labour falling, they would rise; but they would not rise sufficiently to enable the labourer to purchase as many comforts and necessaries as he did before the rise in the price of those commodities.'[14] 'The condition of the labourer will generally decline, and that of the landlord will always be improved.'[15] These, then, were the pessimistic conclusions that Ricardo's theoretical arguments led him to and which seemed completely confirmed by the desperate state of the workers at the start of the 19th century. Because of these dismal conclusions economists of the historical-ethical school upbraided Ricardo for being indifferent to the fate of the working class. The rebuke was highly unjust: Ricardo, with supreme scientific conscientiousness and theoretical intrepidity, was merely revealing what appeared to him as the tendencies inevitably inherent in capitalist economy.

Now, a hundred years after the appearance of Ricardo's work, it is easy to prove that he was wrong in his assessment of these tendencies. The decreasing fertility of the soil, the rising price of corn, the growth of money wages, a fall in profit, the decelerating tempo of capital accumulation, a fall off in the demand for labour, and a decline in real wages—such was the chain of cause and effect that Ricardo had depicted. Many of the links in this logical chain proved weak. The rise in *labour productivity* and the enormous advances made in *technology* and *agronomy* showed his idea of an inevitable and progressive rise in the value of corn to be wrong. Not only money wages, but *real* wages, too, *rose* as a result of rising social needs and the greater social might of the working class, both factors that had been of little import in Ricardo's day. The growth in the productivity of labour outstripped the rise in real wages, and as a result *relative surplus value* (which Ricardo called profit) *increased*, rather than fell. In spite of this *the rate of profit fell* because of the rising organic composition of capital—i.e., precisely because labour productivity rose instead of dropping. In its details Ricardo's effort to explain how the revenues of the different social classes moved proved to be incorrect. Yet this in no way obviates the immense value of the Ricardian theory of distribution, which marked an entire epoch in the history of our science.

Ricardo was the *first* to have posed the *problem of distribution in all its breadth* and to have made it the focal point of his investigation. 'To determine the laws which regulate this distribution, is the principal problem in Political Economy', he writes in the Preface to his *Principles*. In a letter to Malthus, Ricardo counterposes his own conception of political economy as the science concerned with the laws of distribution of products between classes, to the conception of Malthus, which sees political economy as the science of the nature and causes of wealth. While Smith's chapters on distribution remain a collection of disparate facts and observations, Ricardo presents a complete and theoretically reasoned picture of the interdependencies and movements of incomes, which he has constructed upon a single principle. This principle is the principle of labour value. In Smith the theory of value and the theory of distribution remain logically separated: he constantly fluctuates between two viewpoints, sometimes making value his starting point, at other times revenue. Though Ricardo did in one letter express the view that a resolution of the grand problems of political economy—rent, wages, and profit—were not necessarily tied to the theory of value, he in fact based his entire investigation on the *principle of labour value*, upon which he then built his *theory of distribution*.

Ricardo's second great merit is to have given primacy to the problem of the *relative* shares of the different social classes in the *value* of the product, rather than to the distribution between them of *absolute* shares in the *in natura product* (the predominant vantage point found in Smith and in part carried over by Ricardo). Assume, says Ricardo, that the worker receives one and a half times as much food, clothing, and the like as previously. If at the same time the productivity of labour were to double (thus causing the value of products to be halved) we would say that the share (or 'real value') of wages has fallen.

Even though the worker now obtains a greater number of *products in natura*, his *relative* share in the value of the social product has declined. Ricardo was the *first* to have introduced this method of posing the problem into science, and it was subsequently developed by Rodbertus and by Marx, the latter in his so-called 'theory of impoverishment'.

By posing the problem of relative distribution, Ricardo was able to clearly discern *the contradictions of class interests* in capitalist society. In complete accord with the characteristic features of his epoch and with his own social and class position, Ricardo laid special and persistent stress on the conflict between the interests of the landowners and the interests of the remaining classes in society: the fall in agricultural productivity and the rising price of corn lower the rate of profit and hold back the accumulation of capital, cause the position of the workers to deteriorate, and at the same time make the landlords exorbitantly rich. However, along with this basic contradiction, which dominated both the reality of early-19th-century England and his theoretical conceptions, we can find in Ricardo's writings the outlines of the great historical struggle that was beginning to take place between the bourgeoisie and the proletariat. In Smith's schema a rise in wages does not the slightest harm to the capitalists' interests, since it causes the price of the product to go up and is therefore paid for by the consumer. In Ricardo's scheme a rise in wages is not accompanied by a general rise in the product's price, but inevitably brings about a fall in profit: we see reflected in this law the irreconcilable contradiction of class interests between bourgeoisie and proletariat. Admittedly the workers can receive a greater quantity of food, clothing, etc., and thereby improve their lot at the same time as the capitalists grow rich. The apologists for capitalism, Carey and Bastiat, pointed to just this possibility of better conditions for the workers in their polemic against Ricardo's doctrine (see the Chapter on Carey and Bastiat, below). What they ignored, however, was Ricardo's doctrine of *relative* distribution: the working class cannot possibly raise its relative share in the value of the social product unless there is a drop in the relative share going to the capitalists. With Ricardo the Classical school abandoned Smith's naive views on the harmony of interests of different classes and openly acknowledged the existence within capitalist economy of deep class conflicts. But when, in the middle of the 19th century, these class contradictions acquired such force that they began to threaten capitalism's very existence, bourgeois economic science broke with Ricardo's theory. There then began the period of *disintegration of the Classical school.*

1 *Principles*, p. 94.
2 *Ibid*, p. 93.
3 *Ibid*, pp. 96-97.
4 *Ibid*, p. 94.
5 *Ibid*, p. 105.
6 *Ibid*, p. 102.
7 *Ibid*, p. 102.
8 *Ibid*, p. 102 (Rubin's italics).
9 *Ibid*, p. 110.
10 *Ibid*, p. 111.
11 *Ibid*, p. 120 (Rubin's italics).
12 *Ibid*, pp. 120-21.
13 *Ibid*, p. 122.
14 *Ibid*, pp. 101-02.
15 *Ibid*, p. 103.

Part Five

Disintegration of the Classical School

CHAPTER THIRTY-ONE

Malthus and the Law of Population

Thomas Malthus (1766-1834) was born into the gentry, but as the youngest son he did not inherit the family estate and instead entered the ranks of the clergy. A vicar and teacher of political economy, he earned immense fame with his *Essay on the Principle of Population*, the first edition of which appeared in 1798.

Although Malthus was a disciple of Smith and on a number of basic issues sided with the Classical school, he nevertheless holds a special place as the consistent defender of the interests of the landed aristocracy, in opposition to the Classics (Smith, Ricardo, and their followers), who expressed the interests of the industrial bourgeoisie. We find the Classics and Malthus in disagreement first of all on a number of theoretical questions. The Classics advocated a rapid development of the productive forces and a reduction in non-productive consumption; Malthus considers non-productive consumption, and hence also the existence of the landed aristocracy and their household servants, to be essential. For the Classics (Ricardo, James Mill, Say) the possibility of a generalized overproduction of commodities is inconceivable (on this see the chapter on Sismondi, below); Malthus argues that generalized crises are possible. On practical matters, the issue which provoked the greatest controversy between Malthus and the Classics was that of the import duties on corn: the Classics steadfastly demanded their repeal, Malthus defended them as essential. On the other hand, Malthus, along with many other representatives of the English aristocracy of that era, took a more sympathetic attitude towards the rudiments of factory legislation, opposed by the liberal economists of the Classical school, who protested against the state power meddling in relations between capitalists and workers.

Yet if England's landed aristocracy and industrial bourgeoisie waged a bitter struggle against one another during the first half of the 19th century, there was still a wide range of issues over which the two possessing classes shared a community of interest. Thus both the Classics (Ricardo) and Malthus fought with equal zeal for the repeal of the ancient poor laws, which made the upkeep of the local paupers a parish obligation. On theoretical questions, as well, points of harmony existed between Malthus and the Classics. In his debates against the labour theory of value Malthus drew upon ideas put forward by Smith himself—ideas which formed the weakest part of Smith's theory. Malthus chose this weak side of the Classical school as the theoretical support for his reaction against them.

In general, Malthus's theory of population was accepted by the partisans of Classical theory, who used it to explain a number of phenomena, for example wages (see the chapter on Ricardo's theory of wages, above), even though the theory had no essential connection with their main teachings.

Malthus's first work *on population* was a reaction against the bourgeois enlightenment and the socio-political radicalism which closed the 18th century. In 1793 there appeared a book by the Englishman, Godwin, who was a determined partisan of social and political reform and an opponent of private property, which he regarded as the primary cause of the poverty and calamitous state of the lower classes of the population.[1] Godwin had hoped that a reform of social institutions would open up to humanity the possibility of an unlimited improvement and betterment of their lives—an idea being developed at the same time in France by Condorcet. Malthus's *Essay on Population* was a reply to Godwin.[2] Malthus had to show that the true cause of poverty lay not in the inadequacies of the social system, but in the natural, inexorable contradiction between man's unbounded yearning to *multiply* and the limits to the increase in *means of subsistence*. Malthus summed up his ideas in the following three propositions:

'1. Population is necessarily limited by the means of subsistence.

'2. Population invariably increases where the means of subsistence increase, unless prevented by some very powerful and obvious checks.

'3. These checks, and the checks which repress the superior power of the population, and keep its effects on a level with the means of subsistence, are all resolvable into moral restraint, vice and misery.'[3]

Malthus reasons as follows. Suppose that a particular country at a given point in time has a population of one. The amount of means of subsistence in the country are sufficient to feed its existing population, and so also equal one. As shown by the experience of the United States, the population will double approximately every 25 years, i.e., it will grow in a *geometric* progression. Two hundred years hence our country's population will have grown to 256 times its initial figure (1, 2, 4, 8, 16, 32, 64, 128, 256). However, this growing population would have to extract its means of subsistence from the country's same 'limited territory'. Anyone familiar with agriculture knows that each new application of labour to the same plot of land is accompanied by a fall in its productivity (Malthus, in the later editions of his book, therefore couples his law of population with the law of diminishing fertility of the soil). In the best of circumstances, the growth of the means of subsistence over each 25-year period will equal their growth in the preceding twenty-five years. This means that a country's means of subsistence will grow in an *arithmetic* progression, and will, after 200 years, be nine times greater than their original quantity. It is obvious that this amount of means of subsistence will not be able to sustain a population that will have grown 256 times in this period.

There would be a sharp divergence between the number of people and the quantity of means of subsistence unless the rise in population was held back by both *destructive* and *preventive* checks, the former referring to various *calamities* (above all *destitution*), the latter to *abstemiousness* and *vice*, which act to slow down the rate at which people multiply. In the first edition of his book Malthus cited calamities and vice, which he considered would make their inevitable appearance under any social system, even the most perfect, as the only checks on population growth. In later editions Malthus acknowledged that people could avoid the calamities of over-population by consciously abstaining from rapidly increasing their numbers. Only the threat of future poverty and an incapacity to feed a multitudinous progeny could incline people towards abstention. In a socialist society, where there will be no private property, society's members will lose all incentive to abstain. Consequently, social reformers desirous of abolishing private property would, along with it, be doing away with any preventive check on the growth of population and, in a very short while, society would confront the formidable peril of a shortage of means of subsistence for all its members.

Malthus drew from his theory a wide range of *practical* conclusions. The poor have no cause to complain about their poverty, for it is but a necessary consequence of their having bred with excessive rapidity; the poor can improve their lot only by refraining from early marriage. Taxes on behalf of the poor are harmful, since they encourage them to multiply, thus aggravating their poverty in the future. Nor can one lament high grain prices, since the latter merely confirm the fact that the number of people has outstripped the means of subsistence. Tariffs on imported corn and high corn prices are beneficial because they stimulate agriculture and thereby increase the amount of means of subsistence.

Malthus's theory of population engendered an enormous literature:

it was the object of bitter debate, was argued out from economic, religious, and other points of view, and laid the foundation for an entire school, called 'Malthusianism'. In the 1070's this school, under the title of *'neo-malthusianism'*, met with great success. If the old Malthusianism had been harsh in its hositility towards the working class and in deeming the poor as the sole culprits for overpopulation, 'neo-Malthusianism' toned down the reactionary character of its doctrine. In addition, it did not, as had the earlier Malthusianism, propound abstention, but rather the artificial reduction of childbirths through the use of preventive measures. From the 1880's onwards neo-Malthusianism fell into rapid decline, due first of all to the impact of a sharp fall in corn prices (the Malthusians had seen the high price of corn as proof that there was a shortage of means of subsistence compared to the size of the population) and, secondly, to the drop in the birth rate of the European countries that resulted from their rapid industrialization and the growth of their urban populations. By the end of the 19th century there was no longer any fear of absolute overpopulation; on the contrary, anxious voices pointed to the danger of a population decline.

Malthus's enormous popularity and the lively interest in his theory amongst the general public is due not to its theoretical significance, but to its attempt to tackle the topical problems of population, poverty, child-bearing, etc. Even though Malthus's ardent admirers have up to this very day been prepared to proclaim his a work of genius, its theoretical achievements have in fact been far inferior to the influence it has exerted. The substance of Malthus's theory can be reduced to the following propositions: 1) population has a tendency to multiply very rapidly (in geometric progression); 2) there is a far slower (in arithmetical progression) growth in the amount of *means of subsistence*; and 3) the present-day poverty of the broad mass of the population is simultaneously the result of a divergence between the quantity of means of subsistence and the size of the population, and a means (together with vice and abstemiousness) of doing away with this discrepancy.

Malthus's first proposition asserts the existence of a 'natural' law of population, operative at all times and for any social system. In fact the rate and character of population growth will vary depending on a whole range of economic and social conditions. There are a number of social conditions under which the population exhibits a tendency towards very slow multiplication (e.g., in contemporary France and to a lesser extent over Europe as a whole from the end of the 19th century) which lags far behind the increase in means of subsistence.*

Malthus's second thesis is even weaker. It is true that many scholars will say that one ought not take Malthus's words about an arithmetical progression literally. Even so, the onus is on Malthus to show that the means of subsistence cannot expand as fast as the population. Malthus did not prove this, since he ignores in his arguments nothing more nor less than the development of the *productivity of labour* and the *progress* of agricultural technology. With the existence of technological progress Malthus's assertion that a greater and greater number of people will be needed to extract the same quantity of means of subsistence (the so-called 'law of diminishing fertility of the soil', of which Malthus was one of the authors) is no longer valid. On the contrary, the 19th century witnessed a colossal rise in the agricultural output attributable to the individual worker. It is not surprising that Cannan, one of the most recent invesigators, should come to such a conclusion: 'Deprived of the theory that the periodical additions to the average annual produce cannot possibly be increased, or, as Malthus preferred to put it, that subsistence can increase only in arithmetical ratio, the *Essay on the Principle of Population* falls to the ground as an argument, and remains only a chaos of facts collected to illustrate the effect of laws which do not exist. Beyond the arithmetical ratio theory, there is nothing whatever in the *Essay* to show why subsistence for man should not increase as fast as an "unchecked" population.'[4]

Properly speaking it was the *third*, and not the first two of the aforementioned propositions that Malthus saw as the nodal point of his work. The first two had, by the way, been stated repeatedly prior to Malthus, and not even the formula of 'arithmetical' and 'geometric' progressions is his own (one finds it, for instance, in the Italian economist Ortes).[5] The primary aim that Malthus had set himself had not been to prove the law of population in itself, but to investigate its social consequences (above all, the origin and causes of poverty and unemployment). In one passage he expresses himself as follows: 'It has been said, that I have written a quarto volume to prove, that population increases in a geometrical, and food in an arithmetical ratio; but this is not quite true. The first of these propositions I considered as proved the moment the American increase was related, and the second proposition as soon as it was enunciated. The chief object of my work was to inquire what effects these laws,

*This is not even to mention the fact that Malthus had based his assertion that population would double every twenty-five years on unverified factual material.

which I considered as established in the first six pages [of the first chapter of the *Essay on Population*—*I.R.*], had produced and were likely to produce on society.'[6] Thus the most important part of Malthus's theory is his doctrine that poverty results from the absolute overpopulation that ensues from the operation of the first two propositions.

Malthus himself does not in the least conceal that the main task he has set himself is to explain and *justify* the poverty of the working masses in capitalist society. 'That the principal and most permanent cause of poverty has little or no *direct* relation to forms of government, or the unequal division of property; and that, as the rich do not in reality possess the *power* of finding employment and maintenance for the poor, the poor cannot, in the nature of things, possess the *right* to demand them; are important truths flowing from the principle of population, which, when properly explained, would by no means be above the most ordinary comprehensions. And it is evident that every man in the lower classes of society, who became acquainted with these truths, would be disposed to bear the distresses in which he might be involved with more patience; would feel less discontent and irritation at the government and the higher classes of society, on account of his poverty; would be on all occasions less disposed to insubordination and turbulence; and if he received assistance, either from any public institution or from the hand of private charity, he would receive it with more thankfulness, and more justly appreciate its value.'[7]

It would be difficult to find words that more vividly reveal Malthus's *reactionary tendencies* and his desire to prove at all costs the necessity of poverty and unemployment. Yet from a theoretical point of view it is this task of justifying all of capitalism's calamities that Malthus carried out least satisfactorily. Even those economists who are inclined towards full or partial agreement with Malthus's first two propositions grasp the flagrant falsehood of the third. Modern poverty and unemployment result not from any absolute shortage of means of subsistence, but on the contrary, from the colossal growth of the productive forces and machine technology under capitalist conditions. They must be recognized as a product of social, rather than natural conditions, of the 'relative overpopulation' inherent in capitalist economy, and not of any 'absolute overpopulation' flowing from man's own nature. Malthus's attempt to lay responsibility for modern-day poverty on biological and purely technical factors met with total failure.

While Ricardo shared Malthus's theory of population, on the *theory of value* they were determined opponents. It was Malthus's view that the value of commodities is regulated by *supply and demand* and is determined under normal conditions by *costs of production*, i.e., wages plus profit (and rent). Resolutely rejecting the labour theory of value as formulated by Ricardo, Malthus retained the very weakest side of the Smithian theory of value. Like Smith, he pronounced the best *measure* of a commodity's value to be the quantity of labour that it will *purchase* when exchanged. Yet there is a great difference between Smith and Malthus. Smith, who combined together a 'craft' and a 'capitalist' standpoint (see Chapter Twenty-Two), spoke at times of 'purchasable embodied labour' (i.e., products), and at other times of 'purchasable living labour' (i.e., labour power). Malthus consistently adopted the vantage point of the capitalist, and at all times assumed that a commodity is used by its owner to purchase *living* labour, or *labour power*. A capitalist has £100, or commodities to this amount. How do we measure the value of this sum of money or commodities? By the quantity of *living labour* which the capitalist can acquire with them. Suppose that with this sum the capitalist hired ten workers for one week. This means that the value of his money (or commodities) is measured by ten weeks' labour. Suppose now that the product manufactured by these ten in the course of a week is sold by the capitalist at a profit of £20, i.e., for £120. With this sum of money the capitalist can now hire more workers than he could previously, i.e., twelve workers. This means that the value of the product that has been produced is measured by the twelve week's labour which it can purchase when exchanged. The capitalist expends on the product's production a definite quantity of labour (that of his workers) and can purchase a greater quantity of labour when the final product is exchanged. Where does this excess, which forms the profit, come from? To this question Malthus gives no answer. He obviously thinks that profit is a mark up which the capitalist adds onto the value of the commodity to be paid for by the consumer. In this fashion Malthus is taking a step backwards, to the outmoded mercantilist conceptions of 'profit upon alienation'.

In any case, who are these *consumers* who pay more for commodities than their value? The workers are able to purchase only a portion of the commodities produced by their labour since the price of these commodities (£120) is greater than the total wages they have received (£100). Similarly the capitalists, who aspire if possible to bring down their own personal consumption with the aim of accumulating capital, cannot consume the whole of the surplus product. The total product that has been produced cannot be realized without the aid of purchasers—*'third persons'*—who can be none other than landlords, state officials, etc. Thus Malthus arrives at his *theory of markets* and his doctrine of *the usefulness of the non-productive consumer.*

While the aim set for the economy by the Classical school is one of unlimited growth of production, the driving force of which is the class of industrial capitalists, Malthus defends the necessary existence of *unproductive classes* (the landed aristocracy, the bureaucracy, the clergy, etc.). These unproductive classes merely purchase products, but do not sell, they consume but do not produce. In so doing they bring into harmony *production and consumption*, supply and demand, and save the economy from permanent crises of over-production. The Classics considered that general production could not possibly proceed on a basis where each producer is also the consumer of the sum total of his own output. On this question Malthus stands, formally speaking, on firmer terrain by establishing that unlimited capital accumulation and growth of production can lead to *crises*. In the debate over markets and crises Ricardo and Say were arrayed on the one side and Malthus and Sismondi on the other (see Chapter Thirty-Seven on Sismondi, below). Yet the essence of Malthus's argument here is extremely weak and turns on the idea that capitalists and workers are incapable of consuming the entire product, a portion of which has to be sold to unproductive classes. On this Malthus says the following:

'With regard to the capitalists ... they have certainly the power of consuming their profits, or the revenue which they make by the employment of their capitals; and if they were to consume it ... there might be little occasion for unproductive consumers. But such consumption is not consistent with the actual habits of the generality of capitalists. The great object of their lives is to save a fortune, both because it is their duty to make a provision for their families, and because they cannot spend an income with so much comfort to themselves, while they are obliged perhaps to attend a counting-house for seven or eight hours a day ...

'There must therefore be a considerable class of persons who have both the will and power to consume more material wealth than they produce, or the mercantile classes could not continue profitably to produce so much more than they consume. In this class the landlords no doubt stand pre-eminent ...

'... And with regard to [the] workmen, it must be allowed that, if they possessed the will, they have not the power ... But as a great increase of consumption among the working classes must greatly increase the cost of production, it must lower profits, and diminish or destroy the motive to accumulate, before agriculture, manufactures, and commerce have reached any considerable degree of prosperity. If each labourer were actually to consume double the quantity of corn which he does at present, such a demand, instead of giving a stimulus to wealth, would unquestionably throw a great quantity of land out of cultivation, and greatly diminish both internal and external commerce.'[8]

Malthus did not grasp that the capitalists, who indeed cannot put their entire surplus value towards personal consumption, nevertheless prefer not to sell it to landlords, but to accumulate it in the form of new machines, factories, etc., so as to *expand production*. The partisans of the industrial bourgeoisie considered Malthus's 'salutary methods' highly unsuitable. One Ricardian objected to Malthus in the following terms:

'We are continually puzzled, in his speculations, between the object of increasing production and that of checking it. When a man is in want of a *demand*, does Mr. Malthus recommend him to pay some other person to take off his goods?'[9]

The clash between Malthus and his opponents graphically reflected the struggle between the *landed aristocracy and the commerical-industrial bourgeoisie*—a struggle which occupied the whole of England's history during the first half of the 19th century.

1 William Godwin, *An Enquiry concerning Political Justice, and its Influence on General Virtue and Happiness*.

2 'The great error under which Mr. Godwin labours throughout his whole work is, the attributing of almost all the vices and misery that prevail in civil society to human institutions. Political regulations and the established administration of property are, with him, the fruitful sources of all evil, the hotbeds of all the crimes that degrade mankind. Were this really a true state of the case, it would not seem an absolutely hopeless task, to remove evil completely from the world: and reason seems to be the proper and adequate instrument for effecting so great a purpose. But the truth is, that though human institutions appear to be, and indeed often are, the obvious and obstrusive causes of much mischief to society, they are, in reality, light and superficial, in comparison with those deeper-seated causes of evil, which result from the laws of nature and the passions of mankind.' Malthus, *An Essay on the Principle of Population*, reprint of the third edition, (London, Ward, Lock & Co., 1890), pp. 307-08.

3 *Ibid*, p. 14.

4 Edwin Cannan, *A History of the Theories of Production and Distribution in English Political Economy, From 1776-1848* (London, P.S. King & Son, 1924), p. 144.

5 Marx, in Vol. I of *Capital* (Penguin edition, p. 800) refers to Ortes as 'one of the

great economic writers of the eighteenth century [who] regards the antagonism of capitalist production as a universal natural law of social wealth.' He quotes from Ortes's *Della economia nazionale* (1777): 'In the economy of a nation, advantages and evils always balance each other: the abundance of wealth with some people is always equal to the lack of wealth with others ... The great riches of a small number are always accompanied by the absolute deprivation of the essential necessities of life for many others. The wealth of a nation corresponds with its population, and its misery corresponds with its wealth. Diligence in some compels idleness in others. The poor and idle are a necessary consequence of the rich and active ...'.

6 Malthus, *Essay on the Principle of Population*, p. 552, fn.

7 *Ibid*, pp. 541-42; Malthus's italics.

8 Malthus, *The Principles of Political Economy*, facsimile of the 1836 edition, reprinted by the International Economic Circle, Tokyo, in collaboration with the London School of Economics (Tokyo, Kyo Bun Kwan, 1936), pp. 399-405.

9 Cited by Marx in *Theories of Surplus Value*, Part III, p. 60, and attributed to the anonymous author of *An Inquiry into those Principles, respecting the Nature of Demand and the Necessity of Consumption, lately advocated by Mr. Malthus*, London, 1821, p. 55; original italics.

CHAPTER THIRTY-TWO

The Beginning of Vulgar Economy

SAY

The Classical school studied the social forms of things (value, wages, profit, rent) without clearly realizing that these are nothing else but an expression of the social relations of production between people. Hence the duality in the Classical school's conclusions. In as much as they were studying the *social forms* of things, as distinct from the things themselves (e.g., the value of the product as distinct from the product itself as a use value), they regarded them as the result of human *labour* (even though without any clear awareness of the social form in which labour was organized), and in this manner also of human *society*. From this 'labour' point of view the Classical economists reduced wages, profit, and rent to value, and value to labour. They found in labour the deeply-hidden basis of all economic phenomena and, with their labour theory of value, they laid the foundations of political economy as a social science. On the other hand, in as much as the Classical economists studied the social forms of *things*, they were inclined to seek their origin in the natural or material-technical properties of the things themselves. To them it seemed perfectly natural that means of production (machinery, etc.) should have the social form of capital. It seemed no less natural that capital ought to bear a profit. From here is was easy to reach the conclusion that capital in its material-technical form (machinery, etc.) creates the profit that accrues to its owner. Such views were in total harmony with the commonplace, 'vulgar' ideas that reigned within entrepreneurial circles and amongst the general public, which confined itself to the superficial observation of economic phenomena.

The duality between the *'labour'* and *'vulgar'* points of view had left its imprint on Smith's system. Wherever Smith was attempting to employ a theoretical analysis to uncover the motive causes of economic phenomena he identified labour as the source of value and looked upon value as the primary magnitude which is then resolved into wages, profit, and rent. Wherever, on the other hand, Smith confined himself to describing economic phenomena as they presented themselves to superficial observation, he saw value as the result of adding together wages and profit (and rent), the sizes of which were in turn determined by the law of supply and demand. The first point of view led Smith to the labour theory of value, the latter to a vulgar theory of production costs (which rested ultimately on the theory of supply and demand).

Ricardo developed the most valuable side of Smith's doctrine, its 'labour' point of view: he consistently adhered to the labour theory of value and made it the foundation for his theory of distribution. Nevertheless even in the improved version formulated by Ricardo, the labour theory of value found itself in contradiction with the basic fact of capitalist economy, namely that commodities are sold at prices equal to production costs plus average profit. Ricardo's opponents availed themselves of this contradiction: they proposed to completely discard the labour theory of value and to confine themselves to the vulgar theory of costs of production, which merely generalizes the everyday outlook of the capitalist entrepreneur. The entrepreneur reckons that the price of his commodity must at the very least compensate him for all of his outlays on production (the hire of workers and constant capital) plus earn an average profit. Generalizing these views, the 'vulgar' economist says: the value of a commodity is determined by its costs of production plus the customary profit on capital. Where this profit (i.e., the surplus quantum over and above production costs) comes from, why it establishes itself at such-and-such a level, and what determines the size of the costs of production themselves (i.e., the value of the raw materials, machinery, and labour power) are basic questions that earn the attention neither of the entrepreneur, whom they do not in fact concern, nor of the vulgar economist, whose analysis never goes beyond the surface of events.

Just as the Classical school was taking its first steps a *'vulgar'* current, parallel to the main tendency which Ricardo represented, already began to develop, taking as its support the weak side of Smith's theory. We already know that Malthus replaced the labour theory of value with the theory of supply and demand (and the theory of production costs) and endeavoured to develop Smith's mistaken idea about labour—the labour purchasable in exchange for a commodity—as the measure of value. But Malthus's ideas on value were too confused and contradictory to meet with general success. The honour of being the founder of 'vulgar economy' went, therefore, not to him, but to the Frenchman, Jean-Baptiste Say (1767-1832).

Like Malthus, Say had seen his writing appear in print before Ricardo. His *Traité d'économie politique*, which appeared in 1803, ran into several editions and enjoyed a tremendous success. A superficiality of ideas, clarity of construction, and an easy style made the book accessible to a wide circle of readers. Say did much to help diffuse Smith's ideas (granted, in a distorted version) across the continent of Europe and in his own lifetime was considered among the greatest economists of the age. Essentially, however, Say was an extremely superficial scholar, and his sole merit is to have given a systematic and popular exposition of Smith's ideas. It was Say, by the way, who introduced that classification of material now current in bourgeois political economy. He divided up the second edition of his *Traité* as follows: 1) the production of wealth, 2) the distribution of wealth, and 3) the consumption of wealth. In 1821 James Mill, obviously following Say's example, divided up his *Elements of Political Economy* into four sections: 1) production, 2) distribution, 3) exchange, and 4) consumption. This division, which artificially severs the connection between aspects of the ecomomic process that are inseparable, was to become widely practised within science.

Say certainly did not see himself as Smith's popularizer, but pretended to say something scientifically new—the doctrine of three factors of production and productive services, the theory of markets (on his theory of markets see Chapter Thirty-Seven on Sismondi). Not to mention its extreme superficiality, this 'last word' represented a step backward compared to Smith and made its author the father of vulgar economics. If Ricardo developed the positive aspects of Smith's doctrine, Say utilized its weak aspects to vulgarize Classical theory.

We can obtain a clearer idea of Say's doctrine by counterposing it to the basic theses of Ricardo:

	Ricardo		*Say*
1	*Value* is to be fundamentally distinguished from '*riches*' (use value).	1	*Value* is confused with *riches*.
2	Value is created by *labour*.	2	Value is created by *labour, nature, and capital*.
3	The value of a product *resolves itself* into wages and profit. The size of *value* determines the size of revenue.	3	The sum of wages, profit, and rent constitutes a product's value. The magnitude of *revenue* (or the sum of the costs of production) determines the magnitude of *value*.

As is clear from point 1, Say's theory of value diverges sharply from that of Smith and Ricardo. In the tradition of the French school (the Physiocrats, Condillac), he confuses *value* with *use* value. For Ricardo the utility of a product is a necessary condition of its exchange value. For Say this is not enough. According to his view, 'the *utility* of things is the gound-work of their value'—the magnitude of an article's *subjectively recognized utility* determines the magnitude of its objective exchange value. 'Price is the measure of the value of things, and their value the measure of their utility.' 'Exchangeable value, or price, is an index of the recognized utility of a thing.'[1]

Ricardo had argued against this subjective theory of value both in his *Principles* (Chapter XX) and in his correspondence with Say. Why, Ricardo asks Say, do we pay 2000 times more for a pound of gold than for a pound of iron, even though we recognize them as being of equal utility? Say could only answer that 1999/2000 of the iron's utility is given to us gratis by nature, and we only need pay for that share of its utility, 1/2000, which corresponds to the size of the outlays that we had to make to produce it. In so doing Say leaps from the theory of *subjective utility* to the theory of *production costs.*

We need not be surprised then, that having confused value with use value Say rejects the labour theory of value. Products, being use values, can be created by labour only if it is assisted by the forces of nature and means of production (which Say calls capital). 'These three sources are indispensable to the creation of products', which according to Say's theory means the 'creation of utility'.[2] And since utility for Say is indistinguishable from value it is clear that it is not labour alone that creates value, as Smith had taught, but all *three factors of production.*[3] 'Values produced are referable to the agency and concurrence of industry, of capital, and of natural agents; ... no other but these three sources can produce value, or add to human wealth.'[4] Say disagrees with Ricardo when the latter says that 'natural agents, though they add greatly to *value in use,* never add exchangeable value'.[5] No, replies Say, 'that production which is done by nature adds to men's revenues not only value in use, the only value that Smith and Ricardo attribute to it, but an exchangeable value'.[6] How this occurs Say does not show. In precisely identical fashion Say's sole argument for the idea that capital creates value is to adduce the fact that capital yields a special form of revenue, interest.

Thus there exist *three factors of production*: labour, capital, and nature (land). Within the process of production each renders a '*productive service*', for which (with the exception of those services which nature renders free of charge) its owner receives a remuneration or revenue (wages, interest, and rent). This reward is drawn from the product's value: each of these owners receives that share of the value that has been created by the factor of production which belongs to him. *Labour* creates *wages* (i.e., the share of the product's value that equals wages), *capital* creates *interest*, and *land* creates *rent*. When added together the sum of these three revenues determines the size of the value of the entire product. Contrary to the view held by Ricardo, a rise in wages need not necessarily call forth a fall in profit: all talk of a conflict of class interests or of the exploitation of workers by capitalists is out of the question.

Say's doctrine postulating *'three factors of production'* became widely accepted in bourgeois science, and even to this day any textbook will divide up its section on production into the traditional headings of nature, labour, and capital. Say's doctrine of *'productive services'* (or, what is the same thing, his theory of the *'productiveness of capital'*, i.e., that capital creates value) met with equal success. Firstly, in practical terms Say's theory promised to provide a justification for profit and rent as unearned incomes, and to demonstrate the unlawfulness of workers' claims for a share in the product exceeding their wages (since labour creates only the value of its wages and not the whole of the value of the product). True, the aims of the abject apologist were still to a certain extent foreign to Say, but his doctrine was later utilized for precisely these ends. Secondly, Say's 'trinity formula' (labour-wages, capital-interest, land-rent) in effect appeared to provide a very harmonious schema that bound together the phenomena of production, exchange, and distribution: in it labour, nature, and capital simultaneously assume the roles 1) of factors of material production, 2) of creators of value, and 3) of sources of revenue. But this harmony was purchased at the price of confusing value with utility on the one hand, and confusing the process of producing value with the process of producing products on the other. All economic phenomena became fetishized and deprived of any social content: the source of value, the product's social from, was declared to be the material-technical factors as such (machinery, natural agents). Economic theory was reduced to a bare description of the external, material form of economic phenomena. According to Say, capital earns interest; consequently, interest (as a quantum of value) is created by capital (which is made up of the totality of means of production). He thereby ruptures any connection between value and interest, on the one hand, and between human labour and the production relations between people, on the other. Value and interest are created directly by things (capital), while the value of the product is composed of its costs of production, or of revenues (wages, profit, and interest), thinks Say, generalizing the vulgar, every-day notions of the capitalist in the first case and of the manufacturer and merchant in the second. The erroneous idea, sometimes to be found in Smith, that value depends on revenue is taken by Say to its logical conclusion. Yet if the magnitude of a product's value is determined by the size of wages, profit, and rent, what is it that determines the magnitudes of the latter? Here Say can only appeal to the law of supply and demand.

Say knew his greatest success in France. Owing to that country's relative economic backwardness, the traditional failure of French, as opposed to English economists was their inability to work out a clear concept of value and their inclination to replace it with a concept of use value. Yet even in England, the place of its birth, the Classical school, albeit more slowly than in France, entered a period of decline and vulgarization.

1 Jean-Baptiste Say, *A Treatise on Political Economy; or the Production, Distribution, and Consumption of Wealth*, translated from the fourth French edition by C.R. Prinsep, in two volumes (London, 1821), pp. 4-5; Rubin's italics.

2 *Ibid*, p. 40.

3 These are what Say calls the 'productive services' of the three factors of production.

4 Say, *Treatise* (Prinsep translation), Vol. I, pp. 37-38. By 'industry' (*faculté industriëlle*) Say is referring to human labour power; 'labour' is the activity, or productive service rendered by the factor 'industry'.

5 Ricardo, *Principles of Political Economy and Taxation*, Sraffa edition, p. 285 (Ricardo's italics).

6 Say, *Traité d'économie politique*, sixth French edition (Paris, 1841), Vol. I, Book I, chapter 4, p. 72, fn. This note was added in later editions of the *Traité* and does not appear in Prinsep's translation.

CHAPTER THIRTY-THREE

The Debates Surrounding the Ricardian Theory of Value

The theory of *value* forms the touchstone of Ricardo's entire system. It is therefore easy to understand why it was around this that the debates raged with especial intensity during the decade 1820-1830 between Ricardo's opponents and supporters. Ricardo had himself created a breach in his labour theory of value for his opponents to push their way through. He had not known how to square the law of labour value with the law of equalization of the rate of profit (see above, Part 3, Chapter Twenty-Eight). Why should two products manufactured with identical expenditures of labour have unequal values if the capitals advanced for their production circulate for unequal periods of time (or, what comes down to the same thing, if capitals of different sizes are advanced over the same period of time)? This had been a troublesome theoretical question for Ricardo, who constantly, and extremely conscientiously returned to it in his correspondence with Malthus, McCulloch, and others. He frankly admitted that he had despaired of finding a satisfactory solution to this question unaided.

Ricardo had himself indicated that this was the most vulnerable part of his theory, and it was against it that Malthus, Torrens, and Bailey directed their blows. With a single voice they all ruled that the 'exceptions' which Ricardo had admitted deprived his law of labour value of all validity. In Malthus's words, these exceptions 'are both theoretically and practically so considerable as entirely to destroy the position that commodities exchange with each other according to the quantity of labour which has been employed upon them'.[1] This proposition [that products exchange at their labour values—*Ed.*] would, according to Malthus, scarcely apply to one case in five hundred, since the progress of civilization and technology will lead to both a growth in the volume of fixed capital and to differences in the turnover periods of capital—i.e., it will create conditions which violate the exchange of products according to their labour value.

Torrens and Bailey also argued that the law of labour value does not apply within a capitalist economy.

In objecting to Ricardo's theory, what did these critics propose to put in its place? They proposed simply to discard the labour theory of value. In Malthus's opinion the size of a product's value is determined by the relation of supply to demand, while he invited his readers to take the quantity of labour that a product will purchase when exchanged as a measure of its value. He thus proposed to retreat from Ricardo back to Smith's false thesis on the measure of value. Another of Smith's mistaken ideas was resurrected by Torrens (whose main work was his *Essay on the Production of Wealth* (1821)), who argued that while the law of labour value applies to pre-capitalist economy, the only law operative in capitalist economy is the law of production costs, according to which 'when capitals equal in amount ... are employed, the articles produced ... will be equal in exchangeable value'.[2]

Finally there is Bailey,* who advocated renouncing the concept of 'absolute' value and restricting study to 'relative' value, or the proportions in which commodities exchange.

The contradictions which had torn Ricardo's theory to pieces were to be resolved only many years later by Marx through his theory of *prices of production*. Marx showed that in capitalist economy, as distinct from a simple commodity economy, the law of labour value does not assert itself *directly*, but only *indirectly* through the medium of a complex social process of forming the average rate of profit and prices of production. So long as this complex social process went unexamined there would exist an irreconcilable contradiction between the law of labour value and the fact that commodities are sold at their prices of production (equal to production costs plus average profit)—a contradiction which Ricardo's followers and the true bearers of his tradition (James Mill** and McCulloch†) vainly tried to solve. Both of them remained powerless to save from ruin the 'Ricardian' school over which they presided. Mill gave a clear and systematic exposition of Ricardo's theory, but he was no creative thinker and was unable to take economic science forward. Blindly and dogmatically faithful to Ricardo's words, he was prepared to be satisfied with a purely verbal

*His main work was *A Critical Dissertation on the Nature, Measure, and Causes of Value* (1825). [This work, itself highly critical of Malthus, was attacked at length by the latter in his *Definitions—Ed.*]

**Mill's main economic work was his *Elements of Political Economy* (1821). He also wrote works on the history of philosophy.

†A prolific writer, his *Principles of Political Economy* (1825) is one of his most important works.

resolution of the contradictions into which his teacher had got entangled. Still less capable of rescuing Ricardian theory was the presumptuous and flippant McCulloch.

Both Mill and McCulloch endeavoured—to be sure, without any success—to show that despite the 'exceptions' admitted by Ricardo, the law of labour value asserts itself *directly* when commodities are exchanged in capitalist economy. The problem of the exchange of commodities produced by capitals with unequal organic compositions they resolved fairly easily: they simply assumed that the majority of commodities are produced by capitals of average organic composition, and hence are sold at their labour value. But then both Mill and McCulloch scatched their head a great deal over a second exception noted by Ricardo, which arises when capitals are circulated for different periods of time. How do we explain the greater value of a product produced by a capital advanced over a longer period compared to other products containing an identical quantity of labour? In other words, where is the origin of the greater total profit charged on a capital that remains in *circulation for a longer period of time?* This is an extremely difficult problem which touches simultaneously upon the theory of value and the theory of profit. In a letter to McCulloch Ricardo acknowledged that he had completely failed to overcome the difficulty presented by wine which is kept in a cellar for three or four years, or by the oak, which costs two shillings in labour to plant but which is later worth £100. Ricardo, as we know, could find no way out other than to declare these cases 'exceptions' to the law of labour value and to acknowledge that the value of wine or oak (as of any product produced by a capital advanced for a longer period) is determined not simply by the quantity of *labour* necessary for its production, but also by the length of *time* over which the capital is advanced.

This explanation satisfied neither Mill nor McCulloch. 'Time does nothing. How then can it create value?' [3] Mill asked. Labour alone, and not time, creates value—to Mill and McCulloch this was the rule. But how, in that case, was the greater value of old wine to be explained? Obviously there remained no other way out but to assume that the alteration to which the wine was subject during its stay in the cellar is tantamount to an additional expenditure of *human labour*. This was a risky assumption, expressed more cautiously by Mill but developed further by McCulloch.

'Suppose', says McCulloch, 'that a cask of new wine, which cost 50 pounds, is put into a cellar, and that at the end of twelve months it is worth 55 pounds, the question is, whether ought the 5 pounds of additional value given to the wine to be considered as a compensation for *the time* the 50 pounds worth of capital has been locked up, or ought it to be considered as the value of additional labour actually laid out on the wine'. McCulloch answers that the additional value is due to the latter. But how can one show that additional labour has been expended on the wine? Very simply: 'if we keep a commodity, as a cask of wine which has not arrived at maturity, and on which therefore *a change or effect is to be produced*, it will be possessed of additional value at the year's end; whereas, had we kept a cask of wine which had *already arrived at maturity*, and on which no beneficial or desirable effect could be produced for a hundred or a thousand years, it would not have been worth a single additional farthing. This seems to prove incontrovertibly that the additional value acquired by the wine during the period it has been kept in the cellar is ... a compensation for the effect or change that has been produced on it.'[4]

The absurdity of this 'incontrovertible' explanation is obvious. It identifies the action of natural agents, which give an object greater utility or use value, as a source of exchange value and equates it with human labour. What masks this total *renunciation of the labour theory of value* is the naive device of giving the appelation '*labour*' to the operation of *natural forces*. Since machines represent 'accumulated labour' they not only transfer their own value to the product, but in addition create new value. This means that the profit which is charged on fixed capital—and whose origin Ricardo was unable to account for—is created by the machine itself. Does this not contradict the law of labour value? No, answers McCulloch, because 'the profits of capital are only another name for the wages of [the] accumulated labour'[5] contained in the machine. But certainly the labour that in its own time created the machine has long since ceased to function and received its remuneration, and the value of the machine has been fully paid for by the manufacturer who bought it. How is it that in the hands of its new owner the machine not only transfers its value to the product but creates a new value or profit? It is obvious that Mill and McCulloch are acknowledging the ability of dead things (machines) to create value simply by virtue of the fact that these things have at one point been created by human labour.

As we can see, the attempt by Mill and McCulloch to prove the 'direct' applicability of the law of labour value in capitalist economy led to unexpected results. In their desire to remain verbally more faithful and consistent towards the labour theory of value than even Ricardo, they in fact came to completely repudiate its most fundamental idea: that only human labour creates value. By acknowledging the operation of natural forces and machines as a direct source of value, McCulloch (and to a lesser extent Mill), though his words championed the very strictest application of the law of labour value, in fact came very close to Say's 'vulgar economy'. Later on, McCulloch identified labour, completely in the spirit of Say, as 'any sort of action or operation, whether performed by man, the lower animals, machinery, or natural agents'.[6] It would be impossible to think of a greater distortion of Ricardo's theory under the guise of defending it against the attacks of its opponents.

The opponents of the Ricardian theory were perfectly correct to judge Mill's and McCulloch's explanations of the barrel of wine as a *renunciation* of the principle of labour value. Bailey pointed out to Mill that it was impossible to talk about human labour acting on the wine when not a single human being had come anywhere near it the whole time it was in the cellar.[7] Malthus gave an acid chuckle at McCulloch when the latter termed the operation of natural agents as 'labour': 'There is nothing that may not be proved by a new definition. A composition of flour, milk, suet, and stones is a plumb pudding; if by stones be meant plums.'[8] One review of McCulloch's book aptly revealed the basic flaw in his argumentation in these words: 'Extend at a stroke the meaning of the term "labour" to such limits that it embraces, besides human labour, the work of cattle, the operation of machinery, and the processes of nature, and it will become absolute truth that the quantity of labour regulates value; but limit the meaning of the term "labour" to that sense in which it is commonly employed in real life; recognize that the process of fermentation undergone by a liquid in a barrel, or the vegetative process that brings a tree to maturity, are distinct from human labour, and Ricardo's theory of value has the ground cut out from beneath it.'[9]

In reality the theory of value as formulated by Ricardo had 'the ground cut out from beneath it' because it was unable to explain the phenomena of capitalist economy, in particular the tendency for capitals with unequal organic compositions or unequal periods of turnover to have equal rates of profit. The attempts on the part of the orthodox Ricardians to prove that the law of labour value operates *directly* within capitalist economy led in fact to a renunciation of the *labour* theory of value and to capitulation before the vulgar theory of *production costs*.

1 Malthus, *The Measure of Value Stated and Illustrated* (London, 1822), pp. 12-13, fn.: 'The effects of slow or quick returns and of the different proportions of fixed and circulating capitals, are distinctly allowed by Mr. Ricardo; but in his last edition, (the third, p. 32) he has much underrated their amount. They are both theoretically and practically so considerable as entirely to destroy the position that commodities exchange with each other according to the quantity of labour which has been employed upon them; but no one that I am aware of has ever stated that the different quantity of labour employed on commodities is not a much more powerful source of difference in value.'

There is also the much better known passage in Malthus's *Definitions in Political Economy* (London, 1827), pp. 26-27: 'Now this proposition [''that commodities exchange with each other according to the quantity of manual labour worked up in them''—*Ed.*] is contradicted by universal experience. The slightest observation will serve to convince us, that after making all the required allowances for temporary deviations from the natural and ordinary course of things, the class of commodities subject to this law of exchange is most extremely confined, while the classes, not subject to it, embrace the great mass of commodities. Mr. Ricardo, indeed, himself admits of considerable exceptions to his rule; but if we examine the classes which come under his exceptions, that is, where the quantities of fixed capital employed are different and of different degrees of duration, and where the periods of the returns of the circulating capital employed are not the same, we shall find that they are so numerous that the rule may be considered as the exception, and the exceptions the rule.'

2 Torrens, *An Essay on the Production of Wealth* (London, 1821), pp. 28-29, cited by Marx in *Theories of Surplus Value*, Part III, p. 72.

3 Quoted by Marx in *Theories of Surplus Value*, Part III, p. 86.

4 McCulloch, *The Principles of Political Economy* (Edinburgh, 1825), p. 313. Quoted by Malthus in his *Definitions* (1827 edition), pp. 102-103. This passage is from the first edition of McCulloch's book, and was dropped in subsequent editions. The first emphasis is Rubin's, the last two are McCulloch's.

5 McCulloch, *Principles*, first edition, p. 291, quoted in *Theories of Surplus Value*, Part III, p. 185.

6 McCulloch, *Ibid*, p. 75, fn., quoted in *Theories of Surplus Value*, Part III, p. 179.

7 Samuel Bailey, *A Critical Dissertation on the Nature, Measures, and Causes of Value* (London, 1825), pp. 219-20; Marx quotes this same passage from Bailey when discussing Mill's attempt to deduce value from 'time'; see *Theories of Surplus Value*, Part III, pp. 85-88.

8 Malthus, *Definitions in Political Economy* (1827 edition), p. 100.

9 Translated from the Russian.

CHAPTER THIRTY-FOUR

The Wages Fund

As we have seen, even within the close confines of the Ricardian school, James Mill and McCulloch, who looked upon themselves as the true guardians of Ricardo's tradition, in fact vulgarized and distorted the labour theory of value. Even more evident was the process by which the Classical theory became vulgarized during discussions over the problem of *distribution*, a problem more narrowly and immediately tied to the class interests of the bourgeoisie. Here we will look at the fortunes of the theory of *wages* in the post-Ricardian era, in order then to move on to the theory of profit.

It was Ricardo who brought to completion the theory of '*means of subsistence*' (or the iron law of wages), already outlined by the mercantilists and developed further by the Physiocrats (and in part by Smith). He had given a more or less succinct formulation of the quantitative problem of wages, but had not even asked himself to what extent it was possible to reconcile the law of labour value with the fact that the 'value of labour' (i.e., wages) is less than the value that labour creates. Both James Mill and McCulloch were conscious of the difficulties involved in resolving this problem, and they therefore decided to sever once and for all the umbilical cord that for Ricardo had bound, however weakly, the theory of wages to the theory of value. They decided to construct the first of these theories without resort to the second, and put forward a thesis according to which the level of wages is determined exclusively by the relation of the *supply* of labour (i.e., the number of workers) to the *demand* for labour (i.e., the amount of capital earmarked for hiring workers).

The roots of this idea are already to be found in Smith, but it was only after Malthus that they were fully developed. Malthus had taught that in any country there exists a precisely determined and limited *fund of means of subsistence*. If the workers are in receipt of too few means of subsistence this is merely the result of their own, excessively quick multiplication: the nation's fund of means for feeding itself has now to be divided between a growing number of workers. In short the workers are to blame for their own starvation.

Thus at any given moment the fund of means of subsistence

designated for the workers' maintenance is of a strictly determined and limited size, capable of neither increase nor reduction. The economists of the Classical school, however, equated *in natura means of subsistence* as such with the *capital* laid out for hiring labour power (variable capital, in Marx's terminology). Hence they came to the conclusion that the *capital* spent on hiring workers is a strictly determined and *limited* magnitude, which cannot at any given moment be either *increased* or *reduced*. This *'wages fund'* is divided up between all the workers of a given country so that the average wage of the individual worker equals the fraction obtained by dividing the *total wages fund* by the total *number of workers*. A rise in wages is only possible: 1) if the *demand* for labour *grows*, i.e., there is an increase in the total amount of capital spent on hiring workers, or 2) if the *supply* of labour is *reduced*, i.e., there is a fall in the total number of workers. There is only one way that the workers can secure a rise in wages: by heeding the advice of Malthus, delaying having children and thus reducing their own numbers. Strikes, rather than making possible any long-term rise in workers' wages, would only do them harm, since strikes slow down capital accumulation and hence reduce the wages fund. Even if the workers of one group were to acquire higher wages, this would only bring suffering to those other groups of workers who would now be left with a smaller share of the overall wages fund.

The idea of a wages fund was already in the air, so to speak, at the beginning of the 19th century. It was expressed in mild version by Malthus. In one popular book by Mrs. Marcet, published in 1816 (the theories of economists evoked at the time such a lively interest among the general public that they were put forward in light, quasi-*belles lettres* and even taught in female boarding schools), we find two persons engaged in the following conversation:

Caroline: What is it that determines the rate of wages?

Mrs. B: It depends upon the proportion which capital bears to the labouring part of the population of the country.

Caroline: Or in other words, to the proportion which subsistence bears to the number of people to be maintained by it?

Mrs. B: Yes.[1]

The original founders of the wages fund theory were James Mill and McCulloch. According to Mill, the level of wages is determined by the relation of the supply of labour to the demand for labour. 'It thus appears, that, if population increases, without an increase of capital, wages fall; and that, if capital increases, without an increase of population, wages rise.'[2] If the ratio between capital and population stays at its previous level wages will remain the same; if the ratio of capital to population increases, wages will rise; conversely, if it is the ratio of population to capital that rises, then wages will fall.

These same ideas are developed further by McCulloch, who gives the wages fund theory its definitive formulation. 'It is ... on the amount of ... capital, applicable to the payment of wages in its possession, that the power of a country to support and employ labourers must depend ... It is a necessary consequence of this principle, that the amount of subsistence falling to each labourer, or the rate of wages, must depend on the proportion which the whole capital bears to the whole labouring population ... To illustrate this, let it be supposed that the capital of a country appropriated to the payment of wages, would, if reduced to the standard of wheat, form a mass of 10,000,000 quarters: if the number of labourers in that country were *two* millions, it is evident that the wages of each, reducing them all to the same common standard, would be *five* quarters; and it is further evident, that this rate of wages could not be increased unless the amount of capital were increased in a greater proportion than the number of labourers, or the number of labourers diminished more than the quantity of capital.'[3] Here we encounter all the basic ideas of the wages fund theory: an identification of the total volume of *capital* with a known quantity of *means of subsistence*, and the assertion that the wages of the individual worker is the *fractional share* obtained by dividing a country's total number of workers into an already-limited volume of capital.

The wages fund theory as developed in the 1830's and 1840's rapidly gained popularity both in academic circles and among the general public. On the one hand, the most prominent economists, including John Stuart Mill, shared it. On the other, it was readily used by publicists, journalists, and entrepreneurs as a weapon in the struggle against the workers' movement. The economists zealously tried to inculcate into the working class the idea that having *fewer children* and a rapid accumulation of entrepreneurial *capital*—and not *strikes* or forming *trade unions*—were the only means through which the workers could expect to improve their situation. Even McCulloch, who advocated that workers should be free to form combinations, did not believe that they would be of any benefit to the working class: 'It is the extreme of folly to suppose that any combination can maintain wages at an artificial elevation. It is not on the dangerous and generally ruinous resource of combination, but on the forethought, industry, and frugality of work-people, that their wages, and their condition as individuals, must always depend.'[4] Other economists and popularizers of the period propounded these same ideas with even greater self-assurance and dogmatism.

Up until the end of the 1860's the wages fund theory ruled unchallenged in English economic literature, being naively taken as gospel truth by scholars and general public alike. 'There is no use in arguing against any one of the four fundamental rules of arithmetic. The question of wages is a question of division' [5] wrote the economist Perry. It was in Germany that Hermann and Rodbertus raised the first objections to the wages fund theory, but they attracted little attention [6]. With each large-scale strike, with each major conflict, the workers had the theory of the wages fund thrown at them, with its proponents arguing the futility and harm done by their economic struggle. It was for this reason that the theory was employed with such relish in bourgeois circles and provoked such immense hatred on the part of workers and socialists. The successes of the working class's economic struggle and the trade-union movement proved how absurd the wages fund doctrine in fact was. In the 1860's faith in the validity of the theory was undermined even among bourgeois scholars. The works of Longe (1866) and Thornton's book, *On Labour* (1869) dealt it a sharp blow. Soon after Thornton's book appeared, John Stuart Mill, in a special article, declared that he acknowledged the validity of Thornton's argument and would henceforth renounce the wages fund theory. Mill's declaration produced a sensation among bourgeois scholars. And although several among them (Cairnes, for instance) continued to defend this theory, its fate was effectively sealed with Mill's statement and the enormous prestige it carried. Wages fund theory was now jettisoned as being patently false with almost the same unanimity as it has previously been seized upon for its supposed correctness. [7]

The sudden bankruptcy of a theory that for several decades had enjoyed the reputation of being beyond dispute represents, in the words of one economist, one of the most dramatic pages in the history of economic thought. What is astonishing is not that the wages fund doctrine was rejected, but that it was accepted as correct for several decades, despite its obvious lack of theoretical foundation and its denial of reality. At every stage of its existence, capitalist economy provides startling examples of the sudden *expansion* of productive capital (including variable capital) during periods of boom and *contraction* of capital during periods of depression. After this, can one really declare that variable capital is a magnitude that is fixed in advance and strictly *limited?* The idea that the size of the wages fund depends on the size of the stock of *means of subsistence* earmarked for the working class is wrong; on the contrary, it is this latter fund that depends upon the former. If the working class manages by its economic struggle to raise the overall amount of its wages, the workers will demand a greater amount of means of subsistence, and a greater quantity of the latter will then begin to be produced (or imported from abroad in exchange for luxury goods, machinery, and other such commodities of domestic production).

Let us therefore put aside the fund of means of subsistence and investigate whether or not the size of variable capital (the wages fund) is narrowly fixed at any given point in time. This idea was based on two premises, both false. It presumed 1) that the total *capital* employed in production can neither *increase* nor *decrease* at any given moment; and that workers' *wages* are taken out of this *capital.* Rodbertus had already shown that workers' wages are drawn not from the entrepreneur's capital, but from the value of the product that the workers have themselves produced. If all of the capitalists in a given country paid their workers wages totalling 100 million pounds, after the sale of the manufactured product for 150 million pounds they would receive back their entire capital plus, over and above this, a profit of 50 million pounds (assuming here that there is no constant capital). The workers, therefore, have received their wages not out of entrepreneurial capital, which remains intact and unharmed, but out of the *value of the product* created by their own labour. There is absolutely no basis for thinking that at any given moment the workers' share in the national product cannot be raised. It is easy to imagine workers' wages growing to 110 million pounds, with the share of surplus value (or profit) falling to 40 million. In this case the capitalists would have to reduce either their *personal consumption* or their *accumulation of new capital.* Of course the capitalists will here have to *advance* a larger sum to pay for these higher wages, but they can either take this additional sum out of enterprise *reserves* or obtain it from the *banks* by drawing credit.

The theory of the wages fund for long enjoyed scientific recognition not thanks to any theoretical achievements, but in spite of its theoretical bankruptcy. The theory owed its great popularity to the fact that it could be used by the bourgeoisie in order to defend itself against the attacks of the workers. Even bourgeois scholars have acknowledged this fact. 'I would not impeach the scientific impartiality of those who first put forward in distinct form this theory of wages,' wrote Walker, 'but it may fairly be assumed that its progress towards general acceptance was not a little favored by the fact that it afforded a complete justification for the existing order of things respecting wages.' With the help of this theory 'it was an easy task to answer the complaints or remonstrances of the working classes and to demonstrate the futility of trades-unions and strikes as means of increasing wages'.[8]

The history of the wages fund doctrine gives graphic illustration that Classical theory after Ricardo went through a period of disintegration in a *two-fold* sense: first, it became steadily *'vulgarized'*, confining itself to generalizing the surface phenomena of capitalist economy (in the case here, by applying the law of supply and demand to wages) and refraining from making a deeper analysis into their ultimate causes; in the second place, as the class struggle between the bourgoisie and the working class became sharper, economic theory increasingly became an *'apologetic'* tool for defending the interests of the bourgeoisie. Parallel with the decline in the *theoretical* level of Classical doctrine, its *practical* social implications became reactionary. 'Vulgar' economy had become inseparable from bourgeois 'apologetics'. We will find this confirmed no less strikingly in the theory of *profit*.

1 Jane Marcet, *Conversations on Political Economy*, pp. 117-18. Cited in Cannan, *Theories of Production and Distribution in English Political Economy*, p. 242.

2 James Mill, *Elements of Political Economy*, in James Mill, *Selected Economic Writings*, introduced and edited by Donald Winch (Edinburgh, Oliver & Boyd, 1966), p. 230.

3 McCulloch, *Principles of Political Economy*, 1843 edition (Edinburgh, William Tait), pp. 379-80; McCulloch's emphasis.

4 McCulloch, 'Combination by Work-People', *Encyclopaedia Britannica*, Eighth Edition.

5 A.L. Perry, *Elements of Political Economy*, p. 123, quoted in Francis A. Walker, *The Wages Question* (London, Macmillan, 1882), p. 143.

6 Unlike Rodbertus, Hermann's work receives comparatively sparse treatment in marxist histories of economic thought. A good review of his ideas and his book *Staatswirtschaftliche Untersuchungen* (first published in 1832) is to be found in Eugen von Böhm-Bawerk's *Capital and Interest, A Critical History of Economic Theory*, translated by William Smart (London, Macmillan, 1890); Böhm-Bawerk's book is altogether an excellent reference source for most of the economists Rubin discusses in this section.

7 Mill's statement came in a review of Thornton's book that appeared in *Fortnightly Review* (May 1869), part of which is reproduced in the appendix to W. J. Ashley's edition of Mill's *Principles of Political Economy* (London, Longmans, Green & Co., 1921), pp. 992-93. Of these excerpts we offer the most salient passages: 'The price of

labour, instead of being determined by the division of the proceeds between the employer and the labourers, determines it. If he gets his labour cheaper, he can afford to spend more upon himself. If he has to pay more for labour, the additional payment comes out of his own income ... There is no law of nature making it inherently impossible for wages to rise to the point of absorbing not only the funds which he had intended to devote to carrying on his business, but the whole of what he allows for his private expenses, beyond the necessaries of life. The real limit to the rise is the practical consideration, how much would ruin him or drive him to abandon the business: not the inexorable limits of the wages-fund ...

'... The doctrine hitherto taught by all or most economists (including myself), which denied it to be possible that trade combinations can raise wages, or which limited their operations in that respect to the somewhat earlier attainment of a rise which the competition of the market would have produced without them,—this doctrine is deprived of its scientific foundation, and must be thrown aside. The right and wrong of the proceedings of Trade Unions becomes a common question of prudence and social duty, not one which is peremptorily decided by unbending necessities of political economy.'

8 Walker, *The Wages Question*, p. 142.

CHAPTER THIRTY-FIVE

The Theory of Abstinence

SENIOR

Smith and Ricardo, as we know, came very close to conceiving profit (which they often confused with surplus value as a whole) as a portion of the value created by the workers' labour. They can therefore be considered the progenitors of the 'theory of surplus value' (or the so-called '*theory of exploitation*') that Rodbertus and Marx were subsequently to develop with greater consistency. However, Smith's and Ricardo's immediate followers quickly forsook their doctrine of surplus value. Ricardo's pupils, James Mill and McCulloch, though they remained verbally faithful to the labour value doctrine, designated the capitalist's profit as a compensation or wage for the 'accumulated labour' contained in machinery and other means of production. The absurdity of this '*labour theory of profit*' (not to be confused with the labour theory of value) as put forward by James Mill and McCulloch prevented its wide acceptance within bourgeois science. Rather more successful was the '*theory of the productiveness of capital*' that Say had elaborated; this sees profit as stemming from the activity of capital, which stands as an independent factor of production together with labour and the forces of nature. This theory, which was useful as a means of justifying the profit of capital against socialist attacks, did become widely accepted within bourgeois science. No less successful was the '*theory of abstinence*' developed by the Englishman, Senior, and expounded in his book, *Political Economy* (1836).[1]

Senior accepts Say's doctrine of the three independent *factors of production*: labour, natural agents, and capital. However, he introduces one amendment to this division: he replaces 'capital' with the '*abstinence*' of the capitalist. Capital cannot be considered a primary factor of production since it is itself the result of the combined activity of labour, natural agents, and abstinence.

By *abstinence* Senior understands 'the conduct of a person who either abstains from the unproductive use of what he can command, or designedly prefers the production of remote to that of immediate results'. Without the aid of abstinence, in the sense just described, the other two factors of production, labour and natural agents, would not be able to fully exhibit their activity. 'The most laborious population, inhabiting the most fertile territory, if they devoted all their labour to the production of immediate results, and consumed its produce as it arose, would soon find their utmost exertions insufficient to produce even the mere necessaries of existence'. Only where this population 'abstains' from immediately consuming some portion of the produce that it has created and decides to employ it as capital, or 'means of further production' will it be able to draw maximum benefit from the activity of its own labour and the agents of nature.[2]

Modern society owes its immense wealth to the abstinence of preceding generations. 'A carpenter's tools are among the simplest that occur to us. But what a *sacrifice of present enjoyment* must have been undergone by the capitalist who first opened the mine of which the carpenter's nails and hammer are the product! How much *labour directed to distant results* must have been employed by those who formed the intstruments with which that mine was worked! ... We may conclude that there is not a single nail ... which is not to a certain degree the product of *some labour for the purpose of obtaining a distant result*, or in our nomenclature, of some *abstinence* undergone before the Conquest.'[3] It is clear from the words that we have italicized how Senior muddles up the question by identifying 'abstinence' with 'labour for the purpose of obtaining a distant result'. In other passages Senior consistently stresses that abstinence is an 'agent, distinct from labour and the agency of nature'.[4] From Senior's point of view 'labour for the purpose of obtaining a distant result' should be viewed not as abstinence, but as a union of labour with abstinence. The production of capital demanded a double sacrifice on the part of the producer: labour and abstinence. *Abstinence is sacrifice*: 'To abstain from the enjoyment which is in our power, or to seek distant rather than immediate results, are among the most painful exertions of the human will.'[5]

Who is it in contemporary society that makes this sacrifice of 'abstinence'? It is clearly the *capitalists*, who refrain from expending the whole of their 'labour' on immediate consumption, and retain the products of their 'labour' as machines, cotton, and similar 'means of further production'. To Senior's astonishment the workers exhibit no such desire to 'abstain' from spending their pay on 'immediate results', but instead buy bread and potatoes for themselves and their families. Senior is, however, prepared to place the blame for the workers' lack of abstemiousness on their poor education: 'Among the different classes .... those which are the worst educated, are always the most improvident, and consequently the least abstinent.'[6]

Thus abstinence demands from a person the same heavy *sacrifice* as does labour. The capitalists make a sacrifice by engaging in abstinence; for this sacrifice they receive *a reward, in the form of profit on capital* (Senior, like Say, designates entrepreneurial profit as the entrepreneur's wage for the labour of superintending the enterprise), just as the worker receives a wage as his reward for sacrificing his labour.[7] 'Wages and profit are to be considered as the rewards of peculiar sacrifices, the former the remuneration for labour, and the latter for abstinence from immediate enjoyment.'[8] '*Abstinence* ... stands in the same relation to *Profit* as Labour does to *Wages*.'[9] If the worker receives a reward for his sacrifice, the capitalist must have remuneration for the sacrifice that he makes by abstaining. Thus the capitalist from the outset includes the profit on capital as part of the commodity's *production costs*, and it must be paid for out of the latter's price. If the price of a commodity is not high enough to pay a profit on capital the capitalist cuts back production of the commodity in question and thus, by 'limiting supply' raises its price to the level required. Profit, then, in Senior's view is a part of production costs, and not a surplus over and above them—a surplus the origin of which had so persistently puzzled economists.

Senior's doctrine bears the unmistakable mark of an *apologetic*: while serving to *justify* the profit of capital, it does not in the least *explain* its origin. Suppose that a capitalist really does deserve a profit as a reward for his abstinence. Where does he take it from? Senior does not even ask. Certainly value cannot be created passively, by the *purely psychological* fact of abstinence. Senior himself recognizes the weakness of his position: 'It may be said that pure Abstinence, being a mere negation, cannot produce positive effect.' Senior finds no reply to this objection other than to add lamely that the same could apply to 'liberty' or 'intrepedity', and yet these are quite rightly accepted as 'active agents'. Be that as it may, no one up to now has thought to assert that 'intrepedity' can be the source of a product's value. Say, from his own point of view, had been consistent when he examined all three factors of production from their *material-technical* side (labour, nature, and capital in the sense of means of production); Senior, on the other hand, destroys the validity of this schema by placing the *purely psychological* fact of abstinence alongside labour and nature.

Besides being useless as an *explanation* of economic phenomena, the abstinence theory falsely *depicts* both how capitalism came into being and the basic features of this economic system. It assumes that capital was accumulated by industrious and farsighted people abstaining from directly consuming the products created by *their own* labour. We also find this naive 'children's tale' in Smith, and historical science disproved it by showing that the source of primitive capital accumulation was the brazen appropriation by the upper groups in society of the products of *other people's* labour. If 'abstinence' played an insignificant role even during the period of primitive capital accumulation, then it is absurd to see it as a source of profit in a developed capitalist economy. Lasalle (in his book *Kapital und Arbeit*) appraised with acrid sarcasm the true worth of this assertion that only 'the most painful exertions of the human will' restrain the capitalists from the temptation to squander their entire fortunes all at once: 'The profit of capital is the "wage of abstinence". Happy, even priceless expression! The ascetic millionaires of Europe! Like Indian penitents or pillar saints they stand: on one leg, each on his column, with straining arm and pendulous body and pallid looks, holding a plate towards the people to collect the wages of their Abstinence. In their midst, towering up above all his fellows, as head penitent and ascetic, the Baron Rothschild! This is the condition of society! how could I ever so much misunderstand it!'[10]

The ludicrousness of talking about 'abstinence' when dealing with a capitalist inheriting a wharf or a canal worth millions of pounds could not fail to catch the eye even of a Senior. To surmount this difficulty, Senior resorts to a curious sleight of hand: he declares the income of this capitalist to be *not profit, but rent*. 'The revenue arising from a dock, or a wharf, or a canal, is profit in the hands of the *original constructor*. It is the reward of *his* abstinence in having employed capital for the purposes of production instead of for those of enjoyment. But in the hands of his heir it has all the attributes of rent. It is to him the gift of fortune, not the result of a sacrifice.'[11] To be consistent, Senior would have had to acknowledge the income of any capitalist who *inherits* property as *rent*, rather than profit. The fact that such a huge portion of capitalist revenue could not then be acknowledged as profit is itself sufficient to indicate the bankruptcy of a theory which sees profit as a reward for abstinence.

For all its theoretical groundlessness the abstinence doctrine won *widespread scientific acceptance*, and is retained to this day by numerous bourgeois scholars. Even a researcher like Böhm-Bawerk, whom it would be difficult to suspect of any sympathies towards socialist ideas, acknowledges that the success of the abstinence theory is to be explained not so much by its theoretical accomplishments, as by its apologetic character: 'Senior's Abstinence theory has obtained great popularity among those economists who are favourably disposed to interest. It seems to me, however, that this popularity has been due, not so much to its superiority as a theory, as that it came in the nick of time to support interest against the severe attacks that had been made on it.'[12]

Senior was what one might call the economic barrister of the English factory owners, who found him a faithful assistant in their bitter fight against factory legislation. When the Factory Act of 1833 limited the working day of juveniles to twelve hours, Senior came out, in 1837, with a pamphlet against shortening the working day. In it he tried to prove arithmetically (how crudely mistaken his calculations were Marx was to show in Chapter 7 of the first volume of *Capital*)[13] that the whole of the factory owners' profit is contained in the '*last hour*' of the workers' labour; thus to reduce the working day by even a single hour, he argued, would threaten industrialists with total ruin. Fortunately, Senior's sophistical exercizes had just as little effect in holding back the advance of *factory legislation* as did the arguments of the theoreticians of the wages fund in halting the growth of *trade unions*. Just as in his declining years J.S. Mill had been forced to repudiate the theory of the wages fund, so did Senior have to change his position on the Factory Acts and declare himself in support of them. The real-life successes of the workers' movement proved in practice the mistakenness of the apologetic theories of the last plenipotentiaries of the Classical school.

1 Nassau Senior, *An Outline of the Science of Political Economy* (London, Alan & Unwin, 1951)
2 Senior, p. 58.
3 *Ibid*, p. 68, Rubin's italics.
4 *Ibid*, p. 59
5 *Ibid*, p. 60
6 *Ibid*, p. 60.
7 Throughout this discussion Rubin uses the word *protsent*, or 'interest' to refer to profit.
8 Senior, p. 91.
9 *Ibid*, p. 59, Rubin's italics.

10 Quoted in Böhm-Bawerk, *Capital and Interest*, p. 276.
11 Senior, p. 129.
12 Böhm-Bawerk, *Capital and Interest*, p. 286.
13 Chapter 9 of the English edition (Penguin edition, pp. 333-338).

CHAPTER THIRTY-SIX

Harmony of Interests

CAREY AND BASTIAT

Although we have already been led to note examples of bourgeois apologetics when dealing with the works of Ricardo's immediate successors, it was only in *the period 1830-1848* that social conditions in Europe had matured sufficiently for a decisive transformation of economic science to take place. It became a weapon which defended the bourgeoisie directly against the attack of the working class. *The revolution* of 1830 in France and the English *electoral reform* of 1832, had opened the way to political power for the bourgeoisie. The repeal of the English *corn laws* in 1846 signalled the end of a century-long struggle between the industrial bourgeoisie and the landowning class. On the other hand, the *Chartist movement* and the *revolutions* of 1848 showed just how dangerous an enemy the bourgeoisie had in the working class. 'From that time on, the class struggle took on more and more explicit and threatening forms, both in practice and in theory. It sounded the knell of scientific bourgeois economics. It was thenceforth no longer a question whether this or that theorem was true, but whether it was useful to capital or harmful, expedient or inexpedient, in accordance with police regulations or contrary to them. In place of disinterested inquirers there stepped hired prize-fighters; in place of genuine scientific research, the bad conscience and evil intent of apologetics.' (From Marx's preface to the Second Edition of Volume I of *Capital* [Penguin edition, p. 97—*Ed.*])

The working class's revolutionary offensive, on the one hand, and the ideological critique on the part of the socialists, on the other, both accelerated the process leading to the Classical school's decomposition. By the middle of the 19th century it became obvious that there could no longer be a classical theory that would continue to adopt the standpoint of the bourgeoisie and yet carry out the monumental work of making a theoretical investigation into the laws of capitalist economy. Henceforth the epigones of the Classical school would have to make one of two choices: either refrain, in the interests of abject apologetics, from making a sober and disinterested investigation into the laws of capitalism, or attempt to reconcile an outmoded liberalism with the newborn socialism. Carey and Bastiat travelled the first path, John Stuart Mill the second.

The works of the American Henry Carey (1793-1879)* and the Frenchman Frédéric Bastiat (1801-1850)**mark the final stage in the Classical school's dissolution: firstly, because in their works the task of theoretical investigation is wholly relegated to a secondary position in favour of an *apologetic* defence of the capitalist system against attacks by the socialists, and secondly, because their desire to find a justification for capitalism by whatever means forces them to declare total war against *Ricardo's theory*, which was the most mature formulation of Classical doctrine. Both Carey and Bastiat were dilletantes for whom pure theoretical investigation was a secondary objective. Both of them denied the existence of deep-seated class contradictions in capitalist society, a point of view which inevitably compelled them to falsify reality. To the '*pessimistic*' conception of Ricardo and Malthus, the two of them counterposed an '*optimistic*' doctrine which held that the free development of capitalist society necessarily leads to a reconciliation and '*harmony of interests*' among all its component classes. Carey published a book entitled *The Harmony of Interests*, Bastiat a tract called *Economic Harmonies*. The similarities between their two doctrines were sufficiently great to give Carey occasion to accuse Bastiat of plagiarism. In reality Bastiat borrowed little from Carey; the latter, although the theoretical level of his works was extremely low, was nevertheless distinguished by a greater inclination and ability for theoretical investigation than his more tumultuously successful French colleague.[1]

Ricardo had laid bare the basic *class contradictions* of the capitalist system: between the landlords and capitalists, and between capitalists and workers. Though himself an ardent defender of the bourgeois order, Ricardo had in fact forged the theoretical weaponry that the socialists were to make great use of. Carey abhorred Ricardo's doctrine for just this reason. In Carey's words, Ricardo's 'book is the true manual of the demagogue, who seeks power by means of

*His major works are his *Principles of Political Economy* (1837-1840), *The Harmony of Interests* (1851), *The Past, the Present, and the Future* (1848), and *The Principles of Social Science* (1857-1860).

**His major works are *Cobden et la Ligue* (1845), *Sophismes Economiques* (1847), and *Harmonies Economiques* (1850). [Quotations from the latter work are taken from the English translation by Patrick James Stirling which appeared under the title *Harmonies of Political Economy* (London, 1860)—*Ed.*]

agrarianism, war, and plunder'.[2] To slay the revolutionary Hydra, Carey decided first to undermine its theoretical foundations—*Ricardo's theory of income distribution* between the social classes. 'As a harmonist', Marx wrote in a letter to Engels, 'Carey first pointed out that there was no antagonism between capitalist and wage labourer. The second step was to show the harmony between landowner and capitalist'.[3] Let us see, then, how Carey pursues the first of his two apologetic objectives.

Carey was full of optimistic faith in the powerful development of the *productivity of labour.* With every advance in labour productivity the entire stock of accumulated produce will fall in value, since the latter is determined by the quantity of labour necessary to *reproduce* these products, and not by the quantity actually expended on their production. 'The quantity of labour required for *reproducing* existing capital and for further extending the quantity of capital diminish[es] with every stage of progress'.[4] But 'every reduction in the value of existing capital [is] so much added to the value of the *man*',[5] since the latter can now create the same capital with greater facility than before. Thus, as technology advances 'the *labourers of the present* tend to acquire power at the cost of the *accumulations of the past.*'[6] Hence, as the productivity of labour rises, so, too, does the specific weight of living 'labour', or *'man'* himself, in comparison with the accumulated stock of *lifeless things.*

Up to this point Carey has been counterposing to one another abstract, material-technical categories: *'things'* versus *'labour'.* But accumulated stocks of things are, of course, *'capital'*, while labour has the form of *'wage labour'.* By identifying material-technical categories with social categories, Carey arrives at the unexpected conclusion that the specific weight of *wage labour* is constantly growing with respect to *capital.* 'The labourer is rising, as compared with the capitalist [i.e., capital has a diminished command over human labour—*Ed.*], with constantly increasing facility for becoming himself a capitalist.'[7] This means that 'capital [is] thus declining in its power over labour, as labour [is] increased in its power for the *reproduction* of capital'.[8] But under these conditions *the relative share of the worker* in the produce of labour will naturally *increase* at the expense of the relative share going to the capitalist. Carey illustrates his thinking with the following scheme [see table at top of page 329]:

The scheme shows the advance of labour productivity over *four* consecutive periods. As we move from one period to another the gross output of the individual worker doubles at the same time as the

	Total product	*Labourer's Share*	*Capitalist's Share*
First Distribution	4	1	3
Second Distribution	8	2.66	5.33
Third Distribution	16	8	8
Fourth Distribution	32	19.20	12.80

worker's share of the product (both *absolute* and *relative*) is rising. In the first period the worker has received only 1/4 of the product; in the last period he receives 3/5. And though the capitalist's relative share has gradually fallen from 3/4 to 2/5, he finds himself in no way harmed by this and with no cause for complaint: thanks to the rise in the productivity of labour the absolute number of unit products that he gets has grown from 3 to 12.80. 'Both thus profit greatly by the improvements that have been effected. With every further movement in the same direction the same results continue to be obtained—*the proportion of the labourer* increasing with every increase in the productiveness of effort—*the proportion of the capitalist* as steadily diminishing, with constant increase of *quantity* and equally constant tendency towards *equality* among the various portions of which society is composed', i.e., a tendency towards equality of social classes.[9] 'Such is the great law governing the distribution of labour's products. Of all recorded in the book of science, it is perhaps the most beautiful, being, as it is, that one in virtue of which there is established a perfect harmony of real and true interests among the various classes of mankind.'[10] Carey reckons that the *falling rate of profit* graphically confirms the law of a declining capitalist share in production.

One could say that Carey's chain of reasoning contains as many lacunae as it does links. Firstly, the falling value of the individual material components of capital, e.g., an individual machine, is more than compensated for by the rising number of machines; there is an enormous *growth* in the *total volume of capital*, and, with it, in the power of capital over labour. Secondly, while the value of capital (machinery, etc.) is falling by virtue of the fact that less *social labour* is required for its reproduction, this in no way implies a rise in the *value of labour as a commodity*, i.e., that wages take an increasing share of the national product. On the contrary, the decline in the value of the worker's means of subsistence in capitalist economy produces a *fall in the value of labour power*, a rise in *relative* surplus value, and hence also a rise in the *relative* proportion of the product going to the capitalist. Thirdly, that the capitalists' share in the national product should rise along with advances in labour productivity is in no way disproved by the existence of a falling rate of profit (the explanation for which lies in the enormous growth in the total volume of capital that we have just mentioned). Carey is committing a crude blunder in confounding *the rate of profit* with the *capitalists' share of the product* (i.e., with the rate of surplus value).

After having proved that there is a harmony of interests between workers and capitalists, Carey still had to demonstrate a harmony of interests between *capitalists and landlords*. To do this he had to disprove *Ricardo's theory of rent*: Ricardo, of course, had shown that the landowners, while themselves performing no labour, nevertheless appropriate a greater and greater share of the national income, to the detriment of the other classes of the population. Not just socialist thinkers, but such moderate economists as John Stuart Mill deduced from Ricardo's theory of rent the need to nationalize landed property. This revolutionary conclusion disturbed Carey, and he set himself the task of refuting Ricardo's system, according to which 'the interests of the owners of land were constantly opposed to those of all other classes of society' and which 'tends, necessarily, to disturbance of the right of property in land'.[11]

Carey justifiably rejects Ricardo's assertion that the productivity of agricultural labour is constantly falling as a result of farmers ineluctably shifting their cultivation from *good* lands to *poorer* ones. Against this onesided assertion of Ricardo's Carey proposes a contradictory statement that is equally onesided:[12] farmers have always begun by cultivating hilly and less fertile land which was more readily accessible and only later started to take more fertile land, which lay in marshes and bogs, and make it suitable for agriculture. Agriculture is gradually spreading to *more fertile* lands: in addition to this, the quantity of labour needed to make a given plot of land suitable for cultivation will be falling as agricultural technology improves. It therefore follows that no farmer will agree to pay a landlord rent, since he will prefer to occupy a new, more fertile plot of land. If, in spite of this, a farmer expresses a willingness to lease land this is only because the plot that he is renting will have already been made ready for cultivation through the previous application of labour and capital by the landowner or his ancestors. What the landowner receives as a payment on his lease is not, therefore, *ground rent*, but merely *a profit on the capital* that helped improve this particular plot. Land that can be cultivated is just as much a product of labour as any machine; *rent is purely and simply profit on capital*, and the landlord is no different from any capitalist. What is more, the landlord does not even receive a profit on all the capital that he and his forefathers had invested in the land. If the total that they had spent was £1000, then, given the present higher state of technology, the same land could now be improved for an outlay of £500. The value of the capital invested in land (as with that invested in industry) will have fallen from £1000 to £500, so that, assuming an average rate of profit of 5%, the lessee will pay no more than £25 per annum.

It is no accident that Carey's theory was spawned in *America* during the first half of the 19th century—a country where the contradictions of the capitalist system were as yet undeveloped and classes had still not sharply defined themselves with respect to one another; where there was an abundance of free land together with a virtual absence of any rent and a shortage of labour; where high wages and the opportunity to settle on free lands often made it possible for the more industrious workers to become farmers or capitalists. If in *America* the harmony of interests doctrine reflected the immaturity of social relations, in *France* the bourgeoisie sought to employ it to cover over and conceal the severity of the class conflicts which had burst onto the political arena with unprecented force during the revolution of 1848. While Carey's attack was directed against Ricardo and the other ideologists of the more developed English bourgeoisie who had painted a picture of a capitalist system full of contradictions—a picture in which the young American bourgeoisie had no wish to recognize its own future—Bastiat levelled his blows mainly against the socialists.

Prior to the revolution of 1848, Bastiat had feverishly combatted the protectionists in a series of witty pamphlets and *feuilletons* and been a passionate champion of free trade in the mold of the English free traders. The 1848 revolution made a tremendous impression on him and following it he directed his passion against the *socialists*. In Bastiat the kernels of theoretical analysis are completely drowned in a sea of empty phrases and high sounding declamations, and yet his works enjoyed tumultuous success and earned their author a completely undeserved reputation as an eminent economist.

'*All legitimate interests are in harmony*. That is the predominant idea of my work', says Bastiat in his *Economic Harmonies*.[13] Capitalist society is an immense '*natural*' community which, being superior to all the '*artificial*' communities propounded by the socialists, assures people the freedom of co-operation and mutual assistance.

People work for one another and exchange their respective services. The exchange of products is *an exchange of services*. A product's value is determined not by 'the labour *performed* by the person who renders the service', as the Classics had taught, but by 'the labour *saved* to the person who receives it'.[14] 'Value is the *relation of two services exchanged*'—this is Bastiat's law of service-value and he sets great store by it.[15] Relations between capitalist and worker, between landlord and tenant, and between creditor and debter are all subordinated to the law of *service for service*. The right of the capitalist to receive profit is beyond contention. 'Those who are possessed of capital have been put in possession of it only by their labour, or by their privations.' 'On their part, to give away this capital would be to deprive themselves of the special advantage they have in view; it would be to transfer this advantage to others; it would be to render others *a service*. We cannot, then, without abandoning the most simple principles of reason and justice, fail to see that the owners of capital have a perfect right to refuse to make this transfer unless in exchange for another *service*, freely bargained for and voluntarily agreed to.'[16] On such a basis is founded the creditor's right to receive *interest*. The same basis is used to justify the landlord's claim to *rent*, which Bastiat (following Carey) sees merely as a particular form of profit on capital.

Bastiat devotes less attention to the problem of profit. At times he ascribes its origin to the productivity of capital itself, as did Say; more often, however, he follows Senior's doctrine and ascribes it to the capitalist's abstinence. To assuage the workers Bastiat follows Carey and formulates a '*harmonic*' law of distribution: 'In proportion to the increase of Capital, the *absolute* share of the total product falling to the capitalist is augmented, and his *relative* share is diminished; while, on the contrary, the labourers' share is increased both *absolutely* and *relatively*'.[17] On the other hand, the workers also benefit as consumers as products become cheaper with the development of labour productivity. As technology advances, the '*value*' of the product which is created by 'burdensome' labour falls, while the '*utility*' that man acquires '*gratis*' and without any exertion from nature progressively rises. 'Obstacles, formerly onerously combatted by labour, are now gratuitously combatted by nature; and that, be it observed, not for the profit of the capitalist, but for the profit of the community.'[18] *All social classes* benefit from the development of the economy.

Bastiat requests his reader 'to observe' this 'pacifying, consoling, and religious' *law that interests are in harmony*. Bastiat became the zealous advocate of the law of harmony not because he was blind to the class conflicts that were rending society, but because he was too sickened by the force of the shocks which these conflicts engendered. Bastiat had already been witness to the problem which threatened society, this 'ghost of Banquo at the feast of Macbeth', had already sniffed 'the smell of revolutionary gunpowder' and seen 'the pavement of the barricades.'[19] But he hoped that the workers would trust in the law of harmony and refrain from revolutionary struggle. The fear of revolution disturbed the minds of the epigones of the Classical school; it guided their pens and blinded their vision; it compelled them to deny the truths that, through the lips of Smith and Ricardo, the Classical school had pronounced when it was in full flower.

1 If anything, the 'plagiarism' was the other way around. Bastiat was long dead before Carey's major work, his *Principles of Social Science*, was even published. Carey's illustrations in that book, which he uses to try to prove the mutual benefits of economic progress to both capital and labour, are strikingly similar to the illustrations Bastiat had employed in his own *Harmonies Economiques*.

2 Carey, *The Past, the Present, and the Future* (Philapelphia, 1848), pp. 74-75. Carey incorporated virtually this entire chapter (Ch. 1, 'Man and Land') into his latest *Principles of Social Science*; the sentence quoted here re-appears in Vol. iii, p. 154 of that work.

3 Marx, letter to Engels of 26 November 1869, in Karl Marx and Frederick Engels, *Selected Correspondence* (Moscow, 1965) p. 227.

4 Carey, *Principles of Social Science* (Philadelphia, 1858-65), Vol. iii, p. 111.

5 *Ibid*, Vol. iii, p. 111; Carey's italics.

6 *Ibid*, Vol. iii, p. 113; Rubin's italics. Carey's talent for repeating himself and spinning out the same argument, and even the same phraseology in passage after passage is remarkable. As just one of many examples, compare the sentence that Rubin has quoted here with the following on p. 132 of the same volume: 'With each successive stage of improvement, the value of man increases, as compared with capital—present labour acquiring power at the expense of past accumulations.'

7 *Ibid*, Vol. iii, pp. 114-15.

8 *Ibid*, Vol. iii, p. 112 (original italics).

9 *Ibid*, Vol. iii, p. 113 (Rubin's italics).

10 *Ibid*, Vol. iii, p. 113.

11 *Ibid*, Vol. iii, p. 168.

12 Rubin is here merely repeating the gist of Marx's comments in the above-cited letter to Engels (*Selected Correspondence*, p. 228): 'Carey's only merit is that he is just as one-sided in asserting the transition from worse to better lands as Ricardo is in asserting the opposite. In reality however different kinds of land, unequal in degree of fertility, are always cultivated simultaneously ... and it was this which later made the breaking up of the common lands so difficult. However as to the progress of cultivation throughout the course of history, this depending on the circumstances takes place sometimes in both directions, at other times first one tendency prevails

for a period and then the other.'

13 *Harmonies of Political Economy*, p. 1; original emphasis.

14 *Ibid*, p. 114; original italics.

15 *Ibid*, p. 108 (Rubin's italics).

16 *Ibid*, pp. 168-69 (Bastiat's italics).

17 *Ibid*, p. 183; the first emphasis is Bastiat's, the second Rubin's.

18 *Ibid*, p. 181.

19 *Ibid*, p. 9. 'These economists are so much the slaves of their own systems that they shut their eyes to facts for fear of seeing them. In the face of all the poverty, all the injustice, all the oppressions which desolate humanity, they coolly deny the existence of evil. The smell of revolutionary gunpowder does not reach their blunted senses—the pavement of the barricades has no voice for them; and were society to crumble to pieces before their eyes, they would still keep repeating, "all is for the best in the best of worlds."'

CHAPTER THIRTY-SEVEN

Sismondi as a Critic of Capitalism [1]

We have traced out how the *Classical school disintegrated* internally, how its doctrines became vulgarized and distorted in the works of economists who had adopted the standpoint of the *bourgeoisie*. The road from Smith to Bastiat in political economy ran parallel to that traversed at the same time by the industrial bourgeoisie, who at the end of the 18th century had been waging a struggle against the old order and the landlords, but by the middle of the 19th century had switched fronts to face the working class. The outcome of the Classical school's decomposition was not, however, simply vulgar economics and bourgeois apologetics. Just as petty-bourgeois radicalism and proletarian socialism had each emerged as distinct currents out of the general revolutionary movement which had united the whole of the 'third estate' against monarchy and aristocracy at the close of the 18th century, so, too, did tendencies emerge from within the Classical school which stood in principled opposition to the Classics' theories. The petty-bourgeois opposition to Classical doctrine found its representative in Sismondi, the proletarian opposition in the utopian socialists.

Simonde de Sismondi (1773-1842),* almost all of whose life was spent in quiet Switzerland, had been shocked by the contrast between the patriarchal existence of Switzerland's prosperous peasants and craftsmen and the picture that unfolded before him in *England* of capitalism's frenetic development and its accompanying dislocation of peasants, ruin of the manual weavers, and rising pauperism and unemployment. *The crises of 1815 and 1818*, which rocked English industry, brought ruin to the factory owners, and left the workers without a crust of bread, made a deep impression on Sismondi, forcing him to doubt the validity of Classical theory which until then

*His main economic works are *Nouveaux principes d'économie politique* (1819) and *Etudes sur l'économie politique* (1837). In addition he also wrote a number of outstanding historical works: *L'Histoire des Républiques Italiennes dans le moyen âge*, *L'Histoire des Francais*, and others.

he had accepted. In his book *Nouveaux principes d'économie politique* (1819) he made a decisive break with the 'orthodox doctrine' of the Classics and presented an incisive picture of *the capitalist system's contradictions and calamities*.

In the preface to the second edition of his book Sismondi eloquently described the impression that capitalist England had made on him. 'In this astonishing country, which seems to be submitted to a great experiment for the instruction of the rest of the world, I have seen production increasing whilst enjoyments were diminishing. The mass of the nation here, no less than philosophers, seems to forget that the increase of wealth is not the end in political economy, but its instrument in procuring the happiness of all. I sought for this happiness in every class, and I could nowhere find it.' The crises are leaving merchants and factory owners ruined, while the mass of the population are suffering hunger and deprivation. 'The people of England are destitute of comfort now, and of security for the future. There are no longer yeomen, they have been obliged to become day-labourers. In the towns there are scarcely any longer artisans, or independent heads of a small business, but only manufacturers. The *operative*, to employ a word which the system has created, does not know what it is to have a station; he only gains wages, and as these wages cannot suffice for all seasons, he is almost every year reduced to ask alms from the poor-rates.'[2] The above passage characterizes wonderfully Sismondi's own *social position*: he detests capitalism because it is a system which has impoverished the peasantry and craftsmen, created a class differentiation within the homogeneous ranks of the *small independent producers*, and 'made the poor man more poor' and 'the rich man more rich'.[3]

Capitalism has produced a 'false prosperity'. The only way in which the 'immense accumulation of wealth' it has created could contribute to the happiness of all would be if it was 'distributed ... in proportions which cannot be disturbed without extreme danger'—i.e., more or less equally between society's different classes.[4] The reality, however, is that capitalist development has led to an enormous concentration of wealth in the hands of a few and has tended to 'completely separate all kinds of property from all kinds of labour'.[5] According to Sismondi's view the primary contradiction within capitalist economy is that between the feverish *rise in the production* of wealth and the growing *inequality in the way it is distributed*. The contradictoriness of a system which purchases the happiness of the few with the sufferings of the many not only elicits Sismondi's moral indignation —it stirs his 'heart'. He equally wants to show through 'reason' that this contradiction undermines the economy's ability to develop normally and provokes constant shocks, which manifest themselves in the form of crises. To this end Sismondi constructs a new *theory of markets and crises*, as a counter to Classical doctrine.

For Sismondi, the Classics had turned political economy into '*chrematistics*', or the science of increasing wealth.[6] They preached that production should expand without limit, but showed no concern for its balanced and correct distribution. 'M. Ricardo ... has completely abstracted from man, and put forward as the sole aim of science the unlimited increase of wealth'.[7] But, in addition to being oblivious to the just interests of the toiling masses, the Classics also committed a fatal theoretical error: they failed to grasp that a rapid rise in production is impossible when the *purchasing power* (or *income*) of the lower classes *is falling*. Not understanding that *production* depends *upon income*, the Classics 'announced that whatever abundance might be produced, it would always find consumers, and they have encouraged the producers to cause that glut in the markets which at this time occasions the distress of the civilized world'.[8] Sismondi has in mind here *Say's and Ricardo's* renowned *theory of markets*.

The theory of markets as laid down by Say (and James Mill) and as accepted by Ricardo, is reducible to the following, simple proposition. It is impossible to talk about a *general over-production* of commodities or about any overall shortage of demand. If any *additional commodity*, e.g., cloth, appears on the market and has a value of £10,000, this means that simultaneously there is created an *additional demand* of equal amount for other commodities. 'A product', said Say, 'is no sooner created, than it, from that instant, affords a market for other products to the full extent of its own value'.[9] Indeed the factory owner, upon selling his cloth, receives £10,000. Out of this he has a capital of £8,000, which he again advances for the production of cloth, i.e., he hires workers (Say, following Smith's example, ignores outlays on constant capital and assumes that the entire capital is ultimately to be spent on wages) who once in receipt of their £8,000 in wages create a demand for articles of consumption. The factory owner's profit is £2000, which he spends on means of consumption and luxuries. In the end a total demand of £10,000 is created exactly equal to the value of the cloth.

Yet what would happen if the factory owner wanted to accumulate half his profit instead of spending it entirely on personal consumption? Would not demand (£8,000 + £1,000) then be less than supply (£10,000)? The followers of Say and Ricardo would say not, for, by accumulating, the factory owner is adding £1000 to his capital, i.e., he is hiring additional workers who will create a further demand for means of consumption equal to £1000. Total demand as before will equal £10,000 (9000 + 1000), with the difference that the workers' demand for means of consumption will have grown by a thousand pounds, while the factory owner's demand for luxuries will have fallen by the same sum. The character of the demand will have altered, but its total will continue to be determined by the value of the cloth that has been produced, i.e., by the volume of production. In the last instance cloth will have been exchanged for other products: in the words of Say, 'one kind of produce has been exchanged for another', and the increased production of one product will be balanced by a growth in the demand for others. 'The mere circumstance of the *creation of one product* immediately *opens a vent* for other products', says Say.[10] '*Demand* is only limited by *production,*' repeats Ricardo.[11] *Production creates its own market* [or, in common parlance, 'supply creates its own demand'—*Ed.*]; it is therefore foolish to say that the total volume of production can in any way exceed the general level of demand.

But in that case, how are we to explain the outbreak of crises? According to Say's and Ricardo's theory, sales crises arise out of contingent circumstances (e.g., wars, foreign markets being closed off, etc.). Certainly there exists the possibility of *partial over-production* of certain products, but this inevitably means that there is *under-production* of others. If, for example, the owners of factories making luxury articles were to fail to make a proper assessment of the changes in the structure of demand that we described above and again produced 2000 pounds' worth of luxuries, that branch would suffer from over-production. But at the same time an under-production of workers' articles of consumption would be felt. A transfer of capital from the first branch into the second would swiftly eliminate this temporary disproportion in production. Therefore the only crises that are possible are *partial crises*, which arise out of mistakes in the managing of production. *Generalized crises,* where all branches of production suffer from a shortage of demand simultaneously, are, however, *impossible*.

Such was the theory of markets set out by Say and Ricardo; unfortunately, the facts unambiguously refuted it: England periodically went through the shock of generalized crises characterized by inadequate demand and a fall in the prices of every important commodity. This theory was blind to the fundamental contradictions of capitalist economy, depicting it instead as a unified whole distinguished by a perfectly mutual adjustment and harmonious development of all its parts. Say forgot that cloth is not exchanged directly for other products, but must first be sold for money, and what is more, for a definite sum of money which will cover its costs of production plus profit. Say's portrait of the production process mistakenly ignored outlays on means of production. He was therefore unable to grasp the interdependence between the production of means of consumption and the production of means of production or the disparity that exists in their rates of growth. He underestimated the anarchy of production, which renders impossible the balanced and proportional development of all branches of production.

Sismondi's great *merit* was to have rejected the Classical school's theory of markets. From 1819 to 1824 Sismondi took part in three polemical jousts with the best economists of the Classical school: McCulloch, Ricardo, and Say.[12] All three argued that an unlimited growth of production was possible and would not run up against any shortage of demand. Sismondi, like Malthus, argued that 'consumption is not the necessary consequence of production',[13] and that a rapid growth of production will inevitably provoke the outbreak of general crises.

Sismondi's theory is that the volume of *production* is limited by the scale of *consumption*, while the scale of consumption is in its turn limited by the aggregate *income* of society's members. Just as an individual has to balance his consumption with his income, so society must obey the same rule. 'The national income must regulate total production.'[14] Suppose that in the year just ended all of society's classes have received a definite aggregate income, say, five billion pounds. It is clear that the total demand for products that these classes will create in the coming year can amount to no more than five billion pounds. Consequently, the *volume of production* in the year ahead must not exceed the *aggregate income* that has accrued to all classes of society in the year gone by. 'Thus, national income and annual production are mutually in balance and appear as equal quantities.'[15] 'The entire annual income is destined to exchange against the entire annual production.'[16] Should production in the year ahead rise to six billion pounds' worth, one billion worth of commodities would clearly remain unsold. 'If the annual income did not purchase the whole of the year's production a part of this production would remain unsold, piling up in the producers' shops and paralyzing their capitals, and production would come to a halt.'[17]

Sismondi bases his arguments on a monumental theoretical error. Like Smith and Say, Sismondi ignores outlays on *constant capital*. The value of the annual product in reality resolves itself not only into *revenue* (wages, profit, and rent), but also into a portion for the replacement of expended *constant capital*. In consequence, the total annual product must be *greater than* the total annual income. Were Sismondi's statement correct that 'last year's revenue must pay for this year's production'[18] it would mean that the workers and capitalists would consume the whole of the annual product. Sismondi ignores the need to restore used up constant capital and is thus unable to explain how new capital is accumulated.

Sismondi therefore proceeds from the mistaken idea that there can only be equilibrium in capitalist economy if the current year's *production* equals last year's *income*. Yet any equality between them is constantly being disturbed: because the volume of production (owing to the frenzied competition between capitalists) and technical progress (the introduction of machinery) tend to *expand* without limit, there is a relative *fall* in the aggregate income of the broad mass of the population. *The peasantry* become ruined, their purchasing power declines and their demand for industrial products is cut back. The income and purchasing power of the *workers* fall in exactly the same manner, since the introduction of machinery aggravates their unemployment and at the same time offers the capitalists the opportunity to drive down wages. 'Wages almost always fall from a growth of public wealth.'[19] 'Improvements in machinery and economies in human labour lead to a direct reduction in the number of a nation's consumers; because every worker that is ruined was a consumer.'[20]

Profound, hopeless contradictions are thus to be found in the very nature of capitalism: it simultaneously expands production and reduces the income (and consumption) of the broad mass of the population. Hence the *underconsumption* of the mass of the population inevitably provokes constant *crises of overproduction*. In developing his theory of crises, Sismondi did not understand that a fall in workers' demand for articles of consumption goes hand in hand with a stupendous growth in the *production of means of production*. In ignoring what is laid out on constant capital, Sismondi mistakenly thought that the only means by which to compensate for a relative drop in the workers' demand for articles of consumption would be to raise the *capitalists' demand for luxuries*. Sismondi endeavoured to prove that this type of compensation was impossible, but the theoretical flimsiness of the arguments he used was obvious. Sometimes he wrathfully condemned the excessive and 'frivolous enjoyment of luxuries', at others he argued that the capitalists are physically incapable of consuming all the luxury articles they produce, and on still other occasions he pointed out how difficult it was to shift production away from workers' means of consumption towards luxuries for the capitalists. Whatever the case, Sismondi was convinced that the crises rocking capitalist industry had their direct cause in the falling income and purchasing power of the mass of the population, which narrowed that industry's internal market. Capitalism could rapidly develop only if it could sell its commodities on foreign markets. As soon as the capitalist powers had seized all the colonies and foreign markets, crises would occur with ever greater frequency and severity. Sismondi's doctrine that capitalist development is impossible without foreign markets was to be accepted by the Russian Narodniks in the 1870's and 1880's.

For all his theoretical mistakes, it is still Sismondi's great merit to have been the first to pose *the problem of markets and crises* in its full scope. He correctly recognized that crises are the inevitable companion of capitalist economy and, in attempting to find the reasons why they appear, sought them in capitalism's internal structure. Because he based himself on Smith's erroneous theory of reproduction and ignored the increasing importance that the production of means of production acquires in capitalist economy, Sismondi was unable to provide a correct solution to the problem of crises. But it is still his great merit to have posed this problem, to have tried to provide a unitary and reasoned answer to it, and to have thrown light on one aspect of the phenomenon, namely that the development of capitalism depends on the purchasing power of the broad mass of the population. Sismondi *undermined the Classical School's faith* in the possibility of smooth, crisis-free capitalist development. He refuted their optimistic confidence that all classes stand to gain from the functioning of an economic system based on free competition. If Carey and Bastiat had reduced to absurdity Smith's idea of a harmony of interests between all members of society, Sismondi had come to the sad conclusion that 'where all interests are in conflict with one another, injustice will often triumph'.[21]

Sismondi, considering the capitalist system profoundly unjust, came to the conclusion that it could make no pretence to an eternal existence. He correctly explained that the Classics' conviction that capitalism was *the eternal and natural form of economy* was a product of their limited horizons. 'Our eyes have become so accustomed to this new organization of society, to this universal competition which degenerates into hostility between the wealthy and labouring classes that we can no longer conceive of any other mode of existence, not even of *those whose debris* surround us on all sides.'[22] As is clear from Sismondi's last phrase, his incisive critique of the capitalist system stems from the interests and ideals of the *petty bourgeoisie*.

Sismondi sought for capitalism's salvation from calamity in the patriarchal *peasant and handicraft* economies that still flourished in his native Switzerland and whose scattered 'debris' were extant even in England. Sismondi's call is not forward from capitalism, but backwards. He did not intend to counterpose to capitalism a new order, to be founded on collective property: 'Who could ever be capable of conceptualizing an organization that does not yet exist, of visualizing the future, when it is so difficult for us to see the present?'[23] Thus Sismondi emphatically dissociated himself from the utopian socialists of his day (Owen, Fourier, Thompson); in contrast to their schemes for socialist communes he counterposed his own picture of a desirable social system as follows: 'I would like factory, as well as agricultural industry to be divided up into a large number of independent workshops, and not concentrated in the hands of a single entrepreneur superintending hundreds or thousands of workers. I would like industrial capitals to be split up amongst a large number of medium-sized capitalists, and not concentrated in the hands of one person who possesses millions.'[24] This is Sismondi's economic ideal: a society comprised of well-to-do *peasants*, independent *craftsmen*, and *small-scale* merchants.

Once Sismondi had turned his back on any fundamental alternative to capitalist society with its basis in the right to private property, he had no other option but to attempt to smooth over capitalism's disasters via *social reforms*. Sismondi did not accept the Classical doctrine that state intervention into economic life is impermissible. He became one of the first and most ardent champions of social reform, and in this sense is one of the forerunners of the socio-ethical tendency that was to win a wide measure of popularity in the 1870's. We have seen that for Sismondi, capitalism's basic contradiction manifests itself, 1) in an overly rapid *growth of production*, together with a *fall in* the population's *purchasing power*, the latter being a consequence of 2) the destruction of the *petty producers*, especially the peasantry, and 3) the declining standard of living of the *workers*.

Sismondi directed his schemes for social reform at these three targets. To improve the lot of the *workers* he recommended (being in this regard one of the first and most ardent advocates of factory legislation) a series of legislative measures: the right of workers to form combinations; a prohibition on child labour; a mandatory rest day on Sundays; that entrepreneurs be obliged to provide upkeep for their workers during times of sickness and unemployment; etc. In addition, Sismondi wanted to sustain *the small-scale peasant* economy, and eloquently described the latter's advantages over the large-scale estates and latifundia. He did not forget to add that 'a numerous class of peasant proprietors provides a very strong guarantee for the maintenance of the established order'.[25] Sismondi's projects for factory legislation showed the progressive side of his world outlook, but his enthusiastic sympathy for the labouring peasant population was not free from a tinge of conservatism. Finally, his aspirations to *limit the volume of industrial production* are reactionary through and through. It is true that Sismondi rejects the charge levelled at him by his opponents of hostility to industrial progress and technological advance. 'The evil today is not innovation, but the unjust distribution which man makes of its fruits',[26] replies Sismondi to his enemies. But when Sismondi says that there must be slower introduction of machinery so as to avoid displacing craftsmen and workers ('distress', he writes, 'has reached such depths that one could begin to regret the progress of a civilization which ... has only multiplied poverty'[27]) his advice becomes at one and the same time *utopian* and *reactionary*. Ricardo's doctrine that there must be maximum growth of the productive forces bore a more progressive character than Sismondi's lamentings over the excessive growth of production. At a certain point the limitations of the petty bourgeois critique of capitalism which found expression in Sismondi's works became abundantly obvious.

1 There is no English translation of Sismondi's *Nouveaux principes d'économie politique*. Quotations from the Preface to the Second Edition of the *Nouveaux principes* (1827) are from the English translation made by M. Mignet included in the collection of Sismondi's essays which Mignet published under the title *Political Economy and the Philosophy of Government* (London, 1847). The first volume of the second edition, together with three articles replying to McCulloch, Ricardo, and Say, which Sismondi wrote between 1819 and 1824 and later incorporated

into the second edition, have been republished in a modern edition by Calmann-Lévy (Paris, 1971), from which the quotations from Vol. I have been taken. Quotations from Vol. II have been taken from the first, 1819 edition of the *Nouveaux principes*.

2 Preface to the Second Edition, in Mignet, *op cit*, pp. 115-17.
3 *Ibid*, p. 114.
4 *Ibid*, pp. 114-18.
5 *Nouveaux principes*, Vol. I, (1971 edition), p. 356. This is from the reply to Ricardo, 'Sur la balance des consommations avec les productions,' originally published in May 1824, in the *Revue encyclopédique*.
6 The term used by Rubin is *styazhanie*, meaning acquisitiveness.
7 *Nouveaux principes*, Vol. I (1971 edition), p. 322.
8 Preface to the Second Edition, Mignet, pp. 119-20.
9 Say, *Treatise on Political Economy*, Prinsep translation, Vol. I, p. 167.
10 *Ibid*, p. 167, Rubin's italics.
11 Ricardo, *Principles* (Sraffa edition), p. 288.
12 The reply to McCulloch was originally published in 1820, in Rossi's *Annales de Jurisprudence*, and was a response to McCulloch's attack on Sismondi and Owen in the *Edinburgh Review* (October 1819). The reply to Ricardo, cited above, arose out of personal discussions with Ricardo when he visited Sismondi in Geneva, in 1823. In response to Sismondi's article on Ricardo in the *Revue encyclopédique*, Say published his own reply to Sismondi in the July 1824 issue of the same journal (entitled 'Balance des consummations avec les productions'), which Sismondi answered with a brief essay, 'Notes sur l'article de M. Say, intitulé "Balance des consummations avec les productions."' See the Calmann-Lévy republication of Vol. I of *Nouveaux Principes* and note 1 above.
13 Reply to Ricardo, in *Nouveaux principes*, Vol. I (1971 edition), p. 343.
14 *Nouveaux principes*, Vol. I, p. 125. 'The *national income* must regulate *national expenditure*, the latter must absorb, through the fund of consumption, the whole of production.'
15 *Ibid*, p. 120.
16 *Ibid*, p. 121.
17 *Ibid*, p. 121.
18 *Ibid*, p. 129.
19 Reply to McCulloch, in *ibid*, p. 336. 'It is not the worker who gains from the multiplication of the products of labour; his wages are in no way raised by it; M. Ricardo has himself said elsewhere that they must not do so if one does not want the public wealth to stop growing. To the contrary, baleful experience teaches us that wages almost always fall as a result of this multiplication.'
20 *Nouveaux principes*, Vol. II (1819 edition), p. 326.
21 *Nouveaux principes*, Vol. I (1971 edition), p. 289.
22 Reply to Ricardo, in *ibid*, p. 357. Rubin's italics.
23 *Loc cit*, p. 364.
24 Translated from the Russian.
25 This is Sismondi's own index entry to Book III, Chapter 3 of Vol. I. The passage to which it refers appears on p. 160 of the Calmann-Lévy edition: 'The revolution has caused a prodigious increase in the class of peasant proprietors. Today in France one counts more than three million families who are absolute masters of the land that they live on, which means more than 15 million persons. Thus more than half the nation has for its own part an interest in being guaranteed their every right. The multitude and physical force are both on the same side as order.'
26 Reply to Ricardo, in *Nouveaux principes*, Vol I (1971 edition), p. 356.
27 *Nouveaux principes*, Vol. II (1819 edition), p. 328. 'When each day a new machine replaces several families, with no new demand to provide them with employment

and a livelihood, distress has been carried to such depths that one could begin to regret the progress of a civilization which, in gathering an ever larger number of people on the same patch of earth has only multiplied their poverty, whereas in the desert at least, it can only claim a small number of victims.'

CHAPTER THIRTY-EIGHT

The Utopian Socialists

While Sismondi criticized the Classical school from the standpoint of the ruined *petty bourgeoisie*, the early utopian socialists expressed the demands and aspirations of the young *working class*. Their critique was, therefore, deeper and more principled than that of Sismondi. A detailed exposition of the development of socialist ideas is beyond the scope of our task. We will only dwell briefly on a few of the post-Ricardian English socialists who concerned themselves with questions of economic theory and drew socialist conclusions from the Classical school's doctrine.

Among these economists we can distinguish two groups. The first includes Piercy Ravenstone and Thomas Hodgskin.* They both had been at least partially influenced by Godwin's ideas, and each dreamed of replacing the capitalist system, which had brought ruin to the popular masses, with a patriarchal economy of small-scale peasants and craftsmen. They therefore constituted a transitional grouping between the petty-bourgeois critics of capitalism and the socialists. The second group were socialists, and was made up of William Thompson, John Gray, and John Bray.** Combining the economic ideas of Ricardo with the socialist ideas of Owen, these writers were sympathetic to setting up socialist communities. Their socialist ideals, however, were marked by extreme inconsistency, and their works contain innumerable residua of the ideals of the petty bourgeoisie. Like their forefathers, Godwin and Owen, all of these economists made a forceful and incisive critique of capitalist economy, and this is the strongest part of their works. They proved much weaker at

*Ravenstone's main work is his *A few Doubts as to the Correctness of some Opinions Generally Entertained on the Subject of Political Economy*, which appeared in 1821. Thomas Hodgskin was born in 1787 and died in 1869. His main works are *Labour Defended* (1825), *Popular Political Economy* (1827), and *The natural and artificial Right of Property contrasted* (1832).

**William Thompson was born in 1785 and died in 1833. His main works are his *Inquiry into the Principles of the Distribution of Wealth most conducive to Human Happiness* (1824) and *Labour Rewarded* (1827); John Gray was born in 1798 and died in 1850. His main work is his *Social System* (1831); John Bray's main work is his *Labour's Wrongs and Labour's Remedy* (1839).

developing an economic theory. In essence, they all accepted (and at times without any underlying criticism) the basic tenets of Ricardo's theory, merely giving it a different interpretation from the point of view of their own socialist ideals. Wherever the Classics had put a plus, the socialists put a minus; conversely, wherever the Classics had put a minus, they put a plus.

In their *social philosophy* the utopian socialists for the most part shared the ideas of *natural right*. In this regard they did not differ from the early representatives of the Classical school. However, they sharply dissented from the latter in their answer to the question as to just which social system could be considered, '*natural*', rational, and just. Smith had identified the capitalist order as the natural one. The socialists considered that it was based on usurpation (the land and means of production having been seized by the landlords and capitalists), violence, and deceit. They designated it an 'unnatural' system. Is it possible, asks Hodgskin, to acknowledge as 'a natural phenomenon the present distribution of wealth; though it is in all its parts a palpable violation of that natural law which gives wealth to labour and to labour only; and though it is only maintained by an armed force, and by a system of cruel and bloody laws?'[1]

We can already see from this quotation why the socialists hold the capitalist system to be in contradiction with natural law: in their view it is a system which *violates the 'natural' law of labour value*. Here the novel and special meaning that the socialists give to the law of labour value shows itself brilliantly. They accept this law completely, just as Ricardo had formulated it. The socialists persistently repeat after Ricardo, 'labour is the sole source of value', without making any improvement whatsoever on Ricardo's formula. Even Thompson, imputed by many bourgeois historians of economic thought to be Marx's immediate forerunner, failed to differentiate concrete labour from abstract labour and confused exchange value with use value. He identifies labour not only as the sole source of exchange value, but of wealth as well: 'When we value an article of wealth, it is in fact the labour concentrated in its fabrication and in the finding or rearing of its natural material that we estimate.'[2]

Although they took over *in toto* Ricardo's labour value formula, the socialists imparted a different *methodological sense* to it. Ricardo had seen in this formula a law *which actually functions* (albeit with deviations) within capitalist economy. The socialists assumed that in capitalist economy this law *is violated and does not assert itself.* The socialists took what for Ricardo was *a theoretical law of the real phenomena* of capitalist economy and turned it into *a moral postulate* whose realization awaited the future socialist society. They substituted the doctrine of '*the worker's right to the full product of labour*' for the *labour theory of value*. 'Every man', wrote Bray, 'has an undoubted right to all that his honest labour can procure him.'[3]

Methodologically, this new formulation of the labour theory of value signalled a step backward when compared with Ricardo. It calls to mind the 'normative' way that the thinkers of the Middle Ages had posed the problem of value. Whereas for Marx (and to a lesser extent, for Ricardo as well) the labour theory of value served as a tool for comprehending and explaining the phenomena and categories of capitalist economy, the early socialists used the category of labour value as a means for rejecting the extraneity and falseness of other economic categories in the form they possess within capitalist economy (e.g., money, capital, wage labour, etc.).

Indeed, if a commodity's value has to be expressed in *labour*, why then express it in *money*? The early socialists had no understanding that in a commodity economy a product's labour value cannot be expressed in anything other than money. To them it seemed possible to define the value of a commodity directly in *labour units*. 'The natural standard of value is, in principle, human labour', wrote Owen and hence came to the conclusion that 'it has now become absolutely necessary to reduce this principle into immediate practice'.[4] Following Owen's example, the idea of '*labour money*' and '*exchange markets*' became extremely popular among socialists. Gray proposed that a national bank be set up where each producer could hand in his product and in exchange receive a certificate for a determined number of labour units. The owner of the certificate would have the right to obtain any product from the bank's stores valued at the same number of labour units. This type of non-monetary transaction would assure the producer of being able to sell his product at any moment for its *full labour value*. The category of money would be abolished in order to realize more fully the principle of labour value. Regrettably the exchange markets and banks rapidly went bankrupt, and demonstrated for all to see that it is impossible to abolish money so long as the products of labour retain their character as commodities and values. '*Organized exchange*' in the midst of disorganized commodity production proved a utopian undertaking.

Yet if the principle of labour value was to triumph completely it was certainly not just money that would have to be abolished, but all categories inherent in capitalist economy. The exchange of capital for living labour (labour power) sharply contradicts the law of labour value, since the worker receives in this exchange less value than the value of his 'labour'. The merit of the early socialists was to have emphatically underscored this basic contradiction in which Classical theory had got itself entangled. But to resolve this contradiction, i.e., to demonstrate how surplus value rises out of the operation of the law of value, this they did not know how to do. They maintained (and here they were in agreement with Malthus and other critics of the Ricardian theory of value) that the appearance of surplus value contradicts the law of labour value. Hence they concluded that wage labour and capitalism ought to be recognized as harmful and unnatural institutions. If Ricardo had been unable to grasp the historically *transient* character of capitalism, the early socialists could not understand that it was historically *necessary*. In their eyes the capitalist system is nothing more than 'an alienating economic system', 'brazen, albeit legal robbery'. In their justifiable indignation at the inequalities of capitalist economy the early socialists lost sight of the need to gain knowledge of, and to study its real phenomena. Their ethical rejection of capitalism was too readily transformed into a theoretical disregard for its inherent laws. Being overly preoccupied with constructing plans for what *ought* to be, the utopian socialists gave insufficient study to what *is*.

Although this methodological position relaxed their concern for the theoretical study of the phenomena of capitalist economy, when it came to understanding the problem of *surplus value* the proletarian point of view of the early socialists allowed them to make a real advance over the Classics. They understood the mechanism of capitalist exploitation far better and more incisively than did the Classics. The idea that the landlords and capitalists receive their revenue out of the value of the workers' product—an idea that already existed in Smith—was placed emphatically to the fore by the early socialists. 'In this which is at present the case, the labourers must share their produce with unproductive idlers',[5] wrote Hodgskin. Tremendous credit is owed to the author of one socialist pamphlet, published in 1821, for having brought all forms of unearned income together under one category, though it must be admitted that he still called it interest, and not surplus value. 'The interest paid by the capitalists, whether it acquires the form of rent, monetary interest, or entrepreneurial profit, is paid out of the *labour of other people*.'[6] All species of unearned income are here united into surplus value, whose source is recognized as the *'surplus labour'* of the workers. We find virtually the same concept of surplus value in Thompson: 'There can be no other source of profit than the value added to the raw material by the labour, guided by skill, expended on it. The materials, the buildings, the machinery, the wages, can add nothing to their own value. The additional value proceeds from labour alone.'[7] The most important theoretical service of the early socialists was to have understood the nature of surplus value—in this realm they paved the way for Marx.

We have included this review of the economic theories of the early socialists in our section on the break up of the Classical school, not simply because these authors used the Ricardian theory of value as a basis for drawing practical, socialist conclusions which diverged sharply from the doctrine of the Classical school. Even from a purely theoretical point of view the way that we find the early socialists posing economic problems testifies to the collapse of Classical theory. Ricardo was unable to reconcile the law of labour value with the real phenomena of capitalist economy (the exchange of capital for labour power, the equalizing of the rate of profit on capital). So long as no one prior to Marx was able to eliminate this basic contradiction, there remained only two ways out of the difficulty. Either one had to *repudiate the labour theory of value* so as to concentrate total attention on studying the superficial phenomena of capitalist economy; or one could *retain the principle of labour value*, but at the cost of abandoning any theoretical analysis of capitalism's real phenomena. The first tack was taken by Malthus, Torrens, and Ricardo's other critics, who declared the labour theory of value to be a deceptive fiction. Down the second path went the utopian socialists, who declared that the deception and the fiction was the capitalist system itself, based as it was on the 'unequal' exchange of capital for labour. Both directions signalled *the collapse of Classical theory*.

1 Thomas Hodgskin, *Popular Political Economy* (London, 1827), p. 267.
2 William Thompson, *An Inquiry into the Principles of the Distribution of Wealth*, Third Edition (1869), p. 67.
3 John Bray, *Labour's Wrongs and Labour's Remedy; or the Age of Might and the Age of Right* (Leeds, 1839), p. 33; cited by Marx in *The Poverty of Philosophy*, Progress Publishers English edition (Moscow, 1966), p. 61.
4 Robert Owen, *Report to the County of Lanark*, in Robert Owen, *A New View of Society and Other Writings*, Everyman edition (London, 1927), p. 250.
5 Hodgskin, *Popular Political Economy*, p. 245.
6 Translated from the Russian.
7 Thompson, *An Inquiry into the Principles of the Distribution of Wealth*, Third Edition, p. 127.

CHAPTER THIRTY-NINE

The Twilight of the Classical School

JOHN STUART MILL

The Classical school is usually taken to have begun with the appearance of Adam Smith's *Wealth of Nations*, in 1776, and to have ended with the appearance, in 1848, of John Stuart Mill's *Principles of Political Economy with some of their Applications to Social Philosophy*.

Smith's book envisaged vast and optimistic prospects for economic progress. Ricardo's system, which marked the highpoint of the Classical school's development, stood under the banner of the bourgeoisie's struggle against the landowning class and was already displaying presentiments, albeit weakly, of the impending struggle with the working class. After Ricardo, bourgeois economic thought was increasingly directed towards the defence of bourgeois (and landed) property from attack by socialists: the Classical school went through a period of *vulgarization and apologetics*, at the same time encountering more and more forceful opposition from the early socialists. Finally, the Classical school, in the person of John Stuart Mill, in effect gathered together its last and finest forces in order to show itself once again to be in step with the times and to provide an answer to the new problems confronting humanity. It was a *belated* attempt: all it proved was that the Classical school's creative powers had been spent, that the ideas and theories to which it had given birth were already out of date and no longer capable of providing a foundation on which to build a new, all-embracing system of social philosophy.

As the title of his book shows, Mill (1806-1873) dedicated himself to erecting just such a universal system. Mill wanted 'to exhibit the economical phenomena of society in the relation in which they stand to the best *social ideas* of the present time, as he [Adam Smith—*I.R.*] did, with such admirable success, in reference to the philosophy of his century'.[1] It might have seemed that Mill, with his philosophical turn of mind,* his all-round, we might even say, superhuman education at the feet of his father, James Mill, and, finally, with his sensitivity towards the most progressive social currents of his own age, would have been better fitted than other economists for fulfilling this grandiose task. Nevertheless, Mill never succeeded in writing 'a work similar in its object and general conception to that of Adam Smith'.[2] Although his book earned him enormous fame and was considered the best course in political conomy right up to the end of the 19th century, both its social-philosophical and theoretical-economic ideas were shot through with glaring and unresolvable *contradictions*. Adam Smith had expressed the point of view of the most progressive class of his age, the industrial bourgeoisie, from whose social practice he had been able to extract and fuse together his social philosophy and his economic theory. By the middle of the 19th century, the ideas of economic and political *liberalism* that Mill had been brought up on were already *outdated*. The contradictions of the capitalist economy, the destitution of the lower masses of the population, the class struggle of the proletariat, and the critique of the socialist thinkers had already undermined faith in the capitalist system as the bearer of general well-being harmonizing the interests of all members of society. Mill did not remain blind to the signs of the times: he showed ardent compassion for the fate of the Irish peasants, followed sympathetically the successes of the workers' movement, and studied the ideas of the Saint-Simonists and Fourierists with interest. He turned his back on the ideas of bourgeois liberalism that had been so dear to him in his childhood, and in his old age became increasingly inclined towards the ideas of socialism. But Mill never succeeded in shifting completely over to the standpoint of the working class: caught in hesitation and doubt, he stopped halfway between *liberalism and socialism*, and it is from this that the multitude of contradictions in his social philosophy stem.[3]

The basic tone of Mill's social-philosophical reflections is one of profound *disillusionment* with the capitalist system and its inherent *competition and struggle* between individuals and classes. The time when Adam Smith could write that the individual, 'by pursuing his own interest ... frequently promotes that of the society more effectually than when he really intends to promote it' had long since passed.[4] Now Mill was counterposing himself to the 'economists of the old school' and writing: 'I confess I am not charmed with the ideal of life held out by

*Mill was the author of the well-known *System of Logic* and a number of other philosophical works. [*A System of Logic, Ratiocinative and Inductive, being a connected view of the Principles of Evidence, and the Methods of Scientific Investigation*, London, 1843—*Ed.*]

those who think that the normal state of human beings is that of struggling to get on; that the trampling, crushing, elbowing, and treading on each other's heels, which form the existing type of social life, are the most desirable lot of human kind, or any but the disagreeable symptoms of one of the phases of industrial progress.' [5] It was not just Smith's naive faith in the universal harmony of interests that Mill no longer shared; the same applied to Ricardo's more modest hope that the capitalist system, 'by increasing the general mass of productions ... diffuses general benefit.' [6] To point to the mighty growth of the productive forces inherent in capitalism was small comfort: 'Hitherto it is questionable if all the mechanical inventions yet made have lightened the day's toil of any human being. They have enabled a greater population to live the same life of drudgery and imprisonment, and an increased number of manufacturers and others to make fortunes.' [7]

But where was there a way out? Mill had been influenced by the utopian socialists and was not afraid to pose the question either of 'a general reconsideration of all *first principles*' on which the economy was founded or of the possibility of replacing capitalism with *socialism*. Mill rejects those arguments which pretend to prove that a socialist economy is impossible. 'If, therefore, the choice were to be made between Communism with all its chances, and the present state of society with all its sufferings and injustices ... all the difficulties, great or small, of Communism would be but as dust in the balance.' Mill nonetheless does not come out decisively for socialism. Communism is certainly better, he says, than 'the regime of individual property *as it is*', but whether or not it would be preferable to 'private property as it *might be made*' if submitted to thoroughgoing social reforms, we still do not know. The question of the 'comparative merits' of *communism* versus a *reformed capitalism* remains unresolved. 'We are too ignorant either of what individual agency in its best form, or Socialism in its best form, can accomplish, to be qualified to decide which of the two will be the ultimate form of human society.' [8] This being the case the only thing to do is to subject socialism to 'the trial of experience' by setting up a 'modest number' of socialist communities. Meanwhile, for as long as the question of socialism's advantages remains undecided 'the object to be principally aimed at, in the present stage of human improvement, is not the subversion of the system of individual property, but the improvement of it, and the full participation of every member of the community in its benefits.' [9]

While Mill does not, therefore, reject socialism in principle, he nevertheless has as his main objective the implementation of a series of *social reforms* that will improve the lot of the lower classes. He calls for the setting up of workers' *associations* both in and outside production, limitations on the right on *inheritence*, and high taxes on ground rent. Like Sismondi before him, Mill eloquently defends the small-scale *peasant* economy and demands that the lands seized by the landlords be *handed over to those who work them*. On paper and in speech—as a member of parliament—Mill boldly and forthrightly takes up the cause of all the unfortunate, champions the rights of the Irish peasantry, protests against the colonial brutality of the English, and makes a passionate fight for female equality. He shows himself extremely sympathetic toward the advances made by the worker's movement and its growing self-consciousness: no longer do the workers feel 'any deferential awe, or religious principle of obedience, holding them in mental subjection to a class above them'. At the same time, however, Mill is fearful of the sharpness of the class struggle and counsels the workers to 'make themselves rational beings'.[10]

So we see that even in his *social philosophy*, where he most distanced himself from the ideas of his father and other early 19th-century liberals, Mill stopped halfway between liberalism and socialism. Mill, like the early socialists, had posed the problem of socialism in a *utopian* fashion: the object is for the thinker to adjudge the 'relative merits' of capitalism and socialism and to reason out the ideal social system, which ought then to be established by virtue of the perfection of its inherent characteristics. Although Mill had accepted from August Comte the idea of the historical evolution of human society, this did not enable him to comprehend socialism as a necessary phase of human development or the necessary outcome of the development of capitalist economy and the class struggle of the working class. The question for Mill was not of socialism's *necessity*, but of its *desirability and feasibility*. Yet how could one aspire to set in place a socialist system, or even introduce basic social reforms if, as the Classics had taught, the economy was subordinated to immutable *natural* laws? If the way was to be cleared for social reform, Mill would have to overturn the Classical idea that the laws of the economy are eternal and immutable. Yet here, too, he halted half-way when he made his strange division of economic laws into two types: the laws of *production*, and the laws of *distribution*. 'The laws and conditions of the Production of wealth partake of the character of physical truths. There is nothing optional or arbitrary in them.'[11] 'It is not so with the Distribution of wealth. That is a matter of human institution solely. The things once there, mankind individually or collectively, can do with them as they like. They can place them at the disposal of whomsoever they please, and on whatever terms.'[12] Within *production* eternal and inexorable *natural* laws dominate; within *distribution* what dominates is the free will of *human beings*, who can distribute their produce as they see fit and carry out any social reforms.

Mill's error in making this division is obvious. Under any given mode of production definite relations of distribution are established between people which in turn influence that mode of production. Does the introduction of socialism really mean merely a reform of the relations of distribution and not of the mode of production itself? Can people, as participants in production, really dispose of their produce to 'whomsoever they please, and on whatever terms' without by this very act changing the mode of production? Instead of comprehending the economic process as a unified whole that embraces both the production and distribution of products, Mill artificially tore them away from each other. Instead of subordinating both production and distribution to the operation of laws that, while *necessary*, are nevertheless at the same time *historically alterable*, Mill subordinates production to the operation of *eternal* laws, but sees distribution as an *arbitrary* realm within which the different economic forces display no necessary law-determined regularity.

For all its mistakenness, Mill was compelled to divide up economic laws in this way if he was to leave the door open for social reforms and at the same time preserve intact the system of natural economic laws as established by the Classical school. This dualism between the laws of production and the laws of distribution reflects the fundamental dualism in Mill's entire system, the unresolved *contradiction between his social philosophy and his economic theory*. In his social philosophy Mill had left his father far behind, but in his economic theory he was merely repeating and systematizing the ideas of the Ricardian and post-Ricardian economists. The economic theories that he had accepted in the early days of his youth (his father had given him an exposition on political economy together with the works of Smith and Ricardo to read when he was thirteen) remained unaltered right into his old age—despite the thoroughgoing revision that had taken place in his socio-philosophical outlook. This fervent advocate of social reforms at the same time zealously defended the Malthusian law of population, which argued that any reform of the social order was futile. This friend of the trade unions supported (up to 1869) the theory of the wages fund, which argued that it was fruitless and harmful for the workers to wage an economic struggle. This critic of capitalism failed to notice the basic contradictions of the capitalist economy and supported Say's doctrine on the impossibility of general crises.

In *economic theory* Mill was not an original thinker and he broke no new ground. In his early work, *Essays on some Unsettled Questions of Political Economy* (which he wrote in 1830 and published in 1844), Mill did attempt to make some sort of contribution to the development of Classical theory, particularly to the theory of international trade. But in his main and famous work, the *Principles of Political Economy*, leaving aside his socio-philosophical ideas, Mill did nothing more than provide a full, systematic, and lucid exposition of the theory that had been developed by earlier Classical economists. Although he based his book on *Ricardo's system*, it would be difficult to find a single major Ricardian or post-Ricardian economist whose theories Mill did not accept and work into his own system. From Malthus he took the theory of *population*, from Say his doctrine on *crises*. Like Torrens, he turned the labour theory of value into a theory of *production costs*; following Baley, he limited his analysis to the concept of *'relative'* value. From James Mill and McCulloch he accepted the *wages fund* doctrine (which he repudiated in 1869), and from Senior the theory of *abstinence*. Bursting in on this system of ideas developed by the Classics were the critical ideas that Mill had got from that school's opponents. Following Sismondi's example, Mill fervently championed the small-scale *peasant economy*; following in the footsteps of the utopian socialists, he developed a *critique* of the capitalist system.

Thus, where purely theoretical analysis is concerned, Mill opened up no new scientific horizons, but simply summed up what had already gone before. Not only did he prove unable to move beyond the circumference of Classical ideas; he accepted and set out the majority of them in their post-Ricardian version, i.e., when the Classical school was in a state of decay and decline. Even though Mill was absolutely free of the apologetic aims pursued by the epigones of Classical theory, the process of vulgarization that this theory had suffered at their hands nevertheless left its imprint on Mill's presentation. By way of example, let us turn to those central problems of economic theory, *value and profit*: compared to Ricardo, Mill's development of these problems was a genuine step backward.

Mill distinguishes *three categories* of commodities: 1) commodities whose quantity is absolutely limited, e.g., ancient statues; 2) commodities subject to limitless increase in quantity without the production cost per unit going up, for example, manufactured goods; and

3) commodities the quantity of which can be increased, but only with a rise in the production costs of a unit, for example, agricultural produce.

For commodities in the *first* category, value (or more accurately, price) is established on the basis of *the law of supply and demand*, which in the formulation Mill gave it bears some valuable refinements when compared to his predecessors. They mistakenly talked about the *proportion* between supply and demand, whereas one really ought to talk about the *equality* between them. 'Demand and supply, the quantity demanded and the quantity supplied, will be made equal'.[13] The price of a commodity always establishes itself at that level where the quantity of commodities demanded at a given price equals the quantity offered for sale at that same price. Mill was one of the first vigorously to stress that if the price of a commodity depends on the relationship between demand and supply, then, conversely, the levels of demand and supply will change in response to fluctuations in a commodity's price.[14]

Regarding the second category of commodities, supply and demand determine only temporary deviations of prices from value. A 'stable equilibrium' between supply and demand is possible only where a commodity's price coincides with its value. The magnitude of that value is in this case regulated by *the law of production costs*. 'What the production of a thing costs to its producer, or its series of producers, is the labour expended in producing it'. From this it follows that 'the value of commodities, therefore, depends principally ... on the quantity of labour required for their production'. It would seem at first glance that Mill is accepting the law of labour value as formulated by Ricardo. But then he goes on: 'If we consider as the producer the capitalist who makes the advances, the word *labour* may be replaced by the word *wages*: what the produce costs to him, is the wages which he has had to pay.'[15] Thus 'labour' has been imperceptibly slipped in for 'the value of labour' or 'wages'—a confusion that we find in Smith and which Ricardo had criticized and avoided. In place of the formula, *value is determined by labour*, we have the formula, *value* is determined by the amount spent on *wages*, or the size of *production costs* or *advanced capital* (since Mill continues Smith's mistake of ignoring outlays on constant capital and takes the entire capital to be ultimately laid out on wages).

Of course a commodity's value cannot be determined simply by the total amount spent on wages, for if it were the capitalist would receive no profit. 'In our analysis ... of the requisites of production, we found that there is another necessary element in it besides labour. There is also capital; and this being the result of abstinence, the produce, or its value, must be sufficient to remunerate, not only all the labour required, but the abstinence of all the persons by whom the remuneration of the different classes of labourers was advanced. The return for abstinence is Profit.'[16] Consequently the *value* of a commodity is determined by the amount of *wages* spent on its production plus *the average profit* on this sum. Commodities 'naturally and permanently exchange for each other according to the comparative amount of wages which must be paid for producing them, and the comparative amount of profits which must be obtained by the capitalists who pay those wages.'[17] The vulgar theory of production costs had replaced the labour theory of value.

Mill had to effect this replacement if he was to account for those 'exceptions' to the law of labour value that Ricardo had pointed to and over whose explanation James Mill and McCulloch had vainly puzzled. Once the value of a commodity is determined by the sum of wages (or production costs) plus profit it is no longer surprising that the value of wine which has lain for ten years in a cellar should rise: a profit is being charged on the invested capital throughout the ten years and this enters as an independent element into the commodity's value. In a word, if capital is advanced for a longer period in one branch of production than in another (or the complexity of labour or other circumstances give that branch a higher wage rate or profit level) the value of a product produced in the former branch will be higher than the value of a product produced in the second—even though equal quantities of labour were expended on their production. From the point of view of the theory of production costs there is no problem in explaining these 'exceptions'.

What, then, remains of the law of *labour value*? It asserts itself only in *a single*, seldom-encountered circumstance. If two branches of production advance their capitals for *equal* periods of time and have *identical* levels of both wages and profits, then their products will exchange with one another when manufactured with equal expenditures of labour. This, of course, makes perfect sense: given the assumptions, an equality of expended labour means (since the wage levels are equal) that there is equality in total expenditures on wages, and consequently, that there is also equality in the total profit that is charged (since both profit levels and the circulation periods of capital are identical). In essence, what Mill is saying is that commodities will exchange with one another not because they have had equal quantities of *labour* expended on their production, but because they have equal *costs of production* (i.e., total wages) plus *profit*.

Mill, as we see, paid a dear price for his explanation of the 'exceptions' to the law of labour value: a complete (albeit concealed) rejection of this law, which forms the most valuable part of the Smithian and Ricardian legacy. For all the superficial likeness in Ricardo's and Mill's systems, there exists between them a fundamental and principled divergence. Ricardo considers the basic law to be the *law of labour value*. His mistake is to think that in capitalist economy this law must assert itself *directly*. Thus he identifies those cases where commodities sell at *prices of production* deviating from their labour value—cases which form the general rule in capitalist economy—as *exceptions* to the law of value. He cannot explain these exceptions from the point of view of his general law—hence follows the logical collapse of his system. Yet for all the contradictions in his exposition, Ricardo does not *abandon* his basic law of labour value, and because of this leaves the way clear for future scientific progress. Marx was later to show that the law of labour value regulates the phenomena of capitalist economy indirectly, rather than directly and ultimately determines prices of production which in Ricardo's system functioned as exceptions. He demonstrated that it is possible to understand *the law of production costs* only on the basis of *the law of labour value*.[18]

Mill alters Ricardo's doctrine in the reverse direction. Like Ricardo, he sets himself the wrong objective: he wants to discover under what circumstances the law of labour value will *directly* regulate the exchange of commodities. Unlike Ricardo he correctly sees that in capitalist economy this can only occur in *rare cases*. But then Mill takes the *general rule* to be the sale of commodities at their production costs plus average profit, a situation which for Ricardo had figured only as an exception. He acknowledges the *law of production costs* to be the basic law, and in this he is akin to Ricardo's opponents (Torrens, for instance). To preserve the continuity of his own system with Ricardo's theory, however, Mill first separates off those cases where the exchange of commodities is subordinated directly to the *law of labour value* (when conditions in two branches of production are totally equivalent), and secondly identifies the labour necessary for a commodity's production as the '*principal*' element affecting its value. However, neither of these qualifications change the essence of the matter. There are other elements (differences in the turnover periods of capital, differences in wage and profit levels) which, *together with labour*, *independently* (though with less force) determine a commodity's value. We observe the *law of labour value* operating only as a *special case of the law of production costs*, to be found when certain definite conditions combine with one another.

Thus, Ricardo had declared the law of labour value to be the *basic* law, seeing the law of production costs as a theoretically inexplicable *exception*. It was impossible to maintain such a contradictory system. The contradiction between these two theoretical laws could be resolved only by subordinating one to the other. For Marx the *basic* law was the law of labour value from which the law of production costs was to be *derived*. Mill identified the law of production costs as the *basic* law, from which we *derive* the law of labour value as an occasional instance. Mill was able to eliminate Ricardo's contradiction at the price of *repudiating the law of labour value*, which functions as the basic, hidden regulator of commodity-capitalist economy. He refrained from conducting an analysis of the internal laws of capitalist economy, limiting himself to making generalizations about its external phenomena. In this sense he retreated away from Ricardo, over to the camp of the post-Ricardian 'vulgarizers'. Mill purchased the harmony of his model at the expense of its profundity. Mill's formula is at one with the calculations of the manufacturer: the value of a commodity is determined by costs of production plus profit. But how do we determine the level of production costs? Is it not by the value of labour power, raw materials, etc.? But this is to get caught in a vicious circle, explaining the value of one product (the commodity) by the value of others (means of production).

Still more important are the two other questions for which Mill's formula provides no answer: what is *the origin of profit*, and why is it at a particular level? Whereas Ricardo was moving close to the idea of surplus value and regarded profit as a *part of the value* created by the worker's *labour*, profit for Mill emerges as a value *added to the 'value of labour'*(i.e., wages). On this point Mill has not managed to break free from the influence of the post-Ricardian vulgarizers. In one passage he declares, in the spirit of Ricardo, that 'the cause of profit is that labour produces more than is required for its support'.[19] More often, however, he explains profit by citing Senior's theory of abstinence: 'As the wages of the labourer are the remuneration of labour, so the profits of the capitalist are properly, according to Mr. Senior's well-chosen expression, the remuneration of abstinence.'[20]

Mill's work, which he had conceived of as opening a new era in the development of economic thought, was merely the signal that the Classical school was in the final stages of disintegration. It testified to this fact in *two* ways. The part of Mill's work devoted to *social philosophy* made it clear that the ideas of economic liberalism developed by the Classical school had become irrevocably obsolete and were no longer suitable for resolving the great historical task of abolishing a social system based on the exploitation of man by man. The *economic* portion of his work was graphic proof of the fact that Classical theory was powerless to uncover the inherent law-determined regularity of capitalist economy and was going through vulgarization and retrogression even at the hands of its most progressive thinker. The enormous gulf between Mill's social philosophy and his economic theory testified to the fact that bourgeois economic *theory* could no longer serve, as it had previously, as the basis for a progressive social *practice*. Mill's practical activity was superior to his economic theory, and often contradicted it.

Mill was able to bridge the agonizing gulf between theory and practice by giving social problems a *utopian* formulation. He would analyze the pros and cons of social reforms without thinking to ask to what extent these reforms were the necessary product of capitalist society's internal development. It was for just this reason that he was able to content himself with the outmoded theories of his predecessors when analyzing capitalist economy. A utopian social philosophy co-existed with an antiquated economic theory. If economic theory was to have new perspectives opened up for it there had first to be a reformulation of the entire social problem. Once Karl Marx had made the transition from utopian to scientific socialism he set himself the task of demonstrating that socialism is a necessary phase of human history which flows out of capitalist society's own internal development. For Marx to be able to place socialism on a scientific basis he had to uncover the law-governed regularity behind the development of capitalist economy, which forms the basis of the whole of bourgeois society. Marx cleansed economic theory of the vulgar excrescence left by the Classical school's decay: as the starting point of his analysis he took the most vital of Smith's and Ricardo's ideas, thoroughly re-worked them, and incorporated them into a unified and reasoned-out sociological system. In this way Marx developed the most valuable of the Classical school's ideas and at the same time opened up a new era in the development of economic thought. He fulfilled the difficult task that had been beyond Mill's grasp: he presented 'the economical phenomena of society in the relation in which they stand to the best social ideas of the present time.' He synthesized *scientific socialism with economic theory*.

1 John Stuart Mill, *Principles of Political Economy With Some of Their Applications to Social Philosophy*, edited with an introduction by W. J. Ashley (London, Longmans, Green, & Co., 1921), Preface to the First Edition, p. xxviii. Rubin's italics.

2 *Ibid*, p. xxviii.

3 The following observation by Böhm-Bawerk is illustrative: 'Still more striking is the combination of opposed opinions in J.S. Mill. It has often been remarked that Mill takes a middle position between two very strongly diverging tendencies of political economy—the so-called Manchester school on the one side, and Socialism on the other. It is easy to understand that such a compromise cannot, as a rule, be favourable to the construction of a complete and organic system—least of all in that sphere where the chief struggle of socialism and capitalism is being fought out, the theory of interest. The fact is that Mill's theory of interest has got into such a tangle that it would be a serious wrong to this distinguished thinker were we to determine his scientific position in political economy by this very unsuccessful part of his work.' (*Capital and Interest*, pp. 407-08.) Böhm-Bawerk then goes on, as does Rubin later in this chapter, to point to Mill's convoluted attempt to reconcile Ricardo's labour theory of value with both the theory of production costs *and* Senior's theory of abstinence.

4 *Wealth of Nations* (OUP edition), Book IV, Chapter 2, p. 456.

5 Mill, *Principles*, Book IV, Ch. 6, p. 748.

6 Ricardo, *Principles of Political Economy* (Sraffa edition), p. 134.

7 Mill, *Principles*, Book IV, Ch. 6, p. 751.

8 *Ibid*, Book II, Ch. 1, pp. 208-09.

9 *Ibid*, Book II, Ch. 1, pp. 216-17.

10 *Ibid*, Book IV, Ch. 7, pp. 758, 757.

11 *Ibid*, Book II, Ch. 1, p. 199.

12 *Ibid*, Book II, Ch. 1, p. 200.

13 *Ibid*, Book III, Ch. 2, p. 448.

14 'Thus we see that the idea of a *ratio*, as between demand and supply, is out of place and has no concern in the matter: the proper mathematical analogy is that of an *equation*. Demand and supply, the quantity demanded and the quantity supplied, will be made equal. If unequal at any moment, competition equalizes them, and the manner in which this is done is by an adjustment of the value. If the demand increases, the value rises; if the demand diminishes, the value falls: again, if the supply falls, the value rises; and falls if the supply is increased. The rise or the fall continues until the demand and supply are again equal to one another: and the value which a commodity will bring in any market is no other than the value which, in that market, gives a demand just sufficient to carry off the existing or expected supply.' (*Ibid*, p. 448, Mill's italics.)

15 *Ibid*, Book III, Ch. 3, pp. 457-58.

16 *Ibid*, Book III, Ch. 4, pp. 461-62.

17 *Ibid*, Book III, Ch. 6, p. 479.

18 An excellent, and very accessible discussion of the problem of prices of production and their relation to the law of value is Rubin's own essay, 'Value and Production Price,' in his book *Essays on Marx's Theory of Value* (Detroit, 1972).

19 Mill, *Principles*, Book II, Chapter 15, p. 416.

20 *Ibid*, Book II, Ch. 15, p. 405.

Part Six

Conclusion: A Brief Review of the Course

CHAPTER FORTY

Conclusion: A Brief Review of the Course

Modern political economy came into being and developed in parallel with the emergence and growth of capitalist economy, its object of study. In its evolution it reflected the evolution of capitalist economy and that economy's ruling class, the bourgeoisie. Mercantilist literature, for example, clearly expressed the concerns and requirements of merchant capital and the commercial bourgeoisie.

From the middle of the 18th century, when strict state regulation and the monopolies of the trading companies had begun to put a brake on the growth of industrial capitalism, there was wide-spread opposition to mercantilist ideas. In agricultural France it was the Physiocrats who took up the struggle against mercantilism, under the slogan of fostering productive agricultural capital. The efforts of the Physiocrats ended in practical—and, to a lesser extent, theoretical—collapse.

It fell to the English Classical school, which expressed in the first instance the interests of the industrial bourgeoisie, to make the major practical and theoretical advances. In Smith's doctrine the task of waging a struggle against the antiquated restrictions fettering the growth of the capitalist economy managed to conceal and push into the background the conflicting interests of the different classes that make up bourgeois society. Ricardo's doctrine provided the theoretical foundation for the bourgeoisie in its clash of interests with the landowning class, a clash that revealed itself with bitter intensity in England at the beginning of the 19th century.

At the same time Ricardo could not fail to acknowledge that the bourgeoisie and the working class also had divergent interests—an admission which already contained the seeds of the Classical school's disintegration. With the successful conclusion, in the 1830's, of its struggle against the landlords, the bourgeoisie began to feel itself increasingly threatened by the rising working class: the Classical school's decomposition proceeded at an accelerated pace.

1. Mercantilism

Mercantilist policy, which accelerated the break up of the feudal economy and the guild crafts, corresponded to the interests of the commercial bourgeoisie and merchant capital. Its main objective was to foster a rapid growth of foreign trade (together with shipping and such exporting industries as woollen textiles), striving in particular to reinforce the influx of precious metals into the country, which in their turn accelerated the transition from a natural to a money economy. It is therefore understandable that mercantilist literature focused its attention primarily on two, closely inter-related problems: 1) the question of foreign trade and the balance of trade, and 2) the question of regulating the circulation of money. We can distinguish three periods in the way the solution to these problems was approached: a) the early mercantilist period, b) the period of developed mercantilist doctrine, and c) the beginnings of the anti-mercantilist opposition:

a) The early mercantilists devoted their attention mainly to the *circulation of money*, which went through a period of almost total disarray during the 16th and 17th centuries. In part this was due to the 'price revolution' which was taking place at the time, and in part because sovereigns were debasing metal coins.

The debasement of metal coins, the worsening of the currency's rate of exchange, and the outflow of coins of standard value to other countries were severely affecting the interests of the commercial bourgeoisie. The early mercantilists of the 16th and early 17th centuries were advocates of the 'money balance system', and believed that it would be possible to extirpate these evils through *compulsory* governmental *regulation* over the circulation of money. In particular, they demanded an absolute prohibition on the export of metal coins, hoping that by this means the country's 'monetary balance' would improve.

b) The later mercantilists of the 17th century had already come to understand that fluctuations within the sphere of monetary circulation (a deteriorating rate of exchange and the export of metal coins) result from a country's unfavourable balance of trade. They did not believe it possible to regulate the flow of money directly, and so advised those in power to concentrate their energies on regulating the country's balance of trade by stimulating its commodity exports to other countries. In particular they recommended the development of export industries (so that more expensive industrial manufactures could be exported, rather than raw materials) and the transit trade (i.e., the purchase of colonial commodities in oceanic countries, such as India, to be sold in the European countries at higher prices).

In England, the theory of the 'balance of trade' was expressly developed in Mun's work, *England's Treasure by Forraign Trade,* written in the 1630's.

c) With the end of the 17th century an opposition to mercantilism had already begun to appear. North was one of the first free traders. He called on the state to refrain from exercizing compulsory regulation both over the flow of money to and from other countries, and over the circulation of commodities between them. North demanded full freedom of foreign trade and believed it beneficial for both monetary and commodity circulation to be self-regulating.

The economists who debated the problems of the monetary balance and the balance of trade were primarily interested in those practical questions which touched the interests of the commercial bourgeoisie. Alongside this 'merchant' current in mercantilist thought, there appeared, at the end of the 17th century, a 'philosophical' tendency whose representatives (Petty, Locke, Hume) exhibited great interest in working out theoretical problems, first and foremost those of value and money.

As soon as economists directed their thinking toward the theoretical analysis of economic phenomena they found themselves confronted with the problem of value.

In the Middle Ages, when prices were compulsorily fixed by the town and guild authorities, the problem of value had been posed normatively: the Scholastic writers argued over the 'just price' (*justum pretium*) that needed to be compulsorily established to assure the craftsman his customary standard of living.

During the age of merchant capital the formation of prices via regulation gradually ceded to the spontaneous formation of prices through the market. The economists of the 17th century now found themselves facing a new theoretical problem: what were the laws that governed this market formation of prices? Answers to this question were still superficial and undeveloped. John Locke, the well-known philosopher answered that the movement of prices depended on alterations in demand and supply. Barbon, his contemporary, advanced the theory of 'subjective utility': in his words, 'The Value of all Wares, arise from their Use', and depends on the 'Wants and Wishes' of those who consume them.[1] A more profound attempt to find a law-determined regularity to what was at first glance the disorderly and haphazard movement of prices was made by James Steuart, one of the later mercantilists and an advocate of the theory of 'production costs'. In his view a commodity has a 'real value' equal to its costs of production. The price of the commodity cannot be lower than this real value, but is customarily higher, the surplus comprising the industrialist's 'profit'. Profit, therefore, is something added onto the commodity's value and accrues to the industrialist because he has managed to sell it under favourable circumstances—i.e., it is 'profit upon alienation'. The idea that profit is created within the process of circulation is encountered in almost all mercantilist writing and reflects the conditions within the age of merchant capitalism, when there were colossal profits to be made from foreign trade, the colonial trade in particular. From a theoretical point of view, the doctrine of 'profit upon alienation' signified a complete repudiation of any solution to the problem of profit and surplus value in general.

The most sophisticated solution to the problem of value came from William Petty, the ingenious progenitor of the 'labour theory' of value. According to Petty's doctrine, a product's 'natural price' or value is determined by the quantity of labour expended on its production. When a producer exchanges his product he receives a quantity of silver (money) in which there has been embodied as much labour as he himself had expended on producing the product in question. The value of a product, bread, for example, will resolve itself into two components: 1) wages, which equal the worker's necessary minimum of means of subsistence (Petty and other mercantilists were advocates of the 'iron law of wages', in the sense that they recommended limiting the workers' wages to a minimum of means of subsistence in the interests of capitalism's development), and 2) ground rent. Consequently Petty identifies ground rent with surplus value in general, a view which was widely held in a period when capitalism was only just developing, and which later on was explicitly adopted by the Physiocrats.

In making this identification Petty prevented himself from posing the problem of surplus value; yet despite falling into innumerable contradictions in the way he stated it, in his labour theory of value Petty laid the foundation on which the Classics and Marx were later to construct the theory of surplus value. It is safe to say that Petty's theory of value is the most valuable theoretical legacy that mercantilist literature was to bequeath.

The other theoretical problem besides the theory of value that attracted the mercantilists' attention was that of money. The entire body of old mercantilist literature had revolved around the practical problems of monetary circulation: debasement of metal coins, the export of money abroad, etc. By the end of the mercantilist period, however, we already find Hume and Steuart making more or less mature reflections and formulations of the two conflicting theories of money which are still to this day struggling for scientific supremacy. The famous philosopher, David Hume, provided an explicit formulation of the 'quantity theory' of money, according to which the value of a monetary unit depends upon the quantity of money in circulation: the value of money changes inversely to variations in its quantity. The 'quantity theory' had first been formulated as early as the 16th century, under the impact of the 'price revolution' provoked by the inflow of precious metals from America. Hume, however, deepened and refined it. Hume's opponent on this question was the man already mentioned, James Steuart, who argued that the quantity of money in circulation depends on the needs of commodity circulation. Steuart's ideas were later taken over by Thomas Tooke in the first half of the 19th century and subsequently developed by Marx.

2. The Physiocrats

The term 'Physiocrats' came to be applied to a group of economists who had come onto the scene in the 1760's, primarily in France. The head of the school was François Quesnay, who grouped around himself a number of disciples and partisans. After a brief period of brilliant success, Physiocratic doctrine was supplanted by the theories of the new 'Classical' school that had emerged in England and was for a long time regarded with scorn and even mocked. Marx was one of the first to note the Physiocrats' scientific merits, and they were later to gain increasingly widespread scientific recognition.

While mercantilist doctrine reflected English economic conditions during the age of merchant capital. Physiocratic theory corresponded more to the economic and social conditions of mid-18th-century France. This was a time when France was involved in a global struggle with England for naval, commercial, and colonial supremacy and, after protracted wars, had been forced to cede first place to its rival. The mercantilist policy—pursued with especial determination under Colbert's ministry—of encouraging industry, shipping, and trade at state expense had failed to attain its objectives and had devoured enormous resources.

The combined effect of mercantilist policy and feudal survivals had resulted in the devastation of agriculture. A myriad of factors was operating to hold back agricultural growth: a backward agricultural technology, accompanied by poor harvests; the mercantilist policy of forbidding corn exports, which depressed corn prices; and a tax system the entire weight of which fell on the peasantry and spared the gentry. In their programme of economic reforms the Physiocrats strove to eliminate each of these factors. They fervently championed the type of rational agriculture that had met with remarkable success in England. They recommended that land be leased to large-scale farmers with abundant capital. They demanded the repeal of prohibitions on the export of corn, arguing the benefit of high corn prices and low prices on industrial goods. Finally, in order to insulate the farmer from heavy taxes, they called for all taxes to be shifted onto the rent received by the landlords.

The Physiocrats' economic programme, especially their scheme for tax reform, corresponded to the interests of the rural bourgeoisie and was directed against the feudal gentry. However, as they could not rely on any influential social class (the rural bourgeoisie in mid-18th-century France being too small and inconsequential), the Physiocrats pinned their hopes mainly on the crown, whom they expected to carry out the desired reforms. It is therefore quite understandable that the Physiocrats did as much as they could to dull that edge of their programme directed against the feudal gentry and instead sharpened the attack on mercantilist policy. They shunted the bourgeois character of their programme to the rear and emphatically stressed its agrarian nature. The slogan of defending agriculture from the harmful consequences of mercantilist policy became the Physiocrats' favourite watchword.

The mercantilists had maintained that the best means for making a country wealthy was to extensively develop foreign trade. The Physiocrats acknowledged the only source of a nation's wealth to be agriculture. The mercantilists had seen foreign trade as the miraculous source of the flow of precious metals and enormous profits into the country. In order to refute these mercantilist notions, the Physiocrats had to construct a new doctrine of money and surplus value. It was their view that money was nothing more than a convenient aid in the circulation of products: a nation's wealth consisted of products, not money. But since industrial products were nothing more than raw materials obtained from agriculture and refashioned by the labour of the industrial population, and since the latter obtained its means of subsistence likewise from agriculture, a nation's wealth ultimately consisted of agricultural produce, or the material substance which the agricultural population extracted from the bountiful lap of mother nature. Wealth was created only in the process of agricultural production, and not in the process of circulation. Thus the mercantilist policy of one-sidedly and artificially encouraging trade and industry at agriculture's expense was both the height of absurdity and harmful—for such a policy of stringent state regulation and restriction places constraints on individual economic freedom and therefore violates the laws of 'natural right'.

To give their programme of economic policy a more solid foundation the Physiocrats constructed their theoretical system, the central tenets of which were: 1) the doctrine of 'net product', and 2) the theory of the reproduction of social capital.

To demonstrate the need to pump capital out of trade and industry and into agriculture, the Physiocrats advanced the doctrine that only agriculture creates a 'net product', or 'revenue' (i.e., surplus value). In agricultural production nature's bounty provided man with a greater quantity of material substance than was needed merely to provide for the cultivator and to restore his costs of production. This surplus material substance, or 'net product', went to the landlords as rent on their property and formed the basis of the nation's wealth. It constituted the fund that 'fed' the industrial population in the towns and covered the expenditures of the state apparatus. Thus, for the Physiocrats agricultural labour was the only labour that was truly 'productive'; industrial labour was 'sterile' labour, in the sense that it yielded no 'net product' over and above production costs.

To illustrate more clearly the dependence that the landowning and industrial classes had on the class of farmers (which Quesnay viewed as being representative of the entire agricultural population), Quesnay created his famous theory of reproduction, which he set out in his *Tableau Economique* (1758). In the *Tableau* Quesnay showed how the total product of a nation's annual production moved. The entire corn harvest went first of all into the hands of the farming population, which retained part of it for its own provision and payed a part over to the landlord class as rent; a third part of the agricultural produce (raw materials for industrial processing and means of subsistence) passed into the hands of the industrial class, which in turn sent back finished products—partly to the class of cultivators, partly to the landlords. Parallel with the movement of products between the individual social classes, but in the reverse direction, ran the movement of money, which functioned merely in a servicing capacity, to mediate the circulation of commodities. As depicted by Quesnay, the entire process of distributing the social product between the separate social classes was such that in the end all classes in society had their consumption needs met and a new cycle of reproduction was all set to begin.

The *Tableau Economique* represented Quesnay's most important theoretical legacy. It was the first, ingenious attempt to capture the entire process of social reproduction, embracing the production, circulation, distribution, and consumption of the social product within a single scheme. Whilst the mercantilists had occupied themselves with debating isolated, and usually practical problems, Quesnay made a bold attempt to uncover the mechanism of capitalist reproduction as a whole—an attempt which earns him the right to be called the father of contemporary political economy. In his theory of reproduction Quesnay was far ahead of his time. Even the Classical economists proved unable to grasp this theoretical achievement; only Marx was to develop it further.

There is also a valuable theoretical idea in the Physiocratic doctrine of 'net product', although it is hidden beneath a fantastic integument. For the mercantilists the source of profit was trade, while profit was the surplus that remained after covering production costs. The Physiocrats taught that this surplus, or net income, is formed strictly within the process of agricultural production. Consequently they shifted the source of the formation of surplus value out of the circulation process and into the process of production. This was a new formulation of the problem of surplus value and constitutes one of the Physiocrats' great merits. They were unable to solve it, however, because of their naive naturalism, which put the physical productivity of the soil in place of the economic productivity of labour, and the production of material substance in place of the production of value. It was necessary to give a new basis to the theory of value so forcefully advanced by the Physiocrats, namely the labour theory of value set out by the mercantilists, and by Petty in particular. It fell to Adam Smith to carry out that task.

3. Adam Smith

The mercantilists acted as defenders of the interests of merchant capital. But by the 18th century mercantilist policy had already become a brake on the further development of capitalism: it was retarding the transition from the rule of merchant capital to the rule of industrial capital. In France the rural bourgeoisie, for whom the Physiocrats acted as plenipotentiary, was numerically small and had little influence. Hence the Physiocrats were powerless to crush merchant capital's domination. Only the industrial bourgeoisie in the towns had the power to smash the rule of mercantilism; similarly, at the level of economic theory, it was only thanks to the efforts of the Classical school, representing the interests of industrial capitalism, that mercantilism was vanquished as a doctrine. Adam Smith is considered the founding father of the Classical school.

The first half of the 18th century was a transitional period in the history of the English economy. Although the crafts still partially retained their position, they had given way significantly to cottage industry. There was also the more modest spread of the manufactory.

Adam Smith can be called the economist of the manufactory period. The birth of large-scale industrial capitalism, in the form of manufactories based on the division of labour, had made it possible for Smith:

a) to conceive of the whole of society as a gigantic workshop with a division of labour (hence Smith's doctrine on the division of labour);

b) to grasp the importance of industrial labour, together with that of commerce and agriculture (thanks to which Smith overcame the onesidedness of both the mercantilists and the Physiocrats);

c) to conceive of the exchange between different branches of production as an exchange of equivalent products based on equal expenditures of labour (hence the central place that the theory of value occupies in Smith's system);

d) to classify correctly the different forms of revenue (wages, profit, and rent) that go to each of the different social classes.

Smith begins his work by describing the division of labour, which he sees as the best means for raising labour productivity. This view was itself a reflection of the conditions pertaining during the manufactory period, when there was still no widespread application of machinery and the basis of technical progress was above all the division of labour. Since Smith is mainly concerned with the material-technical advantages of the division of labour and not its social form, it is perfectly understandable that he should confuse the social division of labour between individual enterprises with the technical division of labour within the single enterprise. Despite this error, Smith's doctrine on the division of labour is of enormous value. Proceeding from this, Smith conceives of the whole of society as a vast labouring society of people who work for one another and mutually exchange the products of their labour. The conception of society as at one and the same time a *labouring* and an *exchange* society of individuals allowed

Smith to grasp the importance of industry and to accord central place to the labour theory of value.

Smith regarded society as a labouring society of individuals depending on one another by virtue of their productive activity. Unlike the mercantilists, he understood that the exchange of a commodity for money ultimately comes down to an exchange of the products of different producers' labour. On the other hand, he overcame the onesidedness of the Physiocrats, who regarded the movement of commodities as a movement of matter, or the material substance of nature, from the class of cultivators to the other classes of society (i.e., to the landowning and industrial classes). Beneath the exchange of products of labour Smith perceived an exchange of the labouring activities of different producers. If all producers depend on one another, this obviously does away with the privileged position that the mercantilists had accorded to foreign trade and the Physiocrats to agriculture. If industry depends upon agriculture, then the latter must depend upon industry to precisely the same extent. It is absurd to maintain that the farming population 'maintains' the industrial population which in and of itself is 'sterile.' Agriculture and industry are branches of production with equal status: exchange between them is an exchange of equivalents.

Having dispensed with the Physiocrats' error regarding the interrelation between agriculture and industry, Smith was then able to come to a more correct understanding of the productiveness of labour and capital. According to Smith, all labour is productive that yields value or surplus value, independently of whether it is applied to agriculture or to industry (Smith vacillates in his definition: sometimes he defines productive labour as that which gives rise to surplus value, at other times as labour which embodies itself in material products possessing value).

Parallel to extending the concept of productive labour, Smith also expanded the concept of capital. During the mercantilist period what people called capital was usually a sum of money lent out at interest. The Physiocrats maintained that capital (they usually employed the term *les avances*) is not the actual money, but the products employed as means of production. On top of this they had in mind only that capital which is invested in agriculture and, in addition, looked upon capital primarily as a means for increasing the 'net product' (i.e., rent). Smith broadened the concept of capital and extended it equally to industry and commerce. Further, Smith linked the concept of capital closely to the concept of profit, viewing capital as profit-yielding property. By doing this he placed a 'private-economic' concept of capital as a means for extracting profit alongside the 'national-economic' concept of capital (in the sense of produced means of production) that we find with the Physiocrats.

Proceeding from his doctrine on the division of labour, Smith placed the theory of value in a new and central position. The Physiocrats, with their limited, naturalistic vantage point, had confused value with material substance. Smith accepted the ideas that we find already embryonically present among the mercantilists (especially in Petty) and developed the labour theory of value further. Smith's train of thought goes approximately as follows: In a society founded upon the division of labour, each person produces products for other people and, by entering into exchange, receives those products that are necessary for his own subsistence. In acquiring the products of someone else's labour our producer is really disposing over, or 'commanding' the labour of another. But how does our producer determine the value of the product that he himself has produced? By the quantity of other people's labour which he can obtain in exchange for his own product, answers Smith. But how do we determine this quantity of labour? In a simple commodity economy this will equal the quantity of labour that our producer expends on producing his own product. Thus Smith sometimes correctly determines the value of a commodity by the labour expended on its production while at other times he determines it, mistakenly, by the labour which the commodity in question will purchase when exchanged. So long as Smith stays within the bounds of simple commodity economy, this conceptual confusion is of little harm, since these two quantities of labour will coincide. In capitalist economy, however, this coincidence disappears: the capitalist purchases the living labour of the worker (i.e., labour power), for example, eight hours of his labour, in exchange for a product containing a smaller quantity of labour. Being unable to explain the laws of this exchange of capital for labour power, Smith mistakenly concludes that in capitalist economy the value of the product is greater than the quantity of labour expended on its production, and is equal to the sum that the capitalist has laid out to hire workers plus the average profit (in certain circumstances plus rent as well). Consequently, when it comes to capitalist economy Smith denies that the law of labour value operates: here he grounds himself on the vulgar theory of production costs. Because of his vacillations on the theory of value Smith was to become the forebearer of the two currents within economic thought at the beginning of the 19th century: the Classical tendency, which achieved its highest expression in Ricardo, and the vulgar current, represented by Say. The inconsistency of Smith's theory of value impeded him from providing a fully worked out theory of distribution. It is true that he did make a major advance when compared to the Physiocrats. He replaced the Physiocrats' false schema of social classes (landowning class, productive class, and sterile class) with a correct schema of landlords, industrial capitalists, and wage labourers. He correctly enumerated the three forms of revenue that each of these classes receives: wages, profit, and rent. Smith especially deserves credit for having clearly distinguished the category of industrial profit, which the Physiocrats had ignored.

For all the advances that Smith made in the theory of distribution, on the whole his treatment of the latter remained highly incomplete, partially because he did not hold to the standpoint of the labour theory of value, but abandoned it in favour of the theory of production costs. Had Smith held fast to the doctrine that the value of a product is created by human labour and is divided up between the separate social classes, the interdependence of the revenues of the various classes would have lept to his eye and demanded elucidation through a theory of distribution. But so long as Smith grounded himself on the theory of production costs, according to which the product's value is the result of the sum of the various costs of production, or the revenue of those participating in production (wages, profit and rent), these revenues stood out as something prior to value and independent from one another. Instead of regarding the product's value as primary and revenue secondary, Smith looked upon value as a secondary magnitude deriving it from revenue. But if this were the case, the question would immediately arise: how is the size of these revenues—i.e., wages and profit—determined? Smith found no better answer to this question than to make a covert appeal to the theory of supply and demand. In his view the level of profits depends on the abundance of capital or, to be more precise, upon its rate of accumulation: when capital is growing rapidly the rate of profit falls; when a country's total capital declines the rate of profit goes up. But a rise in capital indicates a simultaneous growth in the demand for labour power, and is thus accompanied by a rise in wages. The reverse occurs when a country's total capital is diminishing. Finally, when that capital is in a stationary state both wages and profits establish themselves at a low level. Thus, the movement of the revenues of both capitalists and workers depends upon whether a nation's economy is in a progressive, stationary, or declining state. With such a position,

Smith could hardly be said to have resolved the problem of distribution: he merely gave a factual description, accompanied by a superficial explanation of these facts in the spirit of supply and demand theory.

It was left to the other great economist of the Classical school, David Ricardo, to make a decisive step forward in the theory of distribution.

4. David Ricardo

David Ricardo's life more or less coincides with the age of the English industrial revolution which, by extensively introducing new machinery and rapidly developing factory production, successfully displaced the previous forms of industry (the crafts, cottage industry, and manufactories). If Smith can be called the economist of the manufactory period then Ricardo is the economist of the age of the industrial revolution, the basic characteristics of which were to find their reflection in his theory. In his labour theory of value Ricardo generalized from the multifarious facts associated with the drastic and rapid cheapening of industrial manufactures which resulted from the introduction of new machinery and the rising technical productivity of labour. In his theory of distribution, and most notably in his doctrine on rent, he reflected the condition of sharpening class struggle between bourgeoisie and landlords that went side by side with the first successes of factory industry.

Ricardo's primary merit is to have freed the labour theory of value of the internal contradictions from which it had suffered in Smith's formulation of it, and to have attempted to use this theory to explain the phenomena of distribution.

Smith had failed to make a sufficiently clear distinction between the quantity of labour expended on the production of a product, and the quantity of labour which that product will be able to purchase when exchanged. In keeping with this dualistic standpoint Smith acknowledged that a product's value can change both as a result of changes in the productivity of the labour employed in producing it, and in consequence of alterations in the 'value of labour' (i.e., in the amount of wages or costs of production).

Ricardo took up arms against this error on the part of Smith. He demonstrated clearly that the quantity of labour that can be purchased in exchange for a given commodity cannot serve as an invariable measure of its value, and that to search for such an invariable measure is in general a hopeless undertaking. Ricardo identifies a change in the quantity of labour expended on producing commodities as the sole source (with the exception of the cases noted below) of changes in their value. He therefore makes the *magnitude* of a commodity's *value* depend directly on the development of the *technical productivity of labour*. By adhering consistently to this position Ricardo made a great contribution towards resolving the *quantitative* problem of value, although with his horizons limited (as were Smith's) to capitalist economy he ignored the *qualitative*, or social nature of value as the external expression of a determinate type of production relations between people.

Smith had denied that the law of labour value operates within capitalist economy, where the product's value does not go completely to its producer, but is broken down into wages and profit. To radically disprove this false view of Smith's it would have been necessary to explain the laws by which capital exchanges for labour power. It would only have been possible to explain these laws by analyzing those social relations of production which bind the worker to the capitalist. But the method of analyzing production relations as being relations between people was as unknown to Ricardo as it was to Smith. Ricardo, therefore, had no other recourse but to leave aside the question that Smith had posed. This he did, restricting his investigation on this point to the question of the 'relative' value of commodities. Insofar as it is a question of the 'relative' value of two commodities A and B, it is obvious that any change in workers' wages (a rise, for example) which exerts a uniform influence on the overall production costs of the two commodities will not in the least affect their 'relative' value. The result of a rise in wages is not to increase the product's value, as Smith had thought, but merely to lower the level of profits. No matter how the product's value is distributed between wages and profit, this will not affect the magnitude of the product's value, which in capitalist economy is determined by the quantity of labour necessary to produce it. By taking the position that wages and profit change reciprocally one to the other Ricardo made a decisive stand for the view that profit is a portion of the product's value which the workers have created with their labour and which the capitalist appropriates for himself.

Ricardo in this way rectified Smith's mistake which consisted in denying that the law of labour value operates in capitalist economy. But he did not manage to show how the law of labour value, which does not manifest itself directly in the workings of capitalist economy, nevertheless regulates it indirectly through the medium of prices of production. Ricardo was not successful in explaining the apparent contradiction between the law of labour value and the observable phenomena of capitalist economy. In fact, Ricardo was able to eliminate the influence of wage fluctuations (and the corresponding fluctuations in the rate of profit) on the relative values of the two commodities, A and B, only insofar as wages have approximately the same weight in the costs of production of the two commodities, that is, to the extent that the two branches of production each employ capitals with identical organic compositions. If the capitals that produce commodities A and B have unequal organic compositions (or unequal turnover periods), any rise in wages (or fall in the rate of profit) will more perceptibly affect the commodity produced with the capital of lower organic composition, say, commodity A. In order to preserve the same level of profit in the two branches of production, the relative value of commodity A will have to rise in comparison with commodity B. Thus Ricardo arrives at his famous 'exception' to the law of labour value. The relative values of commodities A and B will change not only with fluctuations in the relative quantities of labour needed for their production, but also with a change in the rate of profit (or with a corresponding change in wages). *Profit on capital* is in effect an independent factor regulating the value of products together with labour.

By allowing for these 'exceptions' to the law of labour value Ricardo opened the way for the vulgar economists (Malthus, James Mill, McCulloch, etc.) to completely abandon the labour theory of value. Ricardo himself, however, considered these 'exceptions' to be of secondary importance compared with the basic principle of labour value—his point of departure for constructing his whole theory of distribution.

Ricardo's *theory of distribution* had two main objects to pursue: firstly, it had to flow from his theory of value and, secondly, it had to account for the real-life phenomena of distribution that Ricardo was observing in England at the beginning of the 19th century. The Smithian theory of distribution had led to a vulgar theory of production costs: the sum of wages, profit, and rent makes up the value of the commodity. We have already seen how Ricardo eliminated the contradiction between the actual existence of profit and the principle of labour value: he regards profit as the portion of the product's value that remains after the deduction of wages (although Ricardo was inclined in his 'exceptions' to treat profit as an independent factor in value formation). Now Ricardo was faced with having to remove the contradiction between this same principle of labour value and the actual existence of rent, which at first view has the appearance of being added onto the commodity's value. Insofar as it was a question of 'differential' rent, Ricardo managed to resolve the contradiction with supreme artistry. Rent arises because different tracts of land have differing productivities of labour. The value of a bushel of corn is determined by the quantity of labour necessary for its production on the most inferior lands then under cultivation. The difference between this social value of corn and its individual value on plots of greater fertility (or on plots situated nearer to their market and thus incurring smaller outlays on transport) makes up the rent that is paid to the landlord. According to Ricardo, the very worst lands under cultivation yield no rent at all (a view which was mistaken, since it assumes that there is no such thing as absolute rent). As people move on to cultivate new, increasingly inferior lands, the value of a bushel of corn will rise. So, too, will ground rent, both in real terms in corn (since the difference in productivity between the best and worst lands will be growing), and even more so nominally, in terms of money (since the value of each bushel of corn will have risen).

By treating rent not as an addition to the social value of corn, but as the difference between this social value and the value of corn on the particular plot of land in question, Ricardo was able to make his theory of rent consistent with the principle of labour value. At the same time he attempted to derive from his theory of rent those logical conclusions that would conform to real events. The age of England's industrial revolution was characterized not only by the tremendous drop in the prices of industrial manufactures that came with the introduction of new machinery, but also, together with this, by an enormous rise in the price of corn. This rise was in fact explainable by the country's rapid industrialization, Napoleon's continental blockade, and by the high duties on corn imports that had been put in place to benefit the English aristocracy. It was a temporary phenomenon, but Ricardo made it into a permanent law of capitalist economy. In his view, the growth of the population would make it increasingly necessary to transfer cultivation to worse and worse plots of land, which would be accompanied by rising corn prices and an upward trend in both real and nominal ground rent. All the advantages of the country's industrialization would accrue to the landlord class. The workers would not share in any of the benefits because though their nominal wages would go up with the rise in corn prices, their real wages would at best remain stationary, i.e., at that minimum level of means of subsistence required by the worker and his family (what Lasalle was to term the 'iron law of wages'). As for profit, this would exhibit a tendency towards an inexorable fall, thanks to the inevitable rise in nominal wages. The fall in profits would dampen the capitalists' drive to accumulate capital, and the nation's economic progress would inevitably slow down, coming nearer and nearer to a total halt.

Ricardo's entire portrait of revenue movements among the different social classes flows from his assumption that corn prices would be necessarily rising. Ricardo underestimated the possibilities for a powerful growth in the productivity of agricultural labour. His doctrine of a necessary and inexorable rise in the price of corn was not born out by events, and neither were the conclusions drawn from it. In spite of this, his theory of distribution represented an enormous scientific advance. It portrayed the vast sweep of the movements in income of all the social classes, and their close inter-connection; it depicted this dynamic as a necessary consequence of the law of labour value; and it clearly revealed the conflicts that exist between the interests of the individual classes.

5. The Disintegration of the Classical School

Ricardo had been courageous enough to acknowledge openly and directly the conflict of interests between capitalists and workers. As the struggle between these two classes flared up and pushed the struggle between capitalists and aristocracy into the background, bourgeois economists increasingly began to switch from a forthright description and explanation of capitalist economy to presenting a justification of it. Bourgeois political economy became increasingly *apologetic* (that is, it set itself the aim of justifying capitalism) and vulgar (i.e., it restricted its investigation to superficially studying phenomena as they might appear to the capitalist, instead of probing into the internal connection between them). Around about the 1830's there began the period of the Classical school's 'disintegration'. Bourgeois economists of that period repudiated the labour theory of value developed by Smith and Ricardo. In order to show that profit is not a part of the value created by the workers' labour, they concocted new theories as to its origin. Say's doctrine was that profit is created because of the productiveness of the means of production belonging to the capitalist (the theory of the 'productiveness of capital'); Senior saw profit as the reward for the 'abstinence' of the capitalist who accumulates capital by refraining from directly satisfying his own personal needs (the theory of 'abstinence'). As bourgeois political economy became apologetic and vulgar, there also began to be opposition to it. Opposing it were the representatives of the landowning class, pushed into the background by the bourgeoisie (Malthus, who taught that only the existence of a wealthy class of landowners could create a market for industrial manufactures), the defenders of the petty bourgeoisie, peasantry, and handicrafts (Sismondi, who argued that capitalism, by bringing ruin to the peasantry and handicrafts, reduces the purchasing power of the population and thereby creates the conditions for constant crises), and, finally, the first defenders of the working class (the utopian socialists).

1 Nicholas Barbon, *A Discourse of Trade* (London, 1690), pp. 13-15. 'The Value of all Wares arise from their Use; Things of no Use, have no Value ... The Use of Things, are to supply the Wants and Necessities of Man: There are Two General Wants that Mankind is born with; the Wants of the Body, and the Wants of the Mind; To supply these two Necessities, all things under the Sun become useful and therefore have a Value ... The Wants of the Mind are infinite, Man naturally Aspires, and as his Mind is elevated, his Senses grow more refined, and more capable of Delight; his desires are inlarged, and his Wants increase with his Wishes ... '.

AFTERWORD
by
Catherine Colliot-Thélène

AFTERWORD

Those acquainted with I.I. Rubin's *Essays on Marx's Theory of Value* are in a position to appreciate the author's remarkable knowledge of Marx's thought. *The History of Economic Thought* is the first English translation of a work which is an important complement to the *Essays*. It is devoted to the study of economic doctrines prior to the writing of *Capital*, doctrines which were therefore known to Marx and discussed by him. It is not, however, just a convenient textbook, taking up perhaps with greater systematicity the elements of analysis scattered throughout *A Contribution to the Critique of Political Economy*, *Capital*, and *Theories of Surplus Value*. On the contrary, it contains original contributions such as the first eight chapters on mercantilism, a doctrine which Marx frequently invokes, which he knew very well, but which never constitutes the object of an ordered study in his texts. Further, and this distinguishes the present work from traditional 'histories of economic thought', Rubin has applied himself to placing the theories he deals with in their respective contexts. Generally with great wisdom, he shows how the particulars of each theory reflect the social and economic state of the country and the period in which it had its day. Of course, this historical perspective is not absent from Marx's texts—witness the remarkable example of the study of the Physiocratic current. (TSV pp. 44-68.)* But Marx usually takes the discussion of his predecessors' theses as a pretext for the development of his own conceptions, so that his approach to those theses pertains rather to 'internal' critique, and his references to the historical context occupy only a subordinate position in comparison.

Rubin's *History* is an original work, then, but it owes much to Marx's analyses and borrows its reading grid from them. Rubin's exceptional knowledge of Marx must not be understood as superior scholarship. For reasons which are perfectly clear to everyone, the work of Marx, like that of no economist before or since and very few philosophers, has attracted a considerable number of honest and hardworking exegetists who have been discouraged neither by the breadth of his writings nor by the frequent technicality of his objects. But the difficulties peculiar to his thought, such as the ambiguity or the awkwardness of some of his formulas, and the weight of the traditional questions of academic political economy, to which it was supposed his theories were meant to be an answer, combine to form an obstacle to the effective appropriation of the very questioning that gives sense and

*The reader is referred to note 3 of the Translator's notes at the end of the Afterword for a list of abbreviations and contractions of the names of works quoted in the text itself.

coherence to the project of *Capital*. Rubin has mastered precisely this questioning. The degree of this mastery is shown by the explanations he proposes in the *Essays* of the fundamental concepts of the Marxist theory of value, concepts which are too often evoked in mysterious or quasi-mythical fashion: content and form of value, equal labour, social labour, abstract labour, socially necessary labour, etc. It is very clear that we find this same mastery in Rubin's perspective on the economic doctrines prior to Marx. That is why the present work is particularly suited to bringing out the main themes of the critique Marx gave of political economy in general and of classical writers in particular, a critique in which Marx's true originality is at work.

It is in Marx's originality, in his own understanding of that originality, and in the eventual displacement between the two, between what Marx thought he was doing and what he effectively did, that we find the network of questions to which recent Marxist exegesis is directed when it raises the problem of the 'object' of *Capital*. What is the relation between Marx's perspective when tackling political economy and the perspective (or perspectives?) of his predecessors? Is it continuity or rupture, partial rupture within a fundamental continuity, rupture incompletely outlined, or what? Of course, Rubin does not pose the question in these terms; these terms were not in common use when he was writing. In his concern to attribute praise and blame to each author on an equitable basis, Rubin seems, above all, to be sensitive to the continuity proclaimed between one author and another, to the progressive looming up of themes which will be organized in a definitive synthesis in the complete system offered in the exposition of *Capital*. With the Physiocrats, he detects the first study of the process of social reproduction as a totality, the emergence of the concept of surplus, and the displacement of the origin of surplus value from the sphere of exchange (where the mercantilists had placed it) to the sphere of production. He attributes to Smith, in spite of the ambiguities in his work, the credit of having seen the exchange of labour behind the exchange of products or as having correctly depicted the class division and forms of incomes characteristic of capitalistic economy. And it hardly needs to be said that the merits he found in Ricardo were without number: constructing his entire inquiry on the concept of labour expended in the production of commodities, making free competition the determining feature of the economic relations he studied, soundly articulating his theory of distribution and the labour theory of value, etc. Briefly, the impression is given that in *Capital* Marx restricts himself to the collection and co-ordination of themes developed by others before him. According to Rubin, even the rational form of Marx's discourse, which allowed this discourse to be considered scientific, had already been acquired with Ricardo. But are we to believe that a work which was significantly baptised 'critique of political economy' by its author had no other aim than the working up of such a synthesis through the correction of the punctual errors committed by this and that author on particular problems? And yet Rubin's generosity with respect to pre-Marxist doctrines only reproduces Marx's own generosity. It will be said, it is true, that that is no justification, because Marx's reflective consciousness may not have been equal to the theoretical revolution he was effecting. Innovators often do their utmost to erase the radical nature of their rupture by inscribing their works in a tradition which legitimates them for their contemporaries. But Marx was not usually afraid of exhibiting the subversive character of his objectives, whether practical or theoretical. His critique of the bourgeois economists, furthermore, is sufficiently complex for us to think twice before challenging it right from the start as hardly pertinent or superficial. And since Rubin has shown in his *Essays* that he has grasped what is at stake in Marx's critique perhaps more clearly than any other commentator, we will follow him for a while in the reading he proposes of the pre-Marxist theoreticians. But we must not put aside the hypothesis that Rubin's ambiguities are directly inherited from Marx and that Marx himself—do we dare formulate this blasphemy?—had effectively shared with his predecessors a fundamental presupposition which at the very least assured the existence of a common theoretical space within which it was possible to maintain a dialogue between his theory and the *economic* theories coming before and, indeed, after his own.

1 The Normative Conception and the Practical Conception of the Object of Political Economy: the Search for an Invariable Measure of Value

Rubin unquestionably contributes to the clarification of the problem of value by pointing out that, according to the authors, two distinct objectives have been identified under this one heading, implying in their turn two completely different conceptions of the finalities inspiring the elaboration of an economic theory.[1] One objective can be qualified as *normative* or as *practical*. It hopes to provide solutions to the problem of equilibrium posed by any system of production; to do that it seeks to determine the reciprocal relations which have to be

1 Cf. above Chapters 22, p. 186, and Chapter 28, pp. 248-249. See also *Essays*, Chapter 13, pp. 125-7, and Chapter 15, p. 167.

respected by elements entering and leaving the global process of reproduction. It therefore proposes an ideal norm of the functioning of economic systems. This approach requires that materially different objects be compared and reciprocally measured, namely, wages, social product and, within the social product, heterogeneous commodities. It can be pursued, therefore, only on condition that an artificial instrument be constituted which makes possible the comparative quantification of the objects in question. This instrument is not to be understood as totally arbitrary but must on the contrary satisfy a certain number of conditions, the principal condition being that the instrument itself should not be subject to the causes of variation for the measurement of which it has itself been adduced. In that perspective, the theory of value becomes the elaboration of an *invariable standard of values*, that is, first, an instrument for the homogenizing of the heterogeneous goods encountered in a process of production and exchange, and second, an instrument for the measurement of the variations undergone by the exchange rates of goods between two successive states of the process of social reproduction. To this first conception of the theoretical task devolved upon political economy is radically opposed the second conception which loudly claims not to be concerned with what ought to be but with what is; i.e., with the study of the objective laws at work in existing systems, or, even more clearly, with disengaging *the order of causes and effects* from within the tangle of numerous processes constituting the real. To the supporters of this second approach, the normative models proposed by the first procedure seem imaginary objects. In particular, the construction of an invariable measure of value, designed to make possible the homogenization of heterogeneous goods, is denounced as an absurd quest. The equalization of commodities, they retort, is *given de facto* in the equivalence of commodities established spontaneously in the process of exchange. The theoretical task is to explain this equivalence, viz. to exhibit the law governing its variations. Authentic theoretical analysis is a study of the effective causes of real phenomena. It is this second interpretation of the object of political economy and, as a result, of the content of the theory of value that Rubin takes to be of sole pertinence. 'However, the theory of value is not concerned with the analysis or search for an *operational standard* of equalization; it seeks a *causal* explanation of the *objective* process of equalization of different forms of labor which actually takes place in a commodity capitalist society'. (*Essays*, p. 169.) That is Rubin's peremptory view in the *Essays*.

The normative conception of the discourse of political economy has dominated twentieth century academic practice and has not encountered serious opposition. The ideal of a mathematical dominance of abstract models was imposed on the scientific community in general in the wake of the brilliant successes brought by the formalization of mathematics and logic and in the wake of the occasionally false hopes which that aroused. The main concern of economists was chanelled into the construction of formally satisfying models, to the detriment of any reflection on the problematic relation these models held with the concrete referent of which, in spite of everything, they claimed to provide an explanation. It is not obvious, however, that the intervention of this new scientific paradigm signified an unexpected mutation of the object of political economy. On the contrary, Rubin shows us that the normative temptation was already latent in much earlier doctrines, precisely in the form of the search for an invariable measure of values. According to him, Smith's conception of value in particular displays an internal tension due to the overlapping of the practical task of determination of an invariable measure of values on the one hand, and of the scientific inquiry into the objective causes of variations in the value of commodities on the other. We can trace back Smith's slips of analysis to this fundamental methodological ambiguity, glosses which are highlighted by all histories of economic doctrines. They include the erroneous identification of the labour expended on the production of a commodity (incorporated labour) with the labour which that commodity can buy (purchasable labour), an identification which Ricardo had already noted and vigorously denounced. According to Rubin, Ricardo's merit lies in having put an end to that ambiguity by unequivocally posing the problem of value in terms of causality, raising economic discourse thereby to an authentically scientific level. All Marx had to do, therefore, was faithfully to welcome the Ricardian heritage. The historical point of view he took on the relations of production studied by economic theories does allow him to make good some lacunae and to correct certain weaknesses in Ricardo's deductions. But the ground of scientific investigation had been clearly delimited by Ricardo; it was no more than a matter of perfecting its internal organization.

The distinction between a normative conception and an objective conception—a term which we prefer to the term 'theoretical' which Rubin uses[2]—seems to us extremely useful. One of its main interests

2 We thus avoid anticipating the form which a discourse ought to take if it is to be recognized as a science. It is possible that a normative approach (or a 'practical' approach, to use Rubin's vocabulary) can on occasion—and under limiting conditions which would, of course, have to be defined—prove useful and even consistent with a search for objective causes. In any case, it is always rash to promote a single type of cognitive approach to the rank of 'scientific' paradigm in general. As we shall see later, Rubin takes certain risks when he identifies 'causal analysis' and *the Theoretical* in general.

is to make it possible to make short work of a traditional objection to the Marxian theory of value advanced by academic critics. We are referring here to the famous problem of the 'reduction of complex labour to simple labour', a reduction which, after Böhm-Bawerk's critique of *Capital*, has been considered one of the thorny issues of any attempt at the determination of value by labour.[3] We are not surprised that academic economists have regularly taken up this objection, because their formation did not predispose them to consider that the object of political economy could itself be disputed. For them the questions a theory ought to answer are supposed to be self-evident. For any theory to be taken seriously, it must be capable of answering these specific kinds of question—such as the rules governing the determination of prices, the magnitude of the surplus in relation to the capital advanced, or the distribution of resources. But the most significant fact about the confusion surrounding the issue of the theory of value is that economists of Marxist persuasion were to be found—and are still to be found—who accept the requirement of 'coefficients of reduction of complex labour to simple labour' as well-founded, that is, who make the validity of the Marxian labour theory of value hang on the possibility of producing such coefficients.[4] Rubin is quite right to emphasize in the *Essays* that such a requirement is in part tied to the normative conception of political economy. As long as one seeks to determine the proportions according to which different goods *ought* to be exchanged independently of any consideration of the effective process of exchange, the necessity is felt for a comparative scale of different labours. (*Essays*, Ch. 15.) For an author who, like Marx, aims to explain the real process, there is no need of recourse to such an artifice. Far from expecting the economist's ingenuity to provide the conditions to bring it about, the reduction of different concrete labours to a homogeneous substratum—what Marx called the

3 Cf. Böhm-Bawerk (1973) pp. 93-6. The economist Joan Robinson makes the same objection when she criticizes Marx for having 'left open the problem of assessing labour of different degrees of skill in terms of a unit of "simple labour".' (Robinson 1972 p. 19.) Right from the start then, in spite of her good intentions, she commits the same error, and it prevents her from understanding the original significance Marx attributes to the 'law of value'.

4 Thus, for example, although the French Marxist economist Carlo Benetti at first shows a relative understanding of the sociological significance of the law of value, he nonetheless ends up by saying that one of the difficulties met by the determination of value by labour time 'consists in the evaluation of labours of different skills used in the production of commodities'. And he adds that 'Marx proposes the reduction of complex labour to simple labour ... on the basis of coefficients determined by the cost of formation of complex labour'. (Benetti 1975, p. 136.) This proves that the correction to CCPE which *Capital* makes on this point (*Capital*, I, p. 51 in n. 2), a correction to which we will refer later in our text, has escaped him.

equalization of labours—can be discovered in the relations of exchange, *as an already realized process*. Nothing is more alien to a normative conception of the theory of value than the first pages of *Capital* (the 'analysis' of exchange value) or, an even better example, the arguments about the value form which invite us to scrutinize the relation of equality constituted by a relation of exchange in order to disengage its implicit significances. The language of commodities is hardly clear, but it is a language all the same, a language in which something is said which the theoretician has to elucidate.[5] This 'said' thing is the abstraction of all concrete characteristics of acts of labour, the reduction of concrete labour to abstract labour, of which the reduction of skilled labour to simple labour is only a subsidiary moment. 'Experience shows that this reduction is constantly being made. A commodity may be the product of the most skilled labour, but its value, by equating it to the product of simple unskilled labour, represents a definite quantity of the latter labour alone.' (*Capital*, I, p. 51.) The academic critique has seen in this assertion the sign of a circularity in Marx's thought. That would indeed be the case if Marx were aiming to provide an instrument or a criterion permitting the *anticipation* of the relations of exchange of commodities. But that is not his project, as emerges from the following quotation from the second German edition in which Marx makes the point more precisely. 'Hence, when we bring the products of our labour into relation with each other as values, it is not because we see in these articles the material receptacles of homogeneous human labour. Quite the contrary: whenever, by an exchange, we equate as values our different products, by that very act, we also equate, as human labour, the different kinds of labour expended upon them'. (*Capital*, I, p. 78.) The result is that it is impossible to determine a priori, that is, before

5 An economist will perhaps find the terminology used here disconcerting, although it simply retraces Marx's own terminology at the beginning of *Capital*. The different versions of the first chapter of *Capital* in fact present themselves as the formulation of phenomena which can be read in reality itself. The approach appears particularly clearly in the study of the simple form of value. (*Capital*, I, p. 48f.) There Marx treats the relation of exchange as a reality signifying something which the theoretician only has to make explicit: 'This shows that when placed in value-relation to the linen, the coat *signifies more* than when out of that relation ... ' (*Capital*, I, p. 58, our emphasis). ' ... all that our analysis of the value of commodities has already told us, *is told us by the linen itself*, so soon as it comes into communication with another commodity ... ' (*Capital*, I, p. 58.) There is no doubt that this is a Hegelian approach, though not a gratuitous flirtation 'with the modes of expression peculiar to him' (*Capital*, I, Afterword to the Second German Edition, p. 29) because the approach is entirely in harmony with one which does not seek to formulate the ideal conditions which the economic system ought to respect in order to function correctly but which seeks to describe the manner in which it really functions.

the exchange is effected, the amount of social labour represented by a determinate mass of whatever products. It is the relations of exchange of commodities, established on the market by the mechanisms of competition, that is, in a process over which no conscious agent has control, that constitute the index of the proportions in which the different concrete labours are equalized.

But our attention should be drawn by something else which explains in part the persistent equivocation on this question to be found both amongst 'bourgeois' critics and amongst Marxist economists themselves. Before arriving at a formulation of the theory of value which absorbs the problem of the reduction of complex labour to simple labour into the more general problem of the process of abstraction of different concrete acts of labour, Marx sought to resolve it by invoking the different value *possessed by* various kinds of labour according to their degree of skill, that is, the differences in wages in so far as they sanction real differences of formation and not simply subjective evaluations. That temptation is manifest in *A Contribution to the Critique of Political Economy* (CCPE) where, side by side with the argument according to which the homogenization of labours of different complexities is a given within daily practice, we find the allusion to 'the laws governing this reduction [sc. from "more complicated labour" to "simple labour"]'. (CCPE p. 31.) The study of these laws is postponed until later. The allusion disappears in the text of *Capital* where Marx even takes the trouble to emphasize in a note that 'we are not speaking here of the wages or value that the labourer gets for a given labour-time, but of the value of the commodity in which that labour-time is materialised'. (*Capital*, I, p. 51 n. 2.) Plainly, the problem of the different exchange values *possessed by* differently skilled labour powers, which the scale of wages is to ratify, is distinct from the problem of the different values *produced by* those labour powers. This restatement is just as much an indication of a clarification finally achieved in Marx's thought as it is a critique of other authors or a precaution against possible misinterpretations. To be completely rigorous Marx ought to have eliminated all mention of this problem from his analysis of value, for it is not a matter of a problem which would receive an original solution in his work but rather a question which, allowing for the new content he gives to the theory of value, has no sense.

In virtue of what are these detailed discussions on the exegesis of Marx's texts of interest to our general objective, that is, the relation between Marx and earlier economic doctrines? As we have said, and on this point we are only repeating Rubin's convincing demonstration, to make the reduction of complex labour to simple labour the condition of the validity of the Marxist labour theory of value is to reveal an incomprehension of the terms in which the problem is posed in Marx. Does the same hold for the Ricardian labour theory of value? It seems that the answer must be negative here. For Marx has inherited both the very question and the temptation to answer it by invoking differences in wages from Ricardo himself. It was Ricardo who, following Smith, had introduced the scale of wages—automatically established, of course, by competition—as a corrective which had to be taken account of in order to support the principle according to which reciprocal values are proportional to the relative quantities required to produce them (*On the Principles of Political Economy and Taxation*, pp. 20-1). And in *The Poverty of Philosophy* Marx showed himself to be an orthodox Ricardian with his statement that 'values can be measured by labour time' but that 'to apply such a measure we must have a comparative scale of the different working days'. (PP p. 46.) But the properly Marxian theory of value elaborated in *Grundrisse*, then in CCPE, and acquiring its definitive form in *Capital*, is the result of putting the Ricardian conception back on the stocks, in order to curb the egalitarian utopias which certain socialist currents (Proudhon, Darimon, etc.) thought they could build on that conception. But this work of remodelling led to more than a simple rectification. In fact it culminated in a radical mutation in the significance of the labour theory of value. So much so that the identity in name and the analogy which some of Marx's formulations offer with Ricardian propositions must be considered quite mystifying. The belated abandonment of the reference to the scale of wages in *Capital* breaks one of the final links which still bound Marx's theory of value to the Ricardian problematic. And, conversely, the presence of that reference in Ricardo's *Principles*, contrary to Rubin's claims, bears witness to the fact that Ricardo's break with a practical conception of the task set for the labour theory of value cannot be sustained without qualification.

It is, however, undeniable that Rubin's eulogies of Ricardo are an echo of Marx's own. The approval Marx gave the Ricardian method is not in the main contradicted between *The Poverty of Philosophy* and *Theories of Surplus Value*. Against Proudhon's claim to build a new world on the basis of the equation: labour time = exchange value, Marx emphasized in PP that in Ricardo the theory of value was not a 'regenerating idea' but 'the theoretical expression of the real movement', 'the scientific interpretation of actual economic life'. (PP p. 43.) The theory of value, then, is the theoretical formulation of a law immanent in the real world. And if, later on, in TSV, Marx notes certain inadequacies in Ricardo's deductions in the *Principles*, they are not in his view sufficient to compromise the solid foundation of their general approach, an approach which presents us with 'the whole bourgeois system of economy as subject to one fundamental law', and which extracts 'the quintessence out of the divergency and diversity of the various phenomena'. (TSV II p. 169.) With respect to the more particular question of the search for an invariable measure of values, Marx certainly did not confirm Ricardo's break with that problematic in the peremptory way Rubin does, but, equally, it must be recognized that his remarks in this respect are equivocal. In fact, Marx, showing here a blindness common to many who pore over the same writings for too long, never grasped what was specifically at stake in the search for an 'invariable measure of value'. He persisted in seeing it as the inadequate expression of a real problem, the very problem which he himself had posed and for the resolution of which he fundamentally subverted the significance of the theory of value, namely, the problem of the nature of value, or again, the problem of the content expressed in the relations of exchange and in their variations. 'The problem of an "invariable measure of value" was simply a spurious name for the quest for the concept, the nature, of *value* itself, the definition of which could not be another value, and consequently could not be subject to variations as value. This was *labour-time*, *social labour*, as it presents itself specifically in commodity production.' (TSV III, pp. 134-5; cf. TSV I, pp. 150-1, TSV II, p. 202.)

Because of this distorted reading, leading Marx to interpret this 'false problem' of the economists as an unskilful approach, Marx never questioned the finalities of the economic theory entailed by the search for an invariable standard. As we have seen, Rubin obviously had his attention caught by the form taken by twentieth-century academic political economy and he was therefore more perspicacious than Marx on this point. But the conviction with which Rubin states that Ricardo 'decisively rejected any and all attempts to find an invariable measure of value, returning time and time again to show that such a measure could not be found' (p. 248 above) leaves more than one contemporary reader perplexed. For Sraffa's efforts in publishing Ricardo's last text, *Absolute Value and Exchange Value*, have brought to light evidence that the determination of an invariable measure of values was far from alien to Ricardo's preoccupations but on the contrary became the main object of his reflections towards the end of his life. Further, the general thrust of his final work (its questions and answers), was already contained in the *Principles*, in particular in Chapter I, Section 6, which has this very title: 'On an Invariable Measure of Values'. Contrary to Rubin's thesis, and Rubin obviously knew this text well, Ricardo does not question at all the *principle* of the search for an invariable standard but he merely underlines the difficulties of the enterprise and poses the conditions such an instrument would have to meet. And the solution he outlines is identical to the one he was to develop later in *Absolute Value and Exchange Value*. He argues that it is not possible to find a perfect instrument of measure, but only 'as near as an approximation to a standard measure of value as can be theoretically conceived'. (*Principles*, p. 45.) He goes on to consider gold as a commodity 'produced with such proportions of the two kinds of capital as approach nearest to the average quantity employed in the production of most commodities' and suggests that those proportions might be 'so nearly equally distant from the two extremes, the one where little fixed capital is used, the other where little labour is employed, as to form a just mean between them'. (*Principles*, pp. 45-6.)

The most commonly accepted interpretation today of Ricardo's problematic has been given by the commentaries and personal labours of Sraffa who edited the publication of Ricardo's complete works. For him, it is self-evident that the search for an invariable standard of values is 'so much at the centre of Ricardo's system' (*Introduction*, *The Works and Correspondence of David Ricardo*, I, p. xlix) that mastery of that standard alone allows him to ground his theory of distribution. It is easy to see that this interpretation is diametrically opposed to Rubin's. It is of interest in that it brings to light a dimension of Ricardian theory of which Marxists, in their haste to assimilate it to Marxian theory, have been generally unaware. But it seems to us to do serious damage to Ricardo's thought. Sraffa believes he can solve the questions Ricardo does not answer by elaborating a 'commodity standard' which no longer involves any reference to labour time but which meets the conditions which derive from the instrumental function assigned to this 'measure'. Now Ricardo was equally concerned to establish *a relation of cause and effect* between the variations in productivity in the different branches of production and variations in prices. It is this aspect of his preoccupations that held Marx's attention and the attention of all Marxist commentators after him; it completely disappears in the Sraffian reading.

How are we to interpret the fact that representatives of both these antagonistic conceptions of political economy claim the Ricardian heritage? It might be simply assumed that the distinction between an 'objective' conception and a 'practical' conception of the theory of value had still not been made by Ricardo and that, in Ricardo as in Smith, the objective conception was encumbered with residues of a prior problematic to which he unfortunately returned at the end of his life. This means that Rubin's judgement must be slightly qualified: the normative and objective conceptions of economic discourse were still mixed in the *Principles*, so that Marx and Sraffa, from their respective stances, might develop one of the virtualities existing in a contradictory way in the thought of the master. But although seductive in its simplicity, this answer would miss an element which is crucial to the understanding of Ricardian theory and its relation with Marxian theory. In truth, the search for an invariable measure of values, whilst distinct from the attempt to assign their cause, is not something extrinsic to the logic of the *Principles*, something independent of the determination of price variations by modifications in the productivity of labour. On the contrary, the search derives from the manner in which Ricardo conceives the relation of causality linking the expenditure of labour to the fixing of exchange values. According to him, relations of exchange between commodities are established in proportion to the respective quantities of labour *effectively* expended in their production, with the proviso of this single corrective: the homogenization of labours of different skills. The variations in productivity due to different causes, the main one of which is technical innovation, affect different branches of production unevenly and consequently involve modifications in the terms of exchange. If at the moment t_2 the production of commodity A requires half the time required at t_1, while the time needed for the production of commodity B remains the same, commodity A will be exchanged at t_2 against commodity B in a relation which is a half of that prevailing at t_1. Whatever the precautions we ought to have taken when stating this principle—if it is to escape the objection that it proposes an excessive schematization of real processes—it remains the case that its acceptance implies that the variation in productivity in a branch of production, all other things being equal, *directly* affects the value of the commodity issuing from that branch, and it alone. Now we must notice that variations in productivity stem from the labour time effectively expended in production, i.e. concrete labour. For Ricardo, it is *the respective durations of concrete labour time* that govern the relations of exchange of commodities. It is true that in one place Marx says that, while classical political economy had never made the distinction between concrete and abstract labour explicit, it nonetheless made it unconsciously, from the moment that it attributed to labour the property of being the source of value. (*Capital*, I, pp. 84-5.) This statement does not endanger our thesis, provided the terminology used is made more precise. It is accurate to say that for Ricardo, when, for example, the product of a day's labour by a jeweller exchanges against the product of a day's labour by a grand couturier, the equivalence of these products is a function of the labour time alone, on the single condition of an identity of skill of the labours concerned. The particular forms these labours assume do not intervene in the determination of the relations of exchange, but it is the equality of the *real* durations of labour—or the durations of *real* labour time—that is the basis of the relation of equivalence of the products. Now this version of the principle of the determination of values by labour time expended on the production of commodities gives it a particular epistemological status: in theory it is *open to empirical verification*. We mean here that in the Ricardian perspective it ought to be possible to show with examples that a commodity A, whose average price is established at double the average price of a commodity B, requires double the time in order to be produced. Similarly, it ought to be possible to show that a modification in the labour time necessary for the production of a given commodity is expressed by a proportional modification in its exchange value. And yet this verification, which nothing rules out in principle, meets a certain number of technical difficulties when one considers its practical implementation. Indeed, when we try to estimate the price variations which follow from changes in the productivity of labour, we meet with a host of obstacles which thwart the possibility of an empirical verification: the different proportions according to which fixed capitals and circulating capitals are distributed according to the branches, as well as the time differences in the turn over of capitals, combined with the identity of the rates of profit which Ricardo accepts as a given within the functioning of the system. And it is in this respect precisely that there arises the need for an instrument capable of isolating the effects of variations in the productivity of labour from the circumstances hiding their manifestations, circumstances which for our author are secondary ones.

For Ricardo, consequently, the theory of value corresponds to two distinct yet related questions. The first is to determine the *causes* of the value proportions between the commodities and their variations. The second question is to provide ourselves with an instrument of measure which, given a certain alteration in the terms of a relation of exchange, would enable us to identify which commodity is 'responsible' for it, and, in the event that each commodity being considered has undergone a change in its conditions of production, to assess what amount of the variation in the exchange relation is to be attributed to one commodity and how much to the other. *These two questions are distinct*. The second is a technical problem included in the conditions of verification of the first. That is why the impossibility of successfully constructing a standard likely to fulfil the role designed for it in a satisfactory way compromises the labour theory of value no more than the difficulties in realizing an experimental device affect the truth value of a physical theory. Supposing that these difficulties receive a solution, we would obviously be in a position to submit the theory to a test which could just as well refute it as corroborate it. But until then it is open to us to develop the implications of the theory containing the proposition which is in principle testable. The distinction between the theoretical question and the practical question is clear from Ricardo's last text, in which his debatable reworking of the definition of an invariable measure of values in no way implies the abandonment of the labour theory of value in so far as it establishes the *cause* of value. Ricardo states that 'in Political Economy we want something more; we desire to know whether it be owing to some new facility in manufacturing cloth that its diminished power in commanding money is owing, or whether it be owing to some new difficulty in producing money'. (AVEV p. 375, semi-colon supplied.) For Ricardo, the *cause* of a modification in the relation of exchange is indeed to be found in transformations of the productivity of labour in the branches concerned and in the ensuing modifications of the labour time required for the production of commodities. What we do not find is the criterion permitting us to distinguish indisputably which commodity it is—clothing or money—whose value has been modified, a criterion which is indistinguishable from the test of empirical verification of the theoretical proposition. *The two questions are nonetheless linked*. If the technical task of elaborating an invariable standard of values is not identical with the formulation of the law governing the relations of exchange, this task nonetheless flows from a particular feature of the form of that law; it is in so far as that law can be, at least in principle, subject to an empirical control, that the requirement of a practical instrument of measure which meets precise conditions is significant. This leads us to suspect that Marx's inability to grasp what specifically is at stake in the search for an invariable standard of values derives from the fact that for him the law of value, although occasionally expressed in formulas comparable to Ricardian formulas, *does not have the status of an empirically verifiable proposition*.

2 The Sociological Significance of the Marxian Labour Theory of Value

Marx himself gives his readers an indication that he considered his highlighting of the dual character of labour, that is, the distinction between concrete labour and abstract labour, to be one of the most original aspects of his thought (*Capital*, I, p. 49 and *Selected Correspondence*, p. 232). Abstract labour, not concrete labour, in his view constitutes the *substance* of value. If we have to speak of cause, it is socially necessary labour time, the quantitative manifestation of abstract labour, that plays this role with respect to the magnitudes of the values of commodities. We shall not spend time here on the naivety of the 'physiological' interpretation of the notion of abstract labour. Its deficiencies and its incompatibility with the general meaning of the Marxian theory of value have been adequately pointed out by Rubin in the *Essays*. (*Essays* Ch. 14 and *Dialektik der Kategorien*, p. 7f.) But Rubin does not, in our view, give enough attention to the epistemological consequences implied by the refusal to accept that caricature of an interpretation. If abstract labour designates no reality which is given to sensible experience, then its duration cannot be the object of a direct measure. That is the reason why the Marxian version of the labour theory of value—the proposition according to which abstract labour is the substance of value, or, and this is the determination of the same phenomenon from a quantitative perspective, commodities exchange proportionately to the socially necessary labour time of their production—is not empirically testable.

To be convinced of this, we must return to what constitutes the keystone of the Marxian labour theory of value, that is, the problematic of fetishism. Here again, Rubin's *Essays*, and in particular the first chapter of that work, provide the irreplacable clue. His interpretation doubtless runs counter to the philosophical interpretation fashionable today. That interpretation seeks to exclude the 'theory of fetishism' from the authentic scientific content of *Capital*, arguing that that theory is concerned with a 'pre-Marxist' theme centered around a humanist critique of the alienation of the individual in the world of things.[6] But the interest of the themes dealt with under the heading

6 We know that this is the position adopted by the French philosopher Louis Althusser and his students. Balibar developed this point of view systematically in *Cinq études du Matérialisme Historique*, 1974. He points out that the text which Marx devotes to the fetishism of the commodity (*Capital*, I, p. 76f.) has often been the point of departure for the elaboration of a 'theory of fetishism', understood as an element of a more general theory of ideologies, even of a theory of knowledge. After an analysis of the text in question, he remarks that these different attempts leaning on the analysis of fetishism have produced only 'philosophies of knowledge or idealist anthropologies' (p. 215), amongst which he includes the works of Lukács in *History and Class Consciousness* (1971) and also some of Rosa Luxemburg's analyses in *Introduction to Political Economy*. According to him, these interpretations are not simple mistakes; they develop something latent in the very problematic of Marx, but latent only in so far as 'this problematic is in the last analysis only a specific variant of a pre-Marxist *philosophical* problematic'. (Balibar 1974, p. 220.) Here we shall not go into the debate on the conditions for the construction of a materialist theory of ideologies. Just one word on this: it cannot be denied that the concept of 'fetishism of the commodity', together with the elaborations which have gone with it as far as

of fetishism is not restricted to providing the (debatable) first draft of a materialist explanation of the forms of bourgeois ideology. In fact, the comparison of commodity production and non-commodity forms of production (Robinson Crusoe on his island, the personal relations of a feudal society, a rural patriarchal society, and a 'union of free men working with communal means of production' in which we can recognize the socialist theory of the future drawn in broad outline), this comparison is developed through the analysis of 'fetishism' (*Capital*, I, p. 76f.), and it alone can disengage the general social function which Marx attributes to the law of value. Had Marx not made that comparison explicitly, his readers would nonetheless have had to acknowledge its tacit presence at the onset of his interrogation and reinterpretation of the law of value. The aims of that comparison is indeed to indicate that the value form of the products of labour performs, in a historically unprecedented way, a necessary function in all human society. The modes of realization may change but the function itself is invariant. That function consists in *the proportional distribution of the totality of disposable labour powers between the different branches of production*, in such a way that the reproduction of all the conditions of production is made possible. Marx's famous letter to Kugelman on 11 July 1868 unambiguously emphasizes that that is indeed the starting point of his study of the relation of exchange. 'Every child knows that a nation which ceased to work, I will not say for a year, but even for a few weeks, would perish. Every child knows, too, that the masses of products corresponding to the different

their gnoseological presuppositions concerning the 'illusion of competition' (*Capital*, III, p. 852f.), were, in Marx's view, meant to explain the mystifications (or the effects of misconception) within which bourgeois political economy in general moves. Now these gnoseological presuppositions, in particular the opposition: appearance/essence of capitalist society, determine the choice Marx makes in the ordering of his exposition of categories in *Capital*, i.e. determine the plan of that work. That is why it is not correct to refute them purely and simply as non-materialist without further concern for the consequences that such a refutation might have for justifications of the demonstrative order of *Capital*.

But there is more to this. As we show here by recalling Rubin's analysis on this subject, only the thematic of fetishism permits the subversion of meaning which Marx makes the law of value undergo. Without that subversion, concepts as important as social labour, general labour or socially necessary labour would be empty of meaning. From this point of view, Balibar seems to us dishonest in the case he mounts against Luxemburg concerning Chapter 4 of her *Introduction to Political Economy*, a chapter in which she imagines a fictitious society where labour is planned and organized, only to be replaced, after the abrupt disappearance of centralized management of production, by commodity exchange. Contrary to what Balibar says, Luxemburg's objective is not to deduce the 'historical necessity' of exchange (Balibar 1974, p. 226) but rather *its functional significance* in a perspective which, as we understand it, is in strict conformity with the logic of Marx's interrogation.

needs require different and quantitatively determined masses of the total labour of society. That this *necessity* of the *distribution* of social labour in definite proportions cannot possibly be done away with by a *particular form* of social production but can only change the *mode* of *appearance*, is self-evident. And the form in which this proportional distribution of labour asserts itself, in a state of society where the interconnection of social labour is manifested in the *private exchange* of the individual products of labour, is precisely the *exchange value* of these products'. (*Selected Correspondence*, pp. 251-2.) In non-commodity societies, these quantitative conditions of the reproduction of the social totality are normally guaranteed by the compulsory assignation of a determinate *concrete* labour to each person. The weight of custom or the authoritarian intervention of institutionalized powers here determines the allocation of producers between branches distinguished by a definite historical state of the social division of labour. In a possible socialist society, it would be for a central management body to effect this indispensable distribution in a rational way. But, one of the most important specific social characteristics of commodity relations of production—of which capitalist relations of production are the most developed form—is the absence of any instance whatever to assure the proportional renewal of the social division of labour. The private character of production signifies the reciprocal independence of economic agents. How can unity of production arise from the conjunction of the numerous, uncoordinated initiatives of economic agents? To understand how, Marx invites us to consider the sole relation which these agents establish amongst themselves: the exchange of commodities. This, of course, is not a matter of an isolated act of exchange but of the relation of exchange in so far as it has become the social form of the process of production, that is, in the way that the whole of production is geared to the creation of exchange values, not to the creation of goods for consumers whose qualitative and quantitative needs have been specified beforehand. The complex phenomena operating in the relation of exchange, for example the abstraction and equalization of different concrete labours, are the means thanks to which the unity of social production (a condition of reproduction) is realized without the producers' knowledge. Through the confrontation of commodities in the act of exchange, therefore, a process takes place which, according to a particular modality, is common to all societies, namely the distribution of social labour between the different branches comprising the totality of production. Questioned from a perspective which Marx has many times emphasized as a crucial characteristic of his approach (cf. PP pp. 28-31, and *Capital*, I, p. 84 n. 1), the labour law of value acquires a sociological significance which it did not possess in the other economic doctrines in which it featured. It is a particular means for realizing a universal social function, effecting social cohesion in a society which no customary or juridico-political instance unifies. And this is not to say that such an instance does not exist but that the specific mode of intervention which it does have consists in not intervening, or in suspending all regulation which might appear to present an obstacle to the freedom of agents-producers to produce or trade.

Thus, Marx seizes on the labour theory of value only because he sees in it the possibility of indicating the presence, behind the structure of the proportions of exchange of commodities, of a second structure on which the first is based, viz. the distribution of socially available labour power between the various branches of production. In his own vocabulary, the exchange of commodities converts private labour into social labour. But this interpretation of the labour law of value really constitutes a total subversion of the meaning and aim of the law in comparison with what it was in the bourgeois doctrines, up to and including what it was in Ricardian theory. And it implies that between labour and value *there exist far more complex structural relations than the mechanical unilateral relations which we have in mind more or less consciously when we talk about a relation of cause and effect*. For the relations of commodity exchange are not simply the result of the proportional distribution of global social labour time; they are also one of the essential moments of the realization of that distribution. By offering his commodity for sale, the producer anticipates in the price he advances, that is, in the equalization of his commodity with the general equivalent (money), the conversion of his effective labour time into a determinate quantity of abstract labour (socially necessary labour time). Failure to sell at the price he counted on[7] will prove him wrong and as a result will make him either modify his production techniques or change the nature of his production, that is, move into another branch of production. It is quite difficult to conceptualize this interconnection of two interlocking structures (the distribution of social labour time and the equations of commodity exchange). Marx plainly designates it a difficulty in CCPE and specifies this difficulty in these terms: 'In the exchange process, the commodity as exchange-value must then become a universal equivalent, materialised general labour-time ... on the other hand, the labour-time of individuals becomes materialised universal labour-time only as the result of the exchange

7 We take the simplest example, that is, the example of an individual estimate which is *higher than* the price imposed by the law of the market. The analysis of the opposite case would require that we take into account the general mobility of labour or the movements of capitals from one branch to another.

process'. (CCPE p. 45.) He tries to solve this difficulty by appealing to the philosophical opposition between the two categories of *actuality* and *virtuality*. (CCPE p. 44.) The process of exchange is considered as the actualization of abstract social labour which exists only potentially, or latently, in commodities before their sale. We shall not discuss the operative efficacy of these categories here. For our purposes it is enough to have shown that the sociological interpretation of the law of value affects the precise form of the causal link formerly established by Ricardo between labour (labour time) and value (magnitude of value). Two important consequences follow. First, the equalization of different concrete labours is effected through the equalization of the products of these labours, and the conversion of concrete labour into abstract labour therefore has no empirical content other than the exchange of commodities against the general equivalent. That is why the conversion cannot be the object of a quantitative evaluation distinct from the one which is spontaneously established in the process of commodity exchange. Secondly, any alteration in the productivity of one branch of production is expressed by a change in the value of the commodity issuing from that branch only if it induces a modification in the distribution of the totality of labour power in a given society. But, in that respect, it works on the whole range of values and not just on the value of the single commodity under consideration. The sociological-functionalist interpretation of the law of value removes any pertinence to questions bearing on the relation between two isolated elements belonging respectively to the structure of average prices (= values) on the one hand and to the structure of production on the other. These two consequences combine (each would be enough by itself) to deprive of any meaning any attempt to test the Marxian labour theory of value empirically.

To what should we attribute the fact that Rubin totally neglects the difference in epistemological status between the Ricardian version and the Marxian version of the determination of value by labour time? First of all, there is a methodological ground for it. The opposition he establishes between a 'practical' determination and a 'theoretical' determination of the goals of political economy, an opposition which has indisputable interest as a preliminary delimitation of objectives, is not enough to exhaust the various implications of the different understandings of the theory of value. In a word, Rubin's epistemological conceptions seem to us to be characterized by an excessive schematism in that he obviously imagines that the concept of 'theory' possesses an unequivocal significance. He endows every approach which aims at the discovery of the causalities at work in reality with the quality of 'theory' or 'science'—both these terms being synonymous for him. He never doubts that the notion of causality could itself be problematic and have different acceptations according to the nature of the objects between which the existence of a causal relation is being posited. He does not suspect, therefore, that causality cannot be the same in a theory which makes concrete labour time the cause of the magnitude of value (the Ricardian theory) and in another which posits abstract labour as the substance of value (the Marxian theory). But this methodological reason is not the only thing which explains Rubin's neglect. His blindness to the irreducible incompatibility between the Marxian theory and the Ricardian theory also derives from the indisputable ambiguity both of Marx's texts and probably of his thought as well. For it cannot be denied that the text of *Capital* is not entirely free of all the elements linked to the Ricardian version, that is, to the concept of a direct causal relation between labour time *actually* expended in the production of a commodity and that commodity's value. We find a clear indication of that in, for example, the lines Marx devotes to 'a few propositions which follow from the reduction of exchange-value to labour-time'. (CCPE p. 37.) Under this heading, Marx has in view the effect of the evolution of the productivity of labour on exchange value. Marx says in substance that if the exchange value of a commodity is effectively determined by the labour time necessary to produce it, the stagnation of the productive forces, and hence of the period of time necessary for the production of a commodity, must result in the stability of its exchange value. An increase in the productivity of labour in the branch under consideration will provoke a fall in the value of commodity-units, while a decline in productivity will, on the contrary, provoke an increase in that value. We find the same form of argumentation, more succinctly put, in *Capital*. (*Capital*, I, pp. 46-7.) Its validity hangs on the meaning of the notion of *socially necessary labour time*, to which we shall return. But the difficulty of the reasoning can already be indicated in simple terms. An increase of or fall in labour productivity directly affects the *actual* duration of concrete labour within a branch of production and within that branch alone. If, therefore, it is the case that 'if the productivity of labour grows' (in a given branch of production), then 'the same use-value will be produced in less time' (CCPE p. 37) (and vice versa), this fact tells us absolutely nothing about the proportional evolution of the total amount of concrete labour assigned to this branch of production in relation to the distribution of the totality of social labour. This latter point of view, however, is the only pertinent one within a sociological interpretation of the value phenomenon. The statement that 'the amount of labour-time contained in a commodity ... is consequently a variable quantity, rising or falling in inverse proportion to the rise or fall of the productivity of labour' (CCPE p. 37) results from an improper identification of abstract (socially necessary) labour time determining value, on the one hand, and concrete labour, the only labour *immediately* affected by a change in the productivity of labour, on the other.

This characteristic example of the persisting influence of the Ricardian problematic on Marx's thought, even in the works where he achieves the greatest lucudity with respect to the implications of his own point of view, is not, however, the only one nor the most significant. Much more interesting to us is the resurgence of Ricardian themes in the texts where Marx tackles the relation between market value and market price. (*Capital*, III, p. 173f., TSV II, Ch. 10 A-5.) This resurgence is obvious in Marx's appeal to the concept of 'individual value' which assumes a direct correlation between *concrete* labour time and value. Here it does indeed appear that the recurrence of Ricardo's conceptions is more than a residue from which Marx's thought can be disengaged with minimal effort. In fact such conceptions crop up—as pseudo-solutions—precisely where Marx's thought stumbles over its own internal limits. And the stubbornness with which Rubin upholds the fundamental continuity between Ricardo and Marx, in spite of his exceptional grasp of the Marxian problematic, becomes altogether symptomatic in this light. Only by paying such a price can he, perhaps unconsciously, hope to shake off the intrinsic difficulties of the 'sociological' approach.

3 The Internal Difficulties of Marx's Conception

This Afterword cannot provide the framework for an exhaustive analysis of all the dimensions of a problem which has been inadequately identified by Marx's critics.[8] We shall therefore evoke only in broad outline the dead-ends characteristic of the Marxian approach concerning the processes of social reproduction. We shall leave for later, or to other critics, the detailed discussion of those attempts at resolving problems presented either by Marx himself or by his exponents. The intrinsic difficulties of the Marxian approach are concentrated in the

8 Contemporary readers are generally not aware of the difficulties arising from the two definitions of the concept of 'socially necessary labour time', which we raise in the pages which follow. Some authors, however, stand out for having caught a glimpse of the existence of a problem in the definition of the concept, for example Rosdolsky, *The making of Marx's 'Capital'* (1968 I chap. 3) and especially Reichelt, *Zur logischen Struktur des Kapitalbegriffs bei Karl Marx* (1974 ch. III, A-4: 'Digression on the concept of socially necessary labour time'.)

definition of the concept which is supposed to explain the mutual relations of exchange of commodities and their variations. We are referring to the concept of *socially necessary labour time*. We have noted that the functionalist-sociological interpretation of the labour theory of value saw the structure of commodity-exchange relations as both the effect *and* the means of realization of a hidden structure, namely, the social division of labour, or, to be more precise, the distribution of the totality of social labour between the different branches of production. In the architecture of Marx's terminology, the concept of *socially necessary labour time* is meant to assure the link between these two structures. Its ambiguities are in proportion to the importance of its role. If they have escaped the majority of commentators, it is because those commentators ordinarily keep to the simplifying formulas employed at the beginning of *Capital*. There, 'socially necessary labour time' is 'that [time] required to produce an article under the normal conditions of production, and with the average degree of skill and intensity prevalent at the time'. (*Capital*, I, p. 47.) But we need only dwell on that definition for a little while—and at the commentaries accompanying it—to notice that it does not permit Marx to use the concept in question in the way he does. The first chapter of *Capital* (or of CCPE) endows the concept of 'socially necessary labour time' with a purely technological content only. It is 'the labour-time required, under the generally prevailing conditions of production, to produce another unit of the *same* commodity'. (CCPE p. 31 our emphasis) i.e. a *technical norm* which competition imposes within each branch. Now the 'normality' of technical conditions of production makes sense only for each branch taken separately, each branch in which techniques are homogeneous. This definition, therefore, gives us no right to invoke the concept which it explicates in order to account for the proportions in which commodities of different natures are exchanged. Yet Marx uses it to this purpose without further justification. 'The value of one commodity is to the value of any other, as the labour-time necessary for the production of the one is to that necessary for the production of the other.' (*Capital*, I, p. 47.) We will read and reread these pages in vain if we want to understand how it is that Marx attributes such operative power to the concept of 'socially necessary labour time'. But we must look elsewhere for the reasons why he accords that concept such importance. By 'elsewhere' we mean in *another* definition of the *same* concept. And this is formulated explicitly only in Volume III of Capital. 'But if the use-value of particular commodities depends on whether they satisfy a particular need then the use-value of the mass of the social product depends on whether it satisfies the quantitatively definite social need for each particular kind of product in an adequate manner, and whether the labour is therefore proportionately distributed among the different spheres in keeping with these social needs, which are quantitatively circumscribed ... The social need, that is, the use-value on a social scale, appears here as determining factors for the amount of total social labour-time which is expended in various specific spheres of production ... *This quantitative limit to the quota of social labour-time available for the various particular spheres of production is but a more developed expression of the law of value in general, although the necessary labour-time assumes a different meaning here*. Only just so much of it is required for the satisfaction of social needs. The limitation occurring here is due to the use-value'. (*Capital*, III, pp. 635-6, our emphasis.)

This passage makes explicit reference to the distribution of social labour between particular branches of production, a distribution which must be proportional to the structure of social needs if reproduction is to proceed without difficulty. From the moment that we adopt the vantage of point of looking at the general process of production, the whole of the labour time spent in a branch of production is recognized as 'socially necessary' only if it 'corresponds' to the existing social need for the commodities of that branch. Of course, Marx also says that 'necessary labour time' has a different meaning here. But one must be clear about this. Are we talking about a second property of the initial concept, a property which can be inferred from the purely technological definition at the beginning of *Capital*? It is quite clear that we are not. Further, and this is the main point, the intelligibility of the beginning of *Capital* is conditional on the second definition of 'socially necessary labour time'. This is the only definition, in fact, that is likely to confer on this concept the role Marx assigns it *from the very beginning of his work*, namely to indicate the social process which, unknown to the parties to the exchange, operates through the value relations of commodities.

It is worth noting that the chapter of Rubin's *Essays* dealing with socially necessary labour time keeps silent on this second definition. (*Essays*, Ch. 16.) Rubin stops at the technological rendering of the notion, that is, he sees in it above all a means of explaining the identity of the unit price of commodities of a given variety, in spite of the diversity of the technical conditions under which they may have been produced. The silence he observes with respect to the wider implications of the notion in question may seem paradoxical on the part of an author who in other places is well aware of the sociological significance of the law of value. But the polemic he develops in this and the following chapter against the supporters of the conception which he dubs 'economist' sheds light on the deep motives of his position. He has indeed grasped perfectly well that to leave any place for references other than technological ones in the definition of the concept is to endanger *the exclusive role of the productivity of labour and the transformations it undergoes in the determination of the values of commodities and their movements*. (Cf. *Essays*, pp. 195-206.) Only a purely technological definition of socially necessary labour time permits one to take the unit value of a product to be a variable independent of demand, and this is the vital part of his demonstration in Chapters 16 and 17 of the *Essays*.

That is what is crucially at stake in the problem we are discussing here, and it is one of the main merits of Rubin's work to bring this to light, even when the solutions he proposes remain unsatisfactory. We have seen that taking account of the second definition of the concept of socially necessary labour time is indispensable if the correlation between that concept and the functional (or sociological) meaning of the law of value is to be assured. At the same time, however, this definition forces us into a confrontation with the most contentious aspects of the internal logic of *Capital*, namely *the exact place occupied by the reference to needs*. The law of value is the immanent law of regulation of a society of private producers, since it stipulates that through the reciprocal adjustments of the exchange values of commodities, a social division of labour adequate to the structure of social needs spontaneously imposes itself. In conformity with this general law, the labour time spent on the production of the whole range of commodities of a given variety is recognized as socially necessary *on the single condition that the total volume of these commodities 'corresponds' to the social need*—to the solvent (and therein lies the rub) social need *which exists for them*. It follows from this that the definition of the fundamental concepts of the Marxian theory of value cannot be made without a preliminary reference to a given structure of social needs or of the demand for commodities.

Now this implicit hypothesis can be considered as compromising Marxist theory from two different points of view. If we tackle it from the angle of its philosophical presuppositions, the problematic of economic discourse seems to be suspended in an ahistorical anthropology. If we focus our attention instead on the internal coherence of the theory, the causal determinations it proposes appear to constitute a circle and as a result the theory can be challenged on grounds of formal incoherence.

Let us look first at the philosophical approach of Marxist theory. The fact that an economic theory deals with the content and organization of needs before even starting the study of the forms of production has often been held to denote a fundamental anthropologism at the root of this theory. (On this point the works of Louis Althusser have simply updated a classical thesis in Marxist exegesis. Cf. *Reading Capital*, Chs. II, VIII, IXI.) This anthropology is suspected of concealing, right from the start, the historical character of economic phenomena, since the needs thus mentioned at the start of the theory seem necessarily to have to receive determinations independent of the conditions of production, i.e., to be defined according to a hypothetical human nature untouched by all historical specification. And it is undeniable that Marx criticized this ideology of 'homo economicus', emphasizing on more than one occasion that the content of men's needs varies according to historical periods, and even that they are determined by the forms of production: as much because these forms of production influence the forms of satisfaction of even natural needs (and create new desires) as because the only economically significant needs, in a commodity-producing society, are ones the satisfaction of which can be paid for. 'Is the entire system of needs founded on estimation or on the whole organization of production?' Marx was already asking in PP. 'More often than not, needs arise directly from production or from a state of affairs based on production'. (PP p. 37.) In particular they depend on the distribution and level of incomes, distribution and level that are themselves a function of production. Now, the way the theoretical exposition of *Capital* was arranged was meant to reproduce the supposed real order of determinations, namely, distribution of products and consumption dependent on production, production being 'the real point of departure and hence also the predominant moment'. (*1857 Introduction*, p. 94.) That is why production must be studied first. But this famous text, in which the preeminence of the moment of production in relation to the other moments of the functioning of the economic system—consumption, distribution and exchange—is confirmed in this way, also contains a number of remarks which rule out a linear conception of the order of causalities at work. In particular it emphasizes that consumption 'mediates production' in the same way that 'production mediates consumption', and it clarifies this formula by means of this commentary: 'a product becomes a real product only by being consumed. For example, a garment becomes a real garment only in the act of being worn; a house where no one lives is in fact not a real house; thus the product, unlike a mere natural object, proves itself to be, *becomes*, a product only through consumption.' (*1857 Introduction*, p. 91.) The difference Marx introduces here between product and natural object, trivial as it may seem, is in fact the nub of the most important structural difficulties in *Capital*. This is because it implies that a product which finds no consumer, that is, in a commodity-producing society, no buyer, is thereby denied its quality of use-value, and as a result its quality of value as well. The actual realization of the sale is the sole condition under which the commodity is said to possess a value (a use value and therefore a value); it follows that the concepts developed in the section on production (*Capital* Vol. I) presuppose the existence of a solvent need for the entire range of commodities, that is, a division of incomes and a series of acts of exchange (the structure of demand) *on which the supply of commodities must be regulated.* One cannot, therefore, understand either the elementary questions of Marx's political economy, or his fundamental concepts, without positing the existence of a human community which, through a specific organization of its productive activity, provides for the satisfaction of an articulated totality of needs in such a way that the conditions of the perpetuation (or expansion) of that productive activity are reproduced. This is the conclusion which we had already reached with the analysis of the concept of socially necessary labour time. The 'anthropological' content of this hypothesis, however, remains pretty thin. True, the existence of needs is posed at the beginning of the analysis but their nature is not in any way specified; nor is that specification necessary so long as we do not try to translate the mode of functioning of the economic system advanced by the theory into a concrete quantitative interpretation. The precise content of our opening presuppositions is simply the following: there exists a given structure of needs composed jointly of the demand of industries (productive consumption) and the demand of private individuals, and, because of an immanent regulation, production tends to adjust to it spontaneously. This structure is itself the result of the prior cycle of production, in the same way that the cycle of production now opening will in its turn determine the distribution of needs for the next cycle. In fact, this *reciprocity of determinations*—the structure of the allocation of labour between branches of production being determined by the structure of demand, and the latter being determined by the former—avoids one from having to resort to an anthropology in the strict sense. From another point of view, however, it does pose a problem. Can we say of a theory which is incapable of isolating, from the totality it studies, one element as the final starting point of the chain of causes implied in its functioning, that it can have an explanatory value?

It is on this interpretation of the internal difficulties of Marx's theory, that 'economist' readings of *Capital* more readily lay stress when, although this is unusual, they are pushed sufficiently far for the functional significance of the law of value to be perceived. An example of this type of interpretation is offered in a recently published collective work by Cutler, Hindess, Hirst and Hussain: *Marx's Capital and Capitalism Today*. This work presents one of the most pertinent critiques of the systematic of *Capital*. Its undeniable superiority over traditional critiques of *Capital* by academic economists is due to the fact that it does not demand of Marx's theory answers to objectives which are alien to it—such as the capacity to provide the immediate basis for a theory of prices—but rather respects the original meaning of the theory, in particular where it relies on Rubin's *Essays*. Following Rubin, they recognize the concept of fetishism of the commodity as the key to the Marxian problematic, and as a result are able to state unambiguously the problem to which the theory of value in its Marxian version claims to be the answer. 'What is the relation between the abstract-labour ratios established in the process of exchange and the distribution of production, and what is the relation of these two to the composition of the product expressed in demand?' (MCCT p. 88.) Better still, they quite clearly perceive that the formulation of the law of value is essentially linked to the normative idea of an equilibrium expressed by the concept of 'socially necessary labour time' (MCCT pp. 81-3), and inevitably come up against the question of the role and determination of social needs in this representation of the economic system. They rightly note that Rubin's position ends up with this difficulty: that value appears to depend on demand, in the sense that it is necessary to invoke a determinate composition of social needs in order to explain the proportions in which commodities exchange. Naturally, Rubin is nervous of this consequence and attempts to avoid it by showing that the proportion of equilibrium is on the contrary exclusively a function of relative productivity between branches of production and between individual enterprises within each branch. But, the authors of MCCT state, Rubin's demonstrations cannot be considered satisfactory in this respect. We must conclude, therefore, that 'the labour theory of value, if it is not to be a simple labour-as-substance theory [that is, a theory of incorporated labour in the Ricardian sense] *must admit the crucial role of demand* if it is (as it must be) combined with a notion of a necessary composition of the social product'. (MCCT p. 93, our emphasis.) But to accord the composition of demand a role in the determination of values totally independent of the productivity of labour endangers the internal coherence of the theory of *Capital*. Now, whatever the difference of objects between *Capital* and 'bourgeois' economic theories may be, internal coherence constitutes an elementary requirement which all theory ought to meet, if it is not to become plain nonsense. It seems to us therefore that it is on this central point—the definition of the state of equilibrium to which the sale of the commodity *at its value* corresponds—that we have identified the fundamental flaw which definitively compromises the validity of the theory of *Capital*.

It must be said immediately that, like the authors of MCCT, we are convinced both that Rubin's interpretation faithfully expresses Marx's thought and that his laborious work is unsuccessful in making demand depend on labour productivity. But we want to tighten up a few points concerning Marx's own awareness of the internal difficulties of his work. One of the weaknesses of its construction lies, as we have indicated above, in the definition of use value. The authors of MCCT claim that Marx makes use of this concept as if it had 'a non-comparative and non-qualitative sense' and that it is in Rubin's interpretation of Marxist theory that we are to see that use value also possesses quantitative determinations in Marxist theory. (MCCT p. 91.) It is correct that in the first chapters of *Capital* Marx makes no mention of any quantity condition in order to attribute a use value to a product. But it is not true any more in Volume III. The texts in which the sociological dimension of socially necessary labour time appears all make it clear that we must admit a quantitative determination of use values.[9] Now as it happens, Marx did see that perhaps that was the weak link in his exposition. We have in mind here a very strange extract in *Grundrisse* where, for a page at any rate, he seems to doubt the appropriateness of the order in which he introduces (or envisages introducing) the different concepts of his theory. He has just tackled the study of the circulation of capital, and he shows that the 'contradiction between use value and exchange value' appears there in a quite original form compared to simple circulation. In fact, the need which exists on a given market for a determinate product 'here [in the circulation of capital] appears as *measure* for it as use value and hence also as *exchange value*. ... What is posited now is that the *measure* of its availability [*seines Vorhandensseins*] is given in its *natural composition* [*seiner natürlichen Beschaffenheit*] itself. In order to be transposed into the general form [money], the use value has to be present in a limited and specific quantity; a *quantity* whose *measure* does not lie in the amount of *labour objectified in it*, but arises from its *nature as use value for others*.' (*Grundrisse*, p. 406.) And a few lines later he adds: 'The indifference of value as such towards use value is thereby brought into just as false a position [*Position*] as are, on the other side, the substance of value and its measure as objectified in labour in general.' (*Grundrisse*, p. 407.)

But, one might ask, why should it matter if Marx did or did not have an intimation of the incoherence which was the inevitable

9 Cf. the quotation above. (*Capital*, III, pp. 635-6.)

stumbling block of his theory of value? In the end, he took no account of it, and throughout the years he did his utmost to construct a theory on foundations whose fragility he had already tested. It is the final product we must judge and not the fast forgotten hesitations of its author. We shall have no difficulty agreeing with that, but it seems to us that it is important to spend a few moments on the reasons why the possible reciprocity of social reproduction, occasionally perceived by Marx, did not lead him to a general questioning of his theses. For such reciprocity is not intrinsically contradictory. It is not self-evident that these lasting changes in the system (modifications in the relations of exchange of products, in the distribution of labour power between branches of production, and in the structure of demand) can in the final analysis always be imputed to a single and unique cause. Earlier we noted certain passages in the *1857 Introduction* which deal with this same concept of use value, in which Marx envisaged that a system including a phase of production, a phase of distribution, a phase of exchange and a phase of consumption could set in motion a plurality of reciprocal causalities. Moreover, at the end of the second part of that work, Marx foresees that this reciprocity of determinations perhaps springs from the implicit scientific paradigm which inspired his reinterpretation of the law of value. This paradigm consists in *likening the sphere of socio-economic relations with an organism* the vital process of which is entirely ordered by imperatives of survival. 'Admittedly, however, *in its one-sided form*, production is itself determined by the other moments. For example, if the market, i.e. the sphere of exchange, expands, then production grows in quantity and the divisions between its different branches become deeper. A change in distribution changes production ... Finally, the needs of consumption determine production. Mutual interaction takes place between the different moments. *This is the case with every organic whole*'. (*1857 Introduction*, pp. 99-100. The second emphasis is ours.) This is not the place to go into the extent to which the organicist paradigm influences Marx's elaboration[10] nor into the question of

10 The influence of biology is noticeable in the metaphors Marx has recourse to in a famous text (TSV II pp. 182-9) in which he compares the respective methods of Smith and Ricardo. In it Smith gets criticized for wavering in constant ambiguity between two heterogeneous plans of analysis, that of the 'appearances of competition' and another, which is situated as it were, beneath the first, in which 'the internal relations of the bourgeois system', its 'hidden structure', its 'internal physiology' or its 'internal organic relations' are articulated. The merit of Ricardo is, on the other hand, clearly to have shown what was the 'true physiology of bourgois society'. The organic metaphor is even more obvious in the review of the Russian critic who describes Marx's method; Marx cites it with approval in the Afterword to the Second German Edition. (*Capital*, I, p. 26.)

how it was able to model his thought. We shall simply note that the influence of the biological concept of organization was widespread in the most diverse areas of research at the end of the eighteenth century and the beginning of the nineteenth, and that the grasp Marx had of Hegel's work would be more than enough to explain the hidden operativeness of this concept in the Marxian approach to the social whole. We well know that biological organicism haunts the metaphors in which Hegel described the process of historical Reason or the process of the Concept. These questions of theoretical filiation could provide ample material for a particular study, but we must put them on one side and confine ourselves here to noting two facts which relate directly to our argument. Firstly, the latent presence of the organic paradigm explains why the Marxian conception of the economic totality can give the impression of being teleological. The authors of MCCT return to this characterization of the system on two occasions. (MCCT pp. 81, 87.) But it is not in the least a question of a particular goal being imposed on the process of social reproduction from outside. The expression 'teleological-functionalist' (MCCT p. 81) is in fact redundant. The functioning of an organism cannot be conceived without an end, and that end is none other than the reproduction of its internal conditions of existence, which is to say, life. One must shed maximum light on this point before putting into question Marx's general principle of interrogation. If we object to this principle in the name of the impasses it seems to lead us into, it is not this or that finality, arbitrarily assigned to the process of social reproduction, that we are disputing but the *operative fruitfulness of the biological paradigm in the domain of the social sciences*. This question is of importance for disciplines which to this day are far from having developed their own methodology and are hence far from able to dispense with the methodological crutches borrowed from sciences foreign to them.

Secondly, it seems that to view society as an organism is precisely what presents an obstacle to the construction of a theory in which the order of causalities is linear. The deficiency of Marxist theory, and the reason for its circularity, lies not so much in the reciprocity of the determinations necessarily implied by its methodological model as in the fact that Marx persists in maintaining a *causal monism* incompatible with that model. By causal monism we mean the desire to attribute to the productivity of labour exclusive responsibility for the equilibrium in the proportions between production and exchange in a capitalist commodity-producing society. For this thesis to which Rubin attaches such great importance unquestionably belongs to the thought of Marx himself. Convincing proof of this can be found by reading, among others, the chapters in *Capital* in which Marx deals with the average rate of profit. (*Capital*, III, Chs. 9-12, 50.) This is why a critique of *Capital* which merely points to the circularity of its approach, however pertinent it may be, does not reach the ground at which Marx's general problematic is engendered nor that point at which the difficulties which finally entangle him are first knotted together. The crucial question which an exhaustive critique of *Capital* must try to answer consists in understanding why, in spite of his likening of the social structure to an organic structure, Marx insisted on maintaining the unitary nature of the principle of equilibrium—and of transformation—of that structure.

To put the question another way, the analyses put forward in MCCT pinpoint very correctly the impasses of *Capital*'s demonstration. They are therefore sufficient *from the point of view of political economy*. But they do not permit us to appreciate the significance of these impasses with regard to Marx's *critical* project. As a theory of equilibrium of the system of production resting on the exchange of commodities, we are told, Marxist theory has deficiencies of internal coherence at least equal to those of the utility theories of value. Agreed, but if we stop at this conclusion, the field remains open to possible attempts at reconstruction on the basis of an amended labour theory of value, or on some other basis. Now we know that these attempts are just what academic research adores. But they are evidence that all the lessons of the failure of *Capital* have not been learned. The full implication of that failure can be wholly appreciated only through a reconsideration of the content Marx gave to his critical project and in particular through an investigation into the role within it of the concept of an equilibrium of social reproduction realized behind the backs of the economic agents through the intermediary of the market.

4 What is at Stake in the 'Critique' of Political Economy

Marx has very clearly designated the absence of a historical perspective in the writings of bourgeois economists as the most fundamental cause of numerous registerable errors or lacunae in their doctrines. The *critique* of political economy, his real ambition, was to restore to economic systems the historicity that conventional economic doctrines had deprived them of. In the mind of its author, this critique had two distinct aspects. First of all, it was to disclose the tacit presuppositions on which the economic theories were built, that is, to show that these theories did not, as they claimed, study economic rationality in general, but the modes of regulation peculiar to a particular form of production and exchange, resting on specific relations of production, namely commodity relations between private owners. But Marx further assumed that once he had brought to light the specificity of the conditions for commodity production, he could show that the laws of internal regulation of that mode of production provide the basis for a particular form of development. This development cannot be expected to be a smooth, indefinitely renewed reproduction of the same system (even on an enlarged basis); it has, on the contrary, to unfold in a chaotic way—disequilibrium being as much a part of its essence as equilibrium—and it has to carry within it the germs of an ineluctable final destructuration.[11] The reproduction of the capitalist mode of production—as the developed form of commodity production—was to be viewed as a necessarily disturbed process jeopardized in the more or less long term by the fractures inscribed in its intrinsic order.

11 Disequilibriums correspond to economic crises, the inevitable final destructuring is foreshadowed in the law of the tendency of the rate of profit to fall, and the means with contradictory effects which the owners of capitals put to work in order to alleviate this falling rate (increased concentration and centralization of capital). Strictly speaking, this law constitutes the law of development of the capitalist system of production. Cyclical crises and the tendency of the rate of profit to decline are, moreover, linked. (*Capital*, III, pp. 256-7.) But the crises can be conceived as moments built into the regulation of the system, since in the end they restore a new equilibrium exactly where they had allowed the collapse of the old one. The phenomenon of the falling rate of profit, on the contrary, gives us an inkling, according to Marx, of the limits of capitalist production. For the 'rate of profit', and not the production of use values, is the 'motive power of capitalist production'. (*Capital*, III, p. 259.) Its fall weakens the principal driving power of accumulation, discourages the constitution of new autonomous capitals, and appears to threaten at some future date the development of the productive forces. (*Capital*, III, pp. 260-1.) That is why, Marx claims, bourgeois economists have a genuine 'horror' when they note this phenomenon. For capitalist production here proves itself not to be the absolute form of development of the productive forces, but a transitory form which will one day curb that development, after having helped it immeasurably compared with earlier relations of production. (*Capital*, III, pp. 241-2, 260-1.)

It is perhaps surprising that, dealing with the Marxian critique of political economy, we focus our attention on the theory of crises rather than on the law of the tendency of the rate of profit to decline. The reason is that we are concerned with the interpretation of *Capital* put forward by Rubin. This interpretation has as its axis the problem of value, so that it rarely deals with the themes of *Capital* Volume III. The theory of crises, however, relates directly to a question which Rubin's interpretation brings to the fore, that is, the role of the hypotheses of equilibrium in the construction of *Capital*—hence the place we assign it here.

With respect to the law of the tendency of the rate of profit to fall, an exhaustive critique of Marx could not, of course, avoid an interrogation of its status and of the validity of the demonstration made about it. We will content ourselves with underlying the following point: Marx's demonstration of the law of the tendency of the rate of profit to fall seems to us to derive from a Ricardian conception of the theory of value. This claim is doubtless paradoxical for people who know that Ricardo

Historical considerations are thus summoned to buttress economic study. They feature at the beginning of the study, in as much as straight away the mode of production is studied *in history* and emphasis is placed on its distinguishing characteristics; they arise spontaneously at the end of the study, once the law of development inherent in the mode of production has been mastered. There is, however, continuity in this study, for the distinctive forms of a mode of organization of production and exchange essentially merge with the laws presiding over its development and its final transformation. It is significant in this respect that Marx links the refusal of bourgeois economists (Ricardo in particular) to accept the possibility of general crises to their failure to recognize the specificity of capitalist production.[12] This implies that, conversely, the reformulation of the law of value on the basis of a correct understanding of its historical significance must lead to an opposite conclusion, namely one which would show that the phenomenon of crises, and the general crisis of production in particular, is the inevitable manifestation of an, as it were, genetic defect in the system of production founded on this law. Thus the deductive chain from the formulation of the concept of value to a highlighting of the conditions wherein the system would become destructured ought to feed itself out in an unbroken series of links.

The first aspect of the critical task was satisfactorily realized in the theory of fetishism. It has not been so with the second. Marx never got to the bottom of his theory of crises. No doubt a remark as peremptory

proposes a quite different demonstration of the same phenomenon from that of Marx. But remember that we are considering as a Ricardian element in Marx's thought the conception according to which productivity of labour, and its variations, is the direct and exclusive cause of the magnitude of the value of the commodity-product, and the changes it undergoes. Now it is indeed *this* conception of the relation labour/value that supports the Marxian demonstration of the falling rate of profit, since 'the progressive tendency of the general rate of profit to fall is, therefore, just *an expression peculiar to the capitalist mode of production* of the progressive development of the social productivity of labour'. (*Capital*, III, p. 213.) On the contrary, a strict adhesion to the sociological significance of the law of value, that is, its properly Marxian significance, which makes it into a law regulating the distribution of social labour between the different branches of production, deprives of meaning any *global* quantification of the value (or surplus value) produced by the totality of the branches in the course of a cycle of production. That Marxian significance consequently puts into question all attempts to estimate the quantitative variations of the mass of surplus value produced, as well as of the rate of profit. This confirms what we were saying earlier, namely that the Ricardian concept of surplus value crops up precisely where Marx's thought comes up against its own internal difficulties.

12 'In order to prove that capitalist production cannot lead to general crises, all its conditions and distinct forms, all its principles and specific features—in short *capitalist production* itself—are denied.' (TSV II p. 501, and cf. TSV II p. 528.)

as that will meet with many a Marxist economist ready to contradict it. The majority are agreed that the theory of crises is incomplete in *Capital* but many nonetheless hold it to be a solid basis for further elaboration. We, however, think that the constitutive hypotheses of *Capital* are an obstacle to that. It is out of the question to entertain the idea that within the limited framework of this Afterword we could tackle the various theses held on this matter and anticipate all the possible objections to our own position. We shall restrict ourselves to indicating the decisive problem that is, in our view, the inevitable stumbling block for all attempts to provide a sequel to *Capital* which would seek to bridge the gulf between the abstract plan within which its analyses are deployed and the study of concrete conjunctures.

It is true that, in *Grundrisse*[13] just as much as in *Capital* or TSV, we find the scattered elements of a study of crises. Part II of TSV even provides us with a synthesis which is sufficiently complete for us to be able to rely on it entirely. Among others, two successive stages in this study are distinguished, as was already the case in *Capital*. (*Capital*, I, p. 145.) The first consists in disclosing, step by step with the increase in understanding of the regularities peculiar to the capitalist system of production, the forms in which the *possibility* of disfunctionings is already inscribed. This amounts to showing how the potentiality of crises is given in the very formulation of the conditions of regulation of the system. This first stage is in its turn separated into different moments ordered according to a principle of increasing concretization. Thus, the following are noted as so many abstract—though to different degrees—forms of possibility of crisis: 1), the separation of sale

13 The first chapter of *Grundrisse* shows with a great deal of clarity that Marx's re-organization of the Ricardian theory of value was motivated by his concern to eliminate the possibility of a utopian interpretation of that theory, an interpretation already propounded by the partisans of labour tokens. The decisive point of this critique was to show that the labour time effectively spent on the production of commodities could in no case serve as the direct measure of values, nor, consequently, as the principle of distribution of social income. The 'genesis' of money—i.e. the demonstration that generalization of the commodity form of the product *implies* the specialization of a particular commodity in the function of general equivalent—acquires meaning within the framework of that polemic. Its realization is sufficient to establish that the success of the act of exchange is contingent on other factors. For the existence of money dissociates the unity of exchange by barter into two distinct acts: purchase and sale, which 'may correspond or not; they may balance or not; they may enter into disproportion with one another. They will of course always attempt to equalize one another; but in the place of the earlier immediate equality there now stands the constant movement of equalization, which evidently presupposes constant non-equivalence. It is now entirely possible that consonance may be reached only by passing through the most extreme dissonance.' (*Grundrisse*, p. 148.) Thus, the most elementary form of crisis is given.

and purchase; 2), the function of money as means of payment; and 3), the autonomization of the phase of circulation in relation to the phase of production of capital.

But, and this is crucial, for Marx *this first stage is insufficient*. It needs to be completed by an analysis of the *reality* of crises, that is, of the conditions which turn a *potential* crisis into an *actual* one. (TSV II p. 515.) Indeed, so long as one restricts oneself to noting the elements which render crises possible, and nothing more, bourgeois economists can continue to give vent to their optimism, since the actuality of crises retains a contingent character. Recognition of the fact that crisis is possible, in so far as reflection stops there, can be rightly considered a fall-back line for bourgeois economists, already convinced that it would be naive to deny any longer the existence of economic crises. 'This shows how insipid the economists are who, when they are no longer able to explain away the phenomenon of over-production and crises, are content to say that these forms contain the possibility of *crises*, that it is therefore *accidental* whether or not crises occur and consequently their occurrence is itself merely a *matter of chance*.' (TSV II p. 512.)

This second stage in the study of crises remained but a draft project in Marx's texts. We shall be told, no doubt, that this can be attributed to the unfinished nature of the work. And we can envisage filling this gap by combining the study of the forms of possibility of crises and the problem of the realization of surplus value, or the law of the tendency of the rate of profit to decline. To estimate such a project's chances of success, it is instructive to remember at what moment in the plan of his work Marx had postponed taking up the reality of crises. It is in the section on competition and credit, in which 'the real movement of capitalist production' is finally to be tackled, that the elements needed for this study were to be pieced together. '... the real crisis can only be educed from the real movement of capitalist production, competition and credit ...'. (TSV II p. 512.) And: 'In so far as crises arise from changes in prices and revolutions in prices which do not coincide with *changes in the values* of commodities, they naturally cannot be investigated during the examination of capital in general, in which the prices of commodities are assumed to be *identical* with the *values* of commodities.' (TSV II p. 515.) We know that the plan of the work has undergone a number of modifications in the course of its writing which was long and laborious. What has become of the section on competition in the final draft of this plan? Rosdolsky has a detailed discussion of this question in his exegetical work devoted to *Grundrisse*. He reaches a conclusion which in our view is a convincing one, namely that at the time Marx was writing *Capital*, in particular Volume III, he had abandoned the main distinction he established in the *1857 Introduction* between the analysis of 'capital in general' and that of competition. (Rosdolsky 1968 I, p. 36 and p. 36 n. 35.) A large part of the themes formerly ranged under this second heading had been absorbed into the analysis of 'capital in general'. Only a few specific problems, among them the 'real movement of market prices', were postponed in this way to a possible future study of competition. But Marx did not consider the necessity of that study crucial to his work. 'The actual movement of competition belongs beyond our scope, and we need present only the inner organization of the capitalist mode of production, in its ideal average, as it were.' (*Capital*, III, p. 831.) It is possible, nonetheless, to concede that Marx confined his initial ambitions and circumscribed his work to the study of 'capital in general' alone, appending to it some themes which did not originally belong to that study. It is also possible to maintain at the same time that there is nothing to stop *us* effecting the study of the 'real movement of market prices' which on Marx's own recognition is indispensable to a complete understanding of the phenomena of crises. But that is exactly where the difficulty lies. Is it possible, on the basis of the Marxian theory of value or of costs of production, to construct a new stage which includes a theory of prices? Or, to formulate it differently, can the process of concretization alluded to in particular at the beginning of Volume III of *Capital* (a process which is brought close to us 'step by step' under the form which it assumes 'on the surface of society, in the action of different capitals upon one another, in competition, and in the ordinary consciousness of the agents of production themselves'), (*Capital*, III, p. 25) can this process be pursued without any break, by adding new parameters and come to explain the causes of price movements? We must look at what those authors who put forward the methodological particularities of the Marxian approach (we mean here 'the method of rising from the abstract to the concrete' (*1857 Introduction*, p. 101)) have in mind when they invoke the new parameters to be taken into consideration for dealing with the problem of market prices at an opportune time. Quite simply, competition, and particularly its effects: constant fluctuations in the relation of supply to demand. They take it as given that the study of 'capital in general' presupposes that fluctuations in supply and demand are bracketed off, as it were—a legitimate theoretical abstraction as a first step, but one that must necessarily be abandoned later. On this point, they have plenty of quotations from Marx to draw on to guarantee the orthodoxy of their interpretation. Thus, in Chapter X of *Capital* Volume III, which is fundamental for the study of this question, Marx reasserts his long held conviction that it is impossible to explain the laws of capitalist production by means of the interaction of supply and demand. He then states the thesis which is ordinarily taken to be the key to his methodology: 'these laws cannot be observed in their pure state, *until supply and demand cease to act*, i.e., *are equated*. In reality, supply and demand never coincide.... But political economy assumes that supply and demand coincide with one another. Why? To be able to study phenomena in their fundamental relations, in the form corresponding to their conception, that is, is to study them independently of the appearances caused by the movement of supply and demand. The other reason is to find the actual tendencies of their movements and to some extent to record them.' (*Capital*, III, pp. 189-90, our emphasis.)[14] Here the cause seems to be understood. The study of 'capital in general', that is, of the laws governing the functioning and the development of the capitalist mode of production, requires that the effects of the supply/demand relation be conceptually suspended, since these effects pertain to deceptive appearance into the trap of which vulgar economists can be so easily led. If, bearing in mind the *critical* objectives of the Marxian theory, we ponder over this statement we can easily see that it raises a considerable difficulty. The analysis of 'capital in general', 'in its ideal average', assumes an identity between the values and prices of commodities, the sale of commodities at their value, or, in a more complex schema, at their costs of production, and this constitutes a sufficient condition for assuring the *equilibrium* of the system of production. As soon as we shelve the factors of disequilibrium (constant discrepancy between supply and demand or, what is the same thing, divergence between price and value) into the realm of 'appearances', isn't the project which claims to demonstrate the essential purtenance of disequilibriums to this system of production condemned to failure right from the start? But in our view, there is an even more fundamental difficulty. While both Ricardo and Marx saw the deviations in the relation of price to value as the effect of the supply/demand relation, Marx, unlike his predecessor, has constantly stressed that those deviations are not accidents or elements disturbing the normal state of equilibrium, but the very process through which value is determined by labour time. In the *Principles*, Ricardo twice returns to the divergence between current price and 'natural' price resulting from fluctuations in supply and demand: in Chapter IV 'On

14 Cf. 'If supply equals demand, they cease to act, and for this very reason commodities are sold at their market-values. Whenever two forces operate equally in opposite directions, they balance one another, exert no outside influence, and any phenomena taking place in these circumstances must be explained by causes other than the effect of these two forces.' (*Capital*, III, p. 189.)

Natural and Market Price' and in Chapter XXX 'On the Influence of Demand and Supply on Prices'. In both chapters, in the end, he puts aside price variations which he considers as 'accidental and temporary' (PPET p. 111.) Marx's position on this point was very different. As early as PP he noted that 'deviations' of prices in relation to values are not accidents but the rule. And if, with Ricardo, he recognized that it is only 'when supply and demand are evenly balanced' that 'the relative value of any product is accurately determined by the quantity of labour embodied in it' (PP p. 52), he added immediately that this equilibrium is never produced, that there is in fact 'no ready-made constituted "proportional relation", but only a constituting movement', and that this 'fluctuating movement' alone makes 'labour the measure of value'. (PP p. 56.)[15] Clearly, the ideal situation or 'ideal average' which we are invited to envisage under the concept of 'capital in general' results from corrections responding automatically to discrepancies existing between supply and demand; so much so that it is just as legitimate to consider the corrections as being the effects peculiar to competition as the discrepancies themselves. The function which competition assumes in relation to the abstraction 'capital in general' thus becomes extremely problematic. In this respect, it is symptomatic that this extract from *Capital*, in which we find expressed the crude and extremely over-simplified opposition between concept and appearance, is immediately followed by a passage in which the ambiguity of the role played by competition appears with full clarity. Indeed, Marx states that the constant discrepancies between supply and demand automatically produce economic processes to compensate them. And it is precisely from the interaction of these contradictory movements that there arises a tendency for an equilibrium situation to be realized. 'Since, therefore, supply and demand never equal one another in any given case, their differences follow one another in such a way—and the result of a deviation in one direction is that it calls forth a deviation in the opposite direction—that supply and demand are always equated when the whole is viewed over a certain period ... ' (*Capital*, III, p. 190). If we follow Marx in this reasoning we are led to invoke the supply/demand relation in order to explain two directly opposed phenomena: on the one hand, the gap between the market

15 Marx was well aware that his position on the question of competition was different from Ricardo's. This emerges from the following remark which is taken from the commentaries which accompany his reading of Ricardo's *Principles*: 'Ricardo made an abstraction of something he considered accidental. It is quite different to represent the real process in which the two aspects—what he calls the accidental movement but which is the constant and real element and its law, the average relation—appear equally important'. (Cf. *Grundrisse*, p. 651 and CCPE p. 174.)

price and market value, hence *the persistent discrepancy between reality and theoretical hypothesis*, and on the other hand, the equally persistent tendency towards the reduction of that gap, hence *the closeness with which the theoretical abstraction accords to reality itself.* Marx comes straight out with the ambiguity of the role given to the supply/demand relation: 'On the one hand, the relation of demand and supply, therefore, only explains the deviations of market-prices from market values. On the other, it explains the tendency to eliminate these deviations, i.e., to eliminate the effect of the relation of demand and supply.' (*Capital*, III, p. 190.)

To put it more rigorously, it is therefore false to say that the effects of competition are provisionally suspended in the analyses of *Capital.* If its hidden action were not in fact assumed, the state of equilibrium to which the 'concept of capital' relates would be a pure product of the imagination. In the absence of some planning body or other, it is this hidden action which constrains the possessors of capital to alter the distribution of their investments in such a way that social labour is distributed among the different branches of production in proportions adequate to the structure of social needs. This hidden action of competition is the indispensable mediation for the realization of the law of value in the sense in which Marx understood that law—an immanent principle of organization of the productive totality. The main point had been made in this connection in *Grundrisse*: 'competition in general is the crucial motor of bourgeois economy; it does not establish its laws but it executes them ... In the case of the economists, to presuppose, as Ricardo does, that unlimited competition exists, is to presuppose the full reality and realization of bourgeois relations of production in their specific difference. Competition, therefore, does not explain these laws; it makes them visible but it does not produce them.' (*Grundrisse*, pp. 649-52.) The means through which the laws peculiar to bourgeois relations of production impose themselves are to be found in the 'free' and individual choices made by the multiple protagonists on the economic chess-board, all these choices being governed by an immanent rationality which ensure their interconnection without the subjects knowing. Marx attributes to competition, and to fluctuations in supply and demand—concrete manifestations of the specificity of the capitalist mode of production—a role which is roughly comparable to that played by human passions in the Hegelian conception of history. They are means in the service of a finality unknown to themselves, but far from impeding it they help in its design.[16]

16 Cf. our interpretation of the role of competition in relation to the laws of the capitalist mode of production with this statement from Hegal: 'This vast congeries

We were wondering earlier whether we could envisage bringing Marx's critical work to completion by demonstrating, as he wanted to, that the conversion of the possibility of crises into reality was inevitable because of the laws governing the capitalist system of production. It seems that the answer must be negative. For competition cannot assume without contradiction the dual role assigned to it by Marx's formulations. If the automatic compensations induced by the mechanisms of competition impose an adjustment of capital distribution which meets the requirements of the qualitative *and* quantitative structure of social needs, it is impossible to introduce the fluctuations in supply and demand as a *new* factor which would permit one to leave the study of 'capital in general' in order to deal with competition. For the existence of these fluctuations has been tacitly presupposed throughout the work on 'capital in general'. The hypothesis of equilibrium on which the formulation of the law of value or of the sale of commodities at their costs of production rests, bars the route, then, to a theory of the conditions of realization of crises. There is, of course, nothing to prevent the carrying out of an empirical study on this question. But the hiatus separating political economy from economic history cannot be abolished. Crises are destined to remain a contingent phenomenon for theory.

5 Conclusion: the Limits of the Marxian Critique of Political Economy

The inability of Marxist theory as presented in *Capital* to account for

---

of volitions, interests and activities, constitute the instruments and means of the World-Spirit for attaining its object; bringing it to consciousness and realizing it. ... But that those manifestations of vitality on the part of individuals and peoples, in which they seek and satisfy their own purposes, are at the same time, the means and instruments of a higher and broader purpose of which they know nothing—which they realize unconsciously—might be made a matter of question.' (Hegel, *Philosophy of History* 1956, p. 25.) The influence that this work of Hegel's, and perhaps even more, the *Logic*, has exercised on the constitution of *Capital* is to this day far from perfectly elucidated, in spite of the numerous research works which the relations between Marx and Hegel have inspired. The reason is that few authors have given systematic attention to the *structures* of the work, save a few rare exceptions amongst which the most important are R. Rosdolsky and H. Reichelt. Reichelt's claim that the law of value is 'a sort of transcendental synthesis, a unifying principle working without our knowledge at the level of social labour' (Reichelt 1970 p. 144) will make more than one Marxist who is persuaded of the materialism of the Marxian gnoseology bristle. But we subscribe to it without hesitation, for reasons which will emerge clearly enough from our interpretation of the particular significance Marx attributes to the 'law of value'.

what Marx called the 'reality of crises' throws new light on a classic problem of the epistemology of political economy. We are talking about the operative character of equilibrium models for the comprehension of the real functioning of the capitalist system of production. Controlable mathematical models, which for the greater part of academic political economy of the twentieth century represent the ideal norm to which, in so far as it is 'scientific', economic theory should tend, have often been the object of critiques, whether emphasis is placed on the illusory character of the understanding they propose or whether, even more radically, the ideological function they perform is stigmatized.[17] By offering the satisfying picture of an economic production all the parameters of which have been mastered, they deny the disorder constitutive of capitalist commodity production. Well before formalist scientific ideology had brought caution to this conception of economic discourse, the nature of the problems posed by 'bourgeois' economic doctrines had prepared the ground which mathematical models have subsequently occupied. Marx had denounced the principle when he criticized as a characteristic sign of the apologetic approach the fact that from the outset it assimilated commodity production to 'social' production, so that 'society, as if according to a plan, distributes its means of production and productive forces in the degree and measure which is required for the fulfilment of the various social needs, so that each sphere of production receives the *quota* of social capital required to satisfy the corresponding need'. (TSV II p. 529.) In so far as he intended his approach to be a critical one, Marx himself hoped to pose the initial question of political economy in a different way. 'On the contrary, the question that has to be answered is: since, on the basis of capitalist production, everyone works for himself and a particular labour must at the same time appear as its opposite, as abstract general labour and in this form as social labour—how is it possible to achieve the necessary balance

17 A critique of the first type can be found in Keynes himself in connection with the specific hypothesis of full employment: 'The classical theorists resemble Euclidean geometers in a non-Euclid world who, discovering that in experience straight lines apparently parallel often meet, rebuke the lines for not keeping straight—as the only remedy for the unfortunate collisions which are occurring'. (Keynes *General Theory* 1974 p. 16.) For the second type of critique, cf. among others Badiou: 'Generally speaking, bourgeois political economy is accomplished in the construction of models of balanced expansion. Here again, the model guards against capitalist "disorder" not by means of an understanding of its cause (i.e. the Marxist science of social formations and its comprehension of the class struggle) but by the integrated *technical image* of the bourgeoisie's class interests ... Models of expansion in equilibrium objectify class objectives under cover of thinking their object (the economy of alleged "industrial societies") ... ' (Badiou *Le concept de modèle* 1969 p. 16.)

and interdependence of the various spheres of production, their dimensions and the proportions between them, except through the constant neutralisation of a constant disharmony?' (TSV II p. 529.) We have seen, however, that representations of the system of production elaborated on the hypothesis of the sale of commodities at their value or at their costs of production are far from fulfilling the programme announced by this question. In fact, those representations, just like bourgeois theories, assume that the distribution of the means of production and labour power or the distribution of social capital among the various branches of production is adequate to the structure of social needs, a condition which must be satisfied, in particular if the concept of *socially* necessary labour time is to have sense. So, Marxist theory seems to constitute direct proof of the independence of the theoretical tool provided by the equilibrium model in economy with respect to every ideology justifying the existing economic system.

Do we have here a sort of demonstration *ad absurdum* forcing us to conclude that the capitalist system of production is really a system in equilibrium and that the upheavals periodically affecting it can be attributed to factors outside its intrinsic logic? In fact, that would be to make the same mistake as those who employ abstract models in political economy, namely to confuse the operative capacities of the theoretical model with the properties of the real object. We think that it is possible, and legitimate, to interpret the impasses of Marxist theory in another way. The centre of interest of that theory is, of course, different from that of the majority of bourgeois economic doctrines: making sense of the conjunctural fluctuations of prices is of little importance to Marxist theory. The law of value, in its Marxian version, is rather, as we have seen, a principle of *organization* in the sense that that concept has for the biology of the first half of the twentieth century—an invisible structure assuring the reciprocal independence of visible structures and guaranteeing the being it animates the very possibility of existence. But, however different the implicit paradigm of the Marxian approach may be in comparison with the formal models inspiring political economy today, it shares a major hypothesis with them. When it is applied to the sphere of socio-economic relations, the Marxian paradigm imposes on them the form of a closed totality endowed with 'natural' laws, that is, laws independent of possible interventions of the political, juridical, or social instances. Marx has taken up, without hesitation, the postulate from bourgeois political economy according to which the world of socio-economic relations is inhabited by a spontaneous dynamic, is regulated by an immanent order.

Rubin's reading is yet again highly instructive here. His faithfulness to Marx, even in what was perhaps a prejudice limiting his critical thought, is shown in the commentaries he devotes to the concepts of 'natural rights' and 'natural law' as used by the Physiocrats and Smith respectively. On this point, the reader should refer to Chapters 11 and 17 above, which tackle the theme of 'natural rights' in the Physiocrats, and to Chapter 20 in which we find very pertinent remarks on the concept of the 'natural' in Smith. In Smith as in the Physiocrats, Rubin notes, the concept of a 'naturalness' peculiar to the field of socio-economic relations merges both theoretical and axiological significances: theoretical in that the concept designates a regularity expressible in scientific laws, a regularity spontaneously discharged by the free play of individual desires and initiatives; axiological because this order, being independent of all institutional intervention, is supposed to bring maximum well-being for the individual and for society alike.[18] But if Smith still juxtaposes a theoretical use and an ideological use of the term 'natural', he does not usually confuse them. According to Rubin, A. Smith was the first to clear the ground for a *purely theoretical* study of natural phenomena. 'Smith's transition from an evaluative to a theoretical understanding of the term "natural" marked *a great step forward for the purely theoretical, scientific-causal study of economic phenomena*'. (See above, p. 175, our emphasis.) It was indeed a deep conviction of Marx himself that it should be possible in principle to clear away the semantic wool surrounding the concept of 'natural' and to isolate a *purely theoretical* meaning of the apologetic interpretations. For him the great merit of classical political economy was to have recognized the '*Naturwüchsigkeit*' of the processes of commodity-producing economy; its main defect was to have ignored its historical character. But we must emphasize this: the '*Naturwüchsigkeit*', the quality peculiar to the phenomena of commodity-producing economy, is the means by which those phenomena can be the subject of reflection within a closed system of scientific laws. Or again, the form which scientific study *ought to take*—the deployment of explanation starting from a single founding law—is determined by the supposed

18 More than a century later, and in a very different historical context, the theoreticians of pure economy have shamelessly taken up these nice optimistic hypotheses once more. For them, as for Smith, the apologetic and *theoretical* significances overlap in their definition of the object of economic theory. The comparison of the different definitions of pure economy proposed by Walras, which he obviously takes to be equivalent, is very eloquent. First he says that pure economy is 'the study of laws in some way natural and necessary according to which exchange, production, capitalization and circulation of social wealth tend to operate under a hypothetical regime of organized free competition'. (Walras 1909.) Later he says that 'pure economy is the science which proves that perfect competition gives the maximum satisfaction of needs ... ' (Walras 1953).

property of the object which Marx calls its '*Naturwüchsigkeit*'. As a result, to acknowledge this property meant the limitation, right from the start, of the possibilities of the *critique* of political economy. Doubtless this critique could reveal the sociological presuppositions of bourgeois economic theory. There was no great difficulty in showing that the behaviours of economic agents, which that theory took to be the expression of natural passions (in particular the search for the maximization of profit), are imposed on them by the specific historical structures of capitalist commodity production. But it was not possible to go beyond the stage of putting into historical perspective the capitalist system of production and the internal arrangements of the economic theory it permitted. In particular, after having posited that the system of capitalist commodity relations was endowed with a law of internal regulation, it was quite out of the question to demonstrate that its destructuring was inevitable.

One wonders whether the supposed autonomy of the field of the Economic, that is, the attribution to the sphere of economic relations of an intrinsic legality which isolates it in an abstract way from the other modalities of social relations (in particular, the juridical and political modalities) is not an excessively narrow interpretation of the theses of historical materialism as they are presented in the first part of *The German Ideology* or in the classic formulation in the *Preface to A Contribution to the Critique of Political Economy*: 'In the social production of their life, men enter into definite relations that are indispensable and independent of their will, relations of production which correspond to a definite stage of development of their material productive forces. The sum total of these relations of production constitutes the economic structure of society, the real foundation, on which rises a legal and political superstructure and to which correspond definite forms of social consciousness.' (*Preface*, p. 181.) Now these relations of production are defined as entailing juridico-political (even ideological) conditions as well as economic ones. The free subjects who meet in the act of exchange of commodities or in the labour contract (a particular form of exchange), are free only through the agency of a legislation and a power which appear in this light to be just as much conditioning as conditioned. Marx in fact is on occasion ready to recognize that certain types of relations of production can have extra-economic constraints as their condition of existence: for example 'in the more or less primitive communal production' in India, or when the State itself owns the land, as in Asia, an example he invokes briefly in the chapter of *Capital* devoted to the 'genesis of capitalist ground rent'. (*Capital*, III, pp. 782-813.) Comparable examples can be found in the long passage of *Grundrisse* which is concerned with 'Forms which precede capitalist production' (*Grundrisse*, p. 471.) No doubt some people will argue that Marx's attitude to *non-commodity* relations of production does not contradict his approach to *commodity* relations of production in *Capital*. Indeed, for him, the autonomy of economic laws was not a general rule, valid for all historical forms of social production, but, on the contrary, a property characteristic of commodity relations of production alone. It is just this property that in his view explained and justified the birth and development of economic theory, a development which follows step for step the appearance of the various elements which became integrated into the fully-fledged capitalist mode of production—commerce, wage labour, manufacturing co-operation, mechanization, etc. These are the conditions of commodity production, and these alone, 'which assert themselves without entering the consciousness of the participants and *can themselves be abstracted from daily practice only through laborious theoretical investigation*; which act, therefore, *like natural laws, as Marx proved to follow necessarily from the nature of commodity-production*'. (Engels, Supplement to *Capital* III, p. 899.) The concept of commodity fetishism, once again, had the task of accounting for the reifying mystification undergone by the relations of production in the commodity producing economy. Yet how are we to understand this mystification? If the 'naturalness' of economic laws is definitely illusory, the critique of political economy must deny the very existence of the object of political economy and not simply its claim to validity for all historical epochs. On the contrary, by according the categories of bourgeois economy an objective validity, even if a relative one (*Capital*, I, pp. 80-1), it may be that Marx himself has in the end fallen into the trap of the ideology secreted by the functioning of capitalist society, an ideology which assigns to the Economic precisely that place where the discourse of *Capital* is still situated.

Catherine Colliot-Thélène, 1979

Bibliography

Primary Sources

K. Marx, *Capital*, Volumes I-III, Lawrence and Wishart, 1974.
K. Marx, *Grundrisse*, Pelican, 1973.
K. Marx, *Theories of Surplus Value*, Parts I-III, Progress Publishers, Moscow, 1969.

K. Marx, *Introduction to A Contribution to the Critique of Political Economy*, in *Grundrisse*, (see above).
K. Marx, *Preface to A Contribution to the Critique of Political Economy*, in Marx/Engels *Selected Works* in one volume, Lawrence and Wishart, 1970.
K. Marx, *The Poverty of Philosophy*, Martin Lawrence, London.
K. Marx and F. Engels, *Selected Correspondence*, Foreign Languages Publishing House, Moscow.
F. Engels, Supplement to *Capital*, Volume III, in *Capital* (see above).
K. Marx, *Grundrisse, de Moscou en langues étrangères*, 2 volumes, 1939-41.

Secondary Sources

L. Althusser and E. Balibar, *Reading Capital*, New Left Books, 1970.
A. Badiou, *Le concept de modèle*, Maspero, Paris, 1969.
E. Balibar, *Cinq études du matérialisme historique*, Maspero, Paris, 1974.
C. Benetti, *Valeur et répartition*, PUG, Maspero, Grenoble, 1974.
E. von Böhm-Bawerk, *Karl Marx and the Close of his System*, Merlin, London, 1975.
A. Cutler, B. Hindess, P. Hirst and A. Hussain, *Marx's Capital and Capitalism Today*, Volume I, Routledge and Kegan Paul, London, 1977.
G.W.F. Hegel, *The Philosophy of History*, Dover, 1956.
G. Lukács, *History and Class Consciousness*, Merlin, 1971.
R. Luxemburg, *Introduction to Political Economy* (not translated).
J.M. Keynes, *The General Theory of Employment, Interest and Money*, Macmillan, London, 1974.
H. Reichelt, *Zur logischen Struktur des Kapitalbegriffs bei Karl Marx*, Europäische Verlagsanstalt, Frankfurt am Main, 1970.
D. Ricardo, *On the Principles of Political Economy and Taxation*, in The Works and Correspondence of David Ricardo, Volumes I-XI, Cambridge 1975.
D. Ricardo, *Abstract Value and Exchange Value*, Works and Correspondence, Volume IV.
J. Robinson, *An Essay on Marxian Economics*, Macmillan, London, 1972.
R. Rosdolsky, *The Making of Marx's 'Capital'*, Pluto, London, 1977.
I.I. Rubin and S.A. Bessonow u.a., *Dialektik der Kategorien*, *Debatte in der USSR 1927-9*, VSA, Westberlin, 1975.
P. Sraffa, *Introduction* to *The Works and Correspondence of David Ricardo*, (see above).

L. Walras, Scientific Jubilee, University of Lausanne, 10 June 1909.
L. Walras, *Abrégé des éléments d'économie politique pure*, Librairie Générale du Droit, Paris and Lausanne, 1953.

Translator's Notes

1 Where there is a discrepancy between Colliot-Thélène's references to the French edition of *Grundrisse* and the Pelican edition, I translate from the French, give her reference, and supply a reference to comparable passages in the Pelican edition.
2 After the first reference to a work in the text I use the following abbreviations and contractions:

TSV	*Theories of Surplus Value*
1857 Introduction	*Introduction to A Contribution to the Critique of Political Economy*
Preface	*Preface to A Contribution to the Critique of Political Economy*
CCPE	*A Contribution to the Critique of Political Economy*
PP	*The Poverty of Philosophy*
MCCT	*Marx's Capital and Capitalism Today*
Principles	*On the Principles of Political Economy and Taxation*
AVEV	*Absolute Value and Exchange Value*
Essays	*Essays on Marx's Theory of Value*
Dialektik	*Dialektik der Kategorien*
History	*History of Economic Thought*
Introduction	*Introduction* to *The Works and Correspondence of David Ricardo*

Name Index

Subject Index

CPSIA information can be obtained
at www.ICGtesting.com
Printed in the USA
BVHW031925201220
595718BV00030B/82